Foundations of Community Policing in Canada

PAUL F. MCKENNA

Public Safety Innovations, Inc.

With a Foreword by Robert W. Runciman, MPP

Prentice Hall Allyn and Bacon Canada

Scarborough, Ontario

Dedicated in loving memory of my mother, Hilda Grace McKenna

Canadian Cataloguing in Publication Data

McKenna, Paul F. (Paul Francis), 1952–
Foundations of community policing in Canada

Includes index.
ISBN 0-13-010691-7

1. Community policing — Canada. 2. Police — Canada. I. Title.

HV7936.C83M34 2000 363.2'3'0971 C99-930923-4

Prentice-Hall, Inc., Upper Saddle River, New Jersey
Prentice-Hall International (UK) Limited, London
Prentice-Hall of Australia, Pty. Limited, Sydney
Prentice-Hall Hispanoamericana, S.A., Mexico City
Prentice-Hall of India Private Limited, New Delhi
Prentice-Hall of Japan, Inc., Tokyo
Simon & Schuster Asia Private Limited, Singapore
Editora Prentice-Hall do Brasil, Ltda., Rio de Janeiro

ISBN 0-13-010691-7

Vice President, Editorial Director: Laura Pearson
Acquisitions Editor: David Stover
Art Director: Mary Opper
Associate Editor: Susan Ratkaj
Production Editor: Matthew Christian
Copy Editor: Jim Zimmerman
Production Coordinator: Peggy Brown
Marketing Manager: Sophia Fortier
Cover Design: Mary Beth MacLean
Cover Image: Digital Vision
Page Layout: Erich Falkenberg

2 3 4 5 04 03 02 01

Printed and bound in Canada.

CONTENTS

FOREWORD

We are in a constant state of change. Worldwide economic structures, political boundaries and societies in general are constantly evolving. An academic examination of the institution of policing within Canada confirms that we are not isolated from the effects of worldwide change. The author of this text examines how we carry out the business of policing in Canada from an historical and contemporary perspective.

The roots of Canadian policing are found entrenched in the British system and, over time, influenced by our neighbours in the United States of America. What has emerged is an understanding that crime prevention, law enforcement, assistance to victims, public order maintenance and emergency response are important to the citizens of our communities. The delivery of these services by our policing institutions has been based on the "community policing" model.

Sir Robert Peel is regarded by many as the father of community policing. His recognition that the police are the public and the public are the police remains true today—only now we say that "we enter into partnerships with our communities to create safe societies."

I congratulate Paul McKenna for his work in this text which brings together a cross-section of the policing community to lend their views to this topic. *Foundations of Community Policing in Canada* is worthwhile reading for those contemplating a career in policing; those practicing the profession; and those who want a better understanding of the concept of community policing.

Robert W. Runciman, MPP
Leeds-Grenville
Solicitor General and
Minister of Correctional Services
May 2, 1999

PREFACE

In approaching the topic of community policing in Canada, it is necessary to state that no single definition will adequately detail its fullest meaning. This textbook is intended to provide exactly what its title suggests; "foundations" for understanding community policing in a Canadian context. Most dictionaries tell us that a "foundation" is the supporting part of a wall, house, or some other structure. A "foundation" may also be the fundamental principle upon which something is based. For example, the foundations of modern science are based on the principles developed by Copernicus, Sir Francis Bacon, and Newton. This textbook is written to support the construction of sound organizational structures based on established, and widely accepted, principles of community policing.

The focus on problem-solving contained in the book reveals the author's position that this approach to community policing is worthy of detailed study and application. While community policing may exist without a commitment to problem-solving, this work will underline the fundamental importance of the theory and practice of problem-solving in any meaningful pursuit of community policing. Because community policing, in its many forms, is relatively new everyone remains a student in this field of endeavour. Of course, some "students" have developed a high degree of sophistication and expertise in the formulation and application of the theories and models which have come into being. However, the genuine pioneers or leaders in the area of community policing will likely be among the first to insist that a great deal of detailed work needs to be done before true mastery is achieved. Therefore, it is hoped that this textbook will provide a useful learning tool for those students who are just entering the field of policing in Canada and who require a detailed briefing on what has gone before in terms of the evolution of community policing. It is also hoped that police practitioners and academics will themselves find this publication useful in so far as it attempts to weave together several strands of theory and practice which constitute the fabric of community policing in the current Canadian context.

While every effort has been made to provide a comprehensive survey of the work being done in the area of community policing across Canada, it is impossible to do justice to the extensive and imaginative applications which are in place in several police services. In creating a textbook that attempts to serve the needs of a wide-ranging audience, it is likelier that the author will provide too little information than too much. It is hoped that this effort will aim high in striking at its target, and, by reaching far, will not fall too short of the mark; which is to present the reader with a solid, practical, and sound understanding of community policing as it has manifested itself in Canada.

There will be substantial reference to important work being done in the United States with regard to this topic, and, while it should not be necessary to apologize for this kind of inclusion, it may be worthwhile to offer some elaboration for the approach taken. Because of its size, influence, and proximity policing in America has a strong gravitational pull for its Canadian counterpart. The fact that American institutions have had a profound impact on their Canadian cousins is not surprising, however, we need to constantly monitor institutional life in the U.S. in order to diagnose symptoms and apply treatments or remedies which may effect our own organizational well-being.

Chapter 1 begins with the seeds of community policing and looks at how those seeds have grown over time. There will be some consideration of both British and American influences. These influences will be placed in a Canadian context, which will remain the key frame of reference for the entire textbook.

Chapter 2 looks at the importance of crime prevention as a precursor to community policing and considers some of the theoretical underpinnings, as well as, the practical applications of crime prevention in several jurisdictions (including Canada, U.S. and Great Britain).

Chapter 3 carefully examines the context within which community policing operates by providing details about the Canadian "character". What are the demographic, social, political, legislative, technological, crime, and global trends that have influenced the growth of community policing in Canada?

Chapter 4 provides some insight into the larger organizational setting of community policing in Canada. Virtually every organization, public or private, is undergoing some process of reexamination or reevaluation which may result in substantial change or, as it is often called, "reengineering". This is certainly true for many police organizations in Canada, and this chapter attempts to provide background information to better understand and anticipate those changes.

Chapter 5 looks at the critical need for creating change within police organizations. The significance of dynamic change is examined in some detail and the unique challenges faced by those who embark on change within a policing context are considered.

Chapter 6 begins our consideration of the problem-solving models of community policing which is central to this textbook. A review of the history and growth of these models is presented along with some current applications from various jurisdictions.

Chapter 7 moves toward a discussion of some the key challenges which confront an organization that seriously attempts to introduce a meaningful approach to community policing. The first challenge relates to the systemic educational demands which are placed on an organization truly committed to community policing. Next are the profound ethical challenges which will contribute in no small measure to the success or failure of the entire community policing enterprise. Finally, this chapter looks at some of the other special challenges that pertain to a community policing mindset.

Chapter 8 attempts to address some of the technological considerations which relate to a mode of operation and management guided by principles of community policing. Are there opportunities for technology to enhance and extend the application of community policing, or, does technology in fact threaten the whole notion of community policing?

Chapter 9 takes stock of the critique presented against community policing. The critics fall into several distinct camps and some effort is devoted to an attempt at understanding and addressing their opposition to the concept and practice of community policing.

Chapter 10 deals at some length with the measurement of community policing. How is it possible to know that community policing is working in addressing the public problems of crime and disorder? How can we measure the effectiveness of individual officers and make determinations about promotion, assignment, transfer, and other personnel matters in the context of community policing? How does the concept of "team policing" impact the department's performance measurement process?

Chapter 11 offers the reader more detail on the actual application of community policing in the Canadian context. Case studies are provided for individual police services across Canada and a range of different approaches begins to emerge for the reader's enlightenment. Of particular interest here is the growing concentration on "community development" as a component of community policing.

Finally, Chapter 12 looks at the future prospects for community policing in the Canadian context and tries to offer some tentative reflections for its further evolution.

Appendix I offers an outline of a sample public opinion/satisfaction survey that can be adapted for use in many jurisdictions.

Appendix II presents the reader with a format for community safety audits, which have been found to be extremely effective in gathering data on a particular community's self-perception.

ACKNOWLEDGEMENTS

A number of people have provided significant assistance to the author in the preparation of this textbook. Throughout many years working in the field of law enforcement, I have been able to learn a great deal by observing individuals who exemplified the highest standards of policing excellence. As a civilian member within a large police organization, I was always able to view the process of policing from an "outsider's" perspective. Within the context "community policing" I found that the best officers were usually the most amenable to the philosophy of proactive, problem-solving policing. The officers, at all levels, who exemplified the highest levels of integrity in their professional lives were usually well-suited to the transformation required for the "new" ways of doing business.

It is virtually impossible to identify all those who have in some manner contributed to my understanding of community policing in Canada. The fact that there are so many such individuals should underline the importance of this theoretical shift in the philosophy of policing. It is certain that I have not done justice to the depth and degree of insight I have received over the last decade. However, I take full responsibility for the presentation contained herein, and hope that the reader unfamiliar with policing in Canada will better understand the quality and talent that reside within police organizations across this country. As policing in Canada advances toward the 21st century there is much to be optimistic about as enlightened leaders have turned their talents to the challenge of problem-oriented policing.

In the first instance, the author wishes to acknowledge the concentrated time and assistance provided by the ten "Leaders in Community Policing" profiled throughout this textbook. These leaders are:

Lenna Bradburn
Chief of Police
Guelph Police Service

Chris Braiden
(retired) Superintendent
Edmonton Police Service

David Cassels
(retired) Chief of Police
Winnipeg Police Service

Bill Closs
Chief of Police
Kingston Police

W. I. James Harding
Deputy Director
Special Investigations Unit
Former Chief of Police
Halton Regional Police Service

Ross Hastings
Professor
University of Ottawa

Robert Lunney
(retired) Chief of Police
Edmonton Police Service &
Peel Regional Police Service

David P. McKinnon
Chief of Police
Halifax Regional Police Service

Christine Silverberg
Chief of Police
Calgary Police Service

Connie Snow
Inspector
Royal Newfoundland Constabulary

Secondly, the author wishes to thank the following organizations and specific personnel who generously offered their assistance in preparing this textbook. While the author has not come close to capturing the full substance of this collective outpouring of resource material and insight, it is hoped that the reader will come to understand the wealth of knowledge and practical expertise that resides in these organizations:

Barrie Police Service
Jack Delcourt, Chief of Police

Belleville Police Service
David Klenavic, Chief of Police (and Chair, Community Policing Committee, Ontario Association of Chiefs of Police)

Brantford Police Service
Alvin W. Barber, (retired) Chief of Police; R.H. Peeling, Deputy Chief

Calgary Police Service
Dr. Yvonne Ko; Dr. Barbara Juschka; Rick Haddow, Inspector; Jim Matthews, (retired) Deputy Chief; Rick Hanson, Deputy Chief; Ms. Debi Perry; Jack Beaton, Superintendent; Murray Stooke, Staff Sergeant; Bill Sherlock, Inspector; Terry Chowanec, Sergeant; Cal Johnston, Staff Sergeant; Ms. Angie Koning; Peter Davison, Sergeant; Cheryl A. Haas; Al Redford, Staff Sergeant; Brian Whitelaw, Sergeant.

Community Policing Advisory Council of Ontario
J.F. (Jim) Campbell, Chair

Cornwall Police Service
Anthony Repa, Chief of Police; Pierre Pilon, Constable

Edmonton (the City of)
His Honour Bill Smith, Mayor & Chair, Safer Cities Initiatives Advisory Committee; Pijush Sarker, Safer Cities Initiatives Advisory Committee

Edmonton Police Service
John Lindsay, Chief of Police; Keith P.J. Duggan, Superintendent; Bill Mowbray, Superintendent;

Bob Montgomery, Staff Sergeant; John Fairweather, Sergeant; A.G. Davis, Constable

Fredericton Police Force
G.M. Carlisle, Chief of Police; Sheldon L. Geldart, Superintendent

Kingston Police Services Board
Jim Clark, Chair; Ms. Donna Harrington, Secretary to the Board

Halton Regional Police Service
Peter Campbell, (retired) Chief of Police; Michael Kingston, Inspector; Dan Okuloski, Inspector

Hayward (California) Police Department
Paul Wallace, Lieutenant; Fred Hoover, Lieutenant; Charley Heitz, Lieutenant; John Espinoza, Sergeant; Raul Valdivia, Sergeant

Midland Police Service
Paul Hamelin, Chief of Police

Moose Jaw Police Service
T.G. Coleman, Chief of Police

National Crime Prevention Centre (Ottawa) Francine Charlebois

Niagara Regional Police Service
J. Grant Waddell, Chief of Police; Gary E. Nicholls, Deputy Chief; Laurie Leonard, Sergeant

Ontario Association of Chiefs of Police
William M. Malpass, Executive Director

Ontario Ministry of the Solicitor General and Correctional Services
Robert W. Runciman, MPP, Leeds-Grenville, Solicitor General and Minister of Correctional Services; Bob Hunter, Special Assistant, Policing; Amanele Mayshak

Ontario Provincial Police
Thomas B. O'Grady (retired) Commissioner; Diane Nagel, Deputy Commissioner; Nancy Caney, Bureau Commander; Phil Duffield, Superintendent; Gary Jewer, Sergeant

Ottawa-Carleton Regional Police Service
Brian Ford, Chief of Police; Garfield Wyche, Superintendent; Dr. Gail Johnson

Owen Sound Police Service
Thomas Kaye, Chief of Police (and former Chair, Community Policing Committee, Ontario Association of Chiefs of Police)

Peel Regional Police Service
Noel Catney, Chief of Police; Ron Bain, Deputy Chief; Fred Shaw, Inspector; Mr. John MacNeil; Steve Lamourie, Constable; Jamie Porter, Constable

Peterborough Community Police
Kevin McAlpine, (former) Chief of Police; Terrence M. McLaren, Deputy Chief; Ray Vandervelde, Inspector

Regina Police Service
Murray Langgard, (former) Chief of Police; L.M. Toupin, Deputy Chief

Reno (Nevada) Police Department
Richard Kirkland, Chief of Police; Bob Galli, Deputy Chief; Ken Bunker, Lieutenant; Phil Galeoto, Lieutenant

Royal Canadian Mounted Police
James Murray, Commissioner; Dwight McCallum, Superintendent, RCMP Ottawa; Gordon Keating, RCMP "A" Division; L.G. Warren, Chief Superintendent, Commanding Officer, "B" Division (St. John's, Nfld.); G.B. Button, Inspector, RCMP, "B"

Division (St. John's, Nfld.); Odillon Emond, Assistant Commissioner, Commanding Officer, "C" Division (Westmount, P.Q.); Staff Sergeant Roger Daigle, RCMP, "C" Division (Westmount, P.Q.); John Moodie, Chief Superintendent, Commanding Officer, RCMP "D" Division (Manitoba); Dick Grattan, Staff Sergeant, RCMP "D" Division (Manitoba); Brian Roberts, Inspector, RCMP, "F" Division (Regina); Rick Wryschreschuk, Sergeant, RCMP , "F" Division (Regina); R.A. Grimmer, Chief Superintendent, Commanding Officer, "G" Division ; Bill Sweeney, Superintendent,"G" Division; Tom Steggles, Sergeant, "G" Division; G.J. Loeppky, Chief Superintendent, Commanding Officer, "J" Division (Fredericton, N.B.); Pat Thibodeau, Constable, RCMP, "J" Division (Fredericton, N.B.); J.P. Dufresnes, Constable RCMP, "J" Division (Fredericton, N.B.); D.N. McDermid, Assistant Commissioner, Commanding Officer, "K" Division (Edmonton, Alta.); Wayne Carroll, Sergeant, "K" Division Alberta; Geoff Elliott, Sergeant, "K" Division Alberta; Al Fraser, Constable, "K" Division Alberta; Jan Lipscombe, Constable, "K" Division Alberta; Dwight Mulally, Sergeant, RCMP, "L" Division (Prince Edward Island); Rick Hamilton, Corporal, RCMP, "O" Division (London); S.A. Duncan, Chief Superintendent, RCMP (Halifax); Dr. Frum Himelfarb

Royal Newfoundland Constabulary
Leonard Power, Chief of Police; Gerry Browne, Deputy Chief

Seattle (Washington) Police Department
Dan W. Fleissner

Six Nations Police
G.M. Lickers, Chief of Police

Sterling Police Service
Frank A. Wallwork, Chief of Police

Sudbury Regional Police
Alex V. McCauley, Chief of Police;
Brian Jarrett, Inspector

Sûreté du Québec
Jacques Labrecque, Inspecteur-chef;
Danielle Corneiller

Sussex Police Force
Mason Johnston, Chief of Police;
Paul Northrup, Constable

Thunder Bay Police
Leo Toneguzzi, Chief of Police;
Peter B. Worrell, Inspector

Timmins Police Service
Denis A. Lavoie, Chief of Police;
Greg Piccolo, Sergeant

Toronto Police Service
David Boothby, Chief of Police;
Chuck Konkel, Staff Sergeant

Trenton Police
Greg MacLellan, Staff Sergeant

University of Reno (Nevada)
Kenneth J. Peak, Professor

Vancouver (City of)
Judy Rogers, Deputy City Manager;
Nancy Cheung, Coordinator, Neighbourhood Integrated Service Teams

Windsor Police Service
John Kousik, Chief of Police;
Jerry Pocock, Inspector

York Regional Police Service
Peter Scott, Acting Chief of Police;
Eric Jolliffe, Inspector

The author would also like to acknowledge the considerable insight provided by Susan Eng, who at the time of writing was Lecturer, Osgoode Hall Law School, York University, and previously, Chair of the Metropolitan Toronto Police Services Board. Her observations on police governance within the context of community policing were singularly enlightening and provocative.

Finally, it is important to thank the following individuals from Prentice Hall Canada for not only making this project a reality, but through their professional expertise, ensuring that it was as much of a pleasure as possible: David Stover, Stefany Dal Mina, Susan Ratkaj, Matthew Christian, and Jim Zimmerman.

The editors would like to thank the following reviewers for their helpful comments on a draft of this manuscript: David MacAlister of Kwantlen University College, Gina Antonacci of Humber College, Robert C. Hays of Conestoga College, Ken Smith of Confederation College, and Len Griffiths of Ontario Police College.

Statistics Canada Information is used with the permission of the Minister of Industry, as minister responsible for Statistics Canada. Information on the availability of the wide range of data from Statistics Canada can be obtained from Statistics Canada's Regional Offices, its World Wide Web site at http://www.statcan.ca, and its toll-free access number 1-800-263-1136.

CHAPTER ONE

> For to root out Fraud, prevent Violence and Oppression, and to preserve Peace and good Order, are the most grateful Pursuits of a good Heart and an ingenious Mind.
>
> John Fielding (1758)

TO BEGIN AT THE BEGINNING

LEARNING OBJECTIVES

1. Identify several definitions of community policing.
2. Describe key assumptions frequently associated with community policing.
3. Describe the main elements of a conceptual framework for community policing.
4. Identify the origins of community policing in traditional police practice.
5. Describe key functions that demonstrate the importance of the office of constable in early policing.
6. Identify the two distinct models of policing developed by Sir Robert Peel.
7. State key components of the American context of policing.
8. List elements of the Canadian context of policing that impact on community policing.

INTRODUCTION

The origins of community policing, however defined, present the first real challenge of this work. Nothing clearly marks the point in time when something called "community policing" came to be applied in a consistent manner. Unlike the introduction of the police cruiser, or bulletproof vest, or DNA testing, which have distinct chronologies, community policing did not burst fully formed onto the scene. Instead, it grew slowly out of several influences, both traditional and modern, which can be examined and understood over a long period of time.

This chapter will attempt to consider a few of the many definitions that have been offered for community policing. It will endeavour to place those, often competing, definitions into a conceptual framework that allows for greater flexibility in understanding this idea. This chapter will look at the origins of community policing in what can be termed traditional police practice. Also, this chapter will look closely at the role of the early constable in order to see how this function incorporated much of what is now being tailored to suit the purposes of community policing. The significant contributions of Sir Robert Peel will be looked at, particularly his role in the formulation of two distinctive, yet divergent, models of policing. Early American influences in the area of community policing will be considered briefly. Finally, the context in which Canadian policing takes place will be summarized in order to better understand the development of community policing in this country and the elements which have fashioned its growth.

WRESTLING WITH DEFINITIONS

Few things are more difficult than the creation of a perfect definition of "community policing." Once someone claims to have arrived at a comprehensive rendering of this term, there will be several others with points of argument regarding this definition. In order to arrive at a better understanding of this term, this chapter will offer several competing definitions from a variety of sources, including academic, analytical, operational, and official. However, this textbook will not give priority to any single definition of community policing. Instead, a simple, conceptual framework will be offered and explained that should be adaptable to a broad range of definitions, while still allowing for considerable flexibility in the precise application of this notion.

It is common in works that have treated the topic of community policing to begin with an attempt at defining the key terms. This effort is commendable in many respects. However, in the final analysis, it is prone to failure because it seeks to impose a degree of precision which cannot be found in something as complex and changeable as policing in modern society. The effort to provide a definition is useful in so far as it puts forward some of the key notions that have been held with regard to community policing. It is also useful because it begins to tell us what community policing is not by explicitly excluding certain elements. In approaching this topic from the perspective of a conceptual framework, it is hoped that there will be enough detail to allow for a relatively clear understanding of the parameters of community policing without constraining its meaning on a Procrustean bed that discards whatever does not fit our overall vision.

What follows is a brief sampling of definitions for community policing that have been offered in earlier efforts. These definitions come from several sources, including individual commentators and academics who have studied this topic, as well as from police organizations that have attempted to outline their own corporate understanding of community policing. It is remarkable that there is considerable variation in the precise terms used to define this concept. However, there are several shared elements that observers have concentrated on in order to penetrate the essential aspects of this idea. Again, in this textbook, there will be no attempt to settle upon a single, preferred definition. Rather, this chapter will provide a quick survey of various perspectives that have been brought to bear on the subject, and then will suggest a conceptual framework that may be useful to the student in making inquiries into the matter. This conceptual framework should also help guide further thinking and research on the topic. While the effort to define is critical to the process of understanding, there is considerable evidence that community policing, by its very nature, is open to substantial variation and adaptability in its scope and application. The very best thinking in this area of police endeavour has recognized this point and has led to the formulation of distinctive meanings for this concept. Commentators often explicitly defend their reluctance to pigeonhole their understanding within a fixed definitive statement about the absolute elements of community policing. As with many discussions that seek to arrive at a shared definition of a concept or idea, the process can become lost in overly scholastic debate that is counterproductive. By touching upon attempts at definition that have been offered by many observers and practitioners of policing, students will become aware of the vast landscape and can view the scenery from their own perspective.

In Chapter 10, the topic of definitions will be revisited to pose some additional questions about the nature and content of the notion of community policing, and to offer some further thoughts about the adequacy of the conceptual framework presented in this textbook. It is not unusual for commentators in this area of police innovation to actually avoid using the phrase "community policing" because overuse has bred a degree of contempt that is not surprising when a term is not only misused in many applications, but persistently and relentlessly overworked (Strauss, 1953). However, when considering the question of success in community policing, and the need for quantitative and qualitative measurements, we must concern ourselves with reasonably precise outcomes and deliverables that will require some clear understanding.

Some have suggested that defining "community policing" is an exercise in the obvious. For example, Chris Braiden has noted that for a police department to say that it is doing community policing is like someone saying he is fishing for fish. However, because so much intense effort and intellect has been invested in the refinement of the notion of community policing, it may be fair to pause in order to see the substance of this effort. Friedmann (1992) has suggested that the topic of community policing can be viewed from three perspectives: the police perspective; the community perspective; and a combined police/community perspective. Each of these perspectives will result in a slightly different understanding of the topic of community policing. In fact, it may be possible to further complicate this picture by including consideration of additional perspectives that may have their own influence on the substance of community policing. If those perspectives that govern both the police and the community may be seen as a distinctive categories, it is possible also to include a perspective that derives from the legislators or decision-makers who guide our society and make determinations about laws and regulations. In any event, these different perspectives set community policing into a series of varying contexts, some of which are open to potential conflict. For example, the police perspective views community policing as a means for increasing their reliance on community resources in order to help them in the control and prevention of crime. The police would like to activate greater levels of public interaction with regard to the generation of intelligence that will assist in the prevention and solving of crime. Furthermore, community policing is seen as a means of increasing their legitimacy within communities. On the other hand, the community perspective focuses on the potential for community policing to improve police services, allow for greater accountability, and provide an opportunity for community members to share in the decision-making processes which relate to public safety and order maintenance.

When these two perspectives are blended it is possible that a better understanding of the nature of crime will be forthcoming. Primarily, this shared perspective should reveal that crime derives from fundamental societal problems that police cannot completely control. Also, the strictly reactive approach to policing, whereby officers respond rapidly to calls for service, is not going to be an effective means of addressing larger public safety issues in society. On the contrary, placing individual officers in closer contact with defined communities in a more proactive framework will contribute to an approach to policing that deals more broadly with "quality of life" considerations. Additionally, concern for human rights and civil liberties consistent with the concerns found in a liberal democracy is also part of this combined police/community perspective (Friedmann, 1992, pp. 3–4).

Friedmann (1992), following a detailed consideration of many sources, offers the following definition:

> Community policing is a policy and a strategy aimed at achieving more effective and efficient crime control, reduced fear of crime, improved quality of life, improved police services and police legitimacy, through a proactive reliance on community resources that seeks to change crime-causing conditions. It assumes a need for greater accountability of police, greater public share in decision-making and greater concern for civil rights and liberties. (p. 4)

The valuable work of Trojanowicz and Bucqueroux (1990) includes this statement on the nature of community policing as an instance in police reform:

> Community Policing, the first major reform in a half-century, changes the way the police think and act. This revolutionary movement broadens the police mandate beyond a narrow focus on fighting crime, to include efforts that also address fear of crime, social and physical disorder, and neighborhood decay. (p. ix)

Trojanowicz and Bucqueroux proceed to develop ten principles of community policing that are extremely useful in comprehending the magnitude of change that would be required for a police department to properly devote itself to its introduction. Generally, these ten principles speak to a complete commitment on the part of the entire police organization to changing its mental model of policing. It means that operational, personal, procedural, and structural changes will be required to effectively and meaningfully introduce the concepts of community policing in ways that will impact on the quality of service provided within the community. From a different angle, Trojanowicz and Bucqueroux (1990) tell us what community policing is not (pp. 20-35). Frequently, it is a struggle for police departments to make sincere and genuine change in the direction of true community policing. This is particularly true when there are observers and critics of the police who are not prepared to allow the latitude and the willing suspension of disbelief needed to introduce a complex philosophical approach to modern policing that requires time, energy, resources, and commitment (from the police, the communities, and the legislators and decision-makers) in order to function properly.

Greene and Mastrofski (1985), for their part, appropriately highlight the fluid nature of the notion of community policing:

> It should be emphasized that community policing means many things to many people. It is at once an ideology, an organizing framework for many police activities, and a set of individual programs. Community policing can be seen as a convergence of several police reform efforts of the past. (p. xiii)

The Crime Prevention Center of the California Department of Justice (1992) offers the following definition:

> Community policing is a philosophy, management style, and organizational strategy that promotes pro-active problem-solving and police-community partnerships to address the causes of crime and fear as well as other community issues. (p. 3)

From a Canadian perspective, the Ontario Ministry of the Solicitor General (1991) provided the following observation:

> Community policing is a multi-dimensional model from which changes in policing strategy, organizational structure and organizational culture can occur. (p. 5)

Seagrave (1997, p. 207) offers a good summary of the broad range of definitions and interpretations of community policing that have been offered in the literature under the following categories: rhetorical; philosophy; program; control; and imprecise. It is clear that many perspectives have been developed with varying degrees of certainty as to the capacity of community policing to address the challenges of modern policing. It is also certain that some of the approaches to the theory and practice of community policing are less rigorous than others, particularly those that present it as simply a program within the department, rather than an overall guiding principle of practice throughout the organization.

In 1993, the Ontario Provincial Police (1993) expressed their policing philosophy as follows:

> Community policing is the fundamental principle under which all our policing services will be delivered. To that end we shall:
>
> - Involve the people of the community in the identification of crime, traffic and social order problems and solutions;
> - Provide policing services that are consistent with the identified concerns, expectations and needs of the community;
> - Participate with other concerned agencies and interest groups to effectively address police and community concerns;
> - Promote public commitment and active involvement in the resolution of crime, traffic and social order problems;
> - Create an open and responsive working environment which encourages commitment and creativity by all members in the design and application of community policing services; and

- Increase our effectiveness through the application of innovative reactive and proactive policing services. (p. 2)

When dealing in the realm of definitions, it is frequently important to begin with a discussion of the operating assumptions that guide the process. Ongoing work funded by the Community Policing Consortium (U.S.) has resulted in a research paper (Oettmeier and Wycoff, 1997) that sets out several assumptions associated with the concept of community policing (Figure 1). An understanding of these common assumptions will provide a useful filter for assessing individual definitions offered for community policing. It will also prove effective in considering the conceptual models and frameworks outlined below.

FIGURE 1

ASSUMPTIONS ASSOCIATED WITH COMMUNITY POLICING

- Continued reliance on random, preventive patrol should be minimized. Random, preventive patrol should only be used as a strategy when police visibility is an issue.
- Citizens will accept a range of response times for different types of calls.
- Differential police response strategies should be implemented to improve the effective management of the dispatch functions.
- Effective management of the patrol function is dependent on intelligent management of the dispatch function.
- Effective management of criminal investigations is indirectly dependent on intelligent management of the dispatch function and directly related to management of the patrol function.
- Case management systems must be developed and implemented to fit the needs of various investigative functions.
- Work demands and resource allocation studies are necessary to ensure equitable deployment of personnel.
- The development of crime and operational analysis procedures is vital in managing patrol and investigative functions.
- The use of directed and self-directed patrol activities for officers should increase when and where appropriate.
- Officers assigned to the patrol function must be actively involved in criminal investigations (e.g., conducting follow-up investigations, recommending early case closures).
- Patrol officers need enhanced status and enriched job responsibilities.
- Attention must be devoted to reassessing the purpose and function of existing beat configurations.
- Initiatives must be undertaken by the police to identify citizen service expectations and to work with citizens in addressing and resolving neighborhood crime and disorder problems.
- To facilitate the development of stronger ties with the community, policies requiring frequent rotation of officers across shifts must be seriously considered.

- Regular public forums should be established so frequent exchanges of information can occur between the police and public, preferably between patrol officers and neighborhood residents.
- Performance measurement systems should serve as a management tool that guides personnel development and facilitates organizational change.
- More meaningful performance evaluation criteria must be developed to reflect the change in officer roles and responsibilities.
- Training curricula must be redesigned so they are more relevant and supportive of patrol and investigative operations.
- Disciplinary processes must become part of a behavioral system that incorporates education, training and counseling as strategies designed to assist officers experiencing behavioral problems.
- Management styles must be more adaptive to varying situations and personalities.
- Managers must begin directing their attention toward the qualitative aspects of service delivery processes and outcomes.
- Police agencies must cultivate leaders who are comfortable and effective working in an environment characterized by constant demands for change.

Source: Oettmeier, Timothy N. and Mary Ann Wycoff (1997). *Personnel performance evaluations in the community policing context.* Washington, D.C.: Police Executive Research Forum. p. 29. Used with permission.

CONCEPTUAL FRAMEWORKS

The Ontario Association of Chiefs of Police, in collaboration with the Ontario Ministry of the Solicitor General and Correctional Services, has developed a framework for community policing that provides a good basis for the version that has been developed for presentation in this textbook:

A Framework for Community Policing

(Source: Ontario Association of Chiefs of Police & Ontario Ministry of the Solicitor General and Correctional Services)

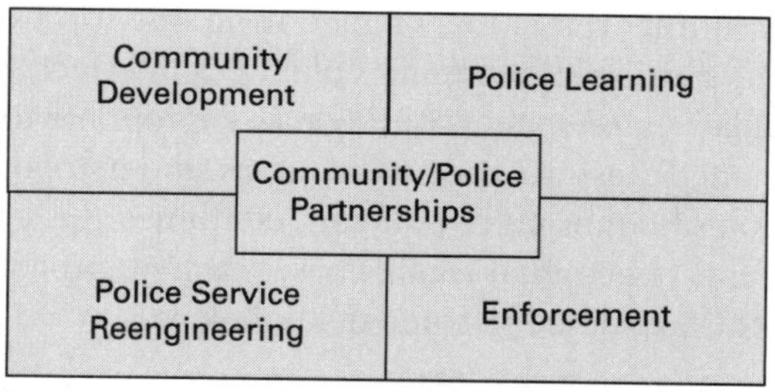

While the framework above presented important elements that can be considered in the creation of an overall organizational approach to community policing, this textbook

A Conceptual Framework for Community Policing:
Transformative Policing

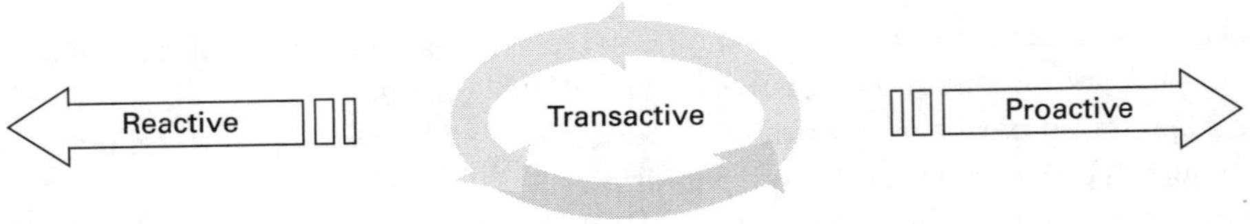

- Emergency Response
- Public Order Maintenance
- Rapid Response
- Centralized Services
- Management
- Command & Control
- Hierarchy
- Training
- Discipline, Public Complaints & Inquiries

- Law Enforcement
- Assistance to Victims of Crime
- Problem-oriented
- Organizational Learning
- Partnerships
- Accountability
- Cooperation
- Empowerment
- "Quality of Life"
- Multi-agency Cooperation
- Continuous Improvement
- Coproduction of Order
- Assistance to Victims of Crime

- Crime Prevention
- Community Consultation & Surveys
- Community Catalyst
- Education & Development
- Decentralized Services
- Leadership
- Civilianization
- Organizational Change & Reengineering
- Performance Measurement & Evaluation
- Team Policing
- Innovation & Benchmarking
- Tiered or Differential Response
- Research

offers the following adaptation of several sources of information in order to provide a strategically-oriented conceptual framework that blends several perspectives:

By referring to this model as a "framework," it is intended that much of the content will be suggestive, rather than exhaustive. In order to distinguish this model from other versions that are tied to a specific definition of "community policing," the adjective "transformative" is applied. In conventional usage, the word "transform" means to change the form or outward appearance of something. It also may imply some change in character or personality, and this is highly appropriate within the context of this textbook's understanding of the need to alter not only individual, but also the organizational character of police departments that are planning to pursue some version of community policing. In physics, the word "transform" refers to a change from one form of energy to another. Because the fundamental aim of community policing is in effect, to transform the philosophy and practice of law enforcement, it is seen as appropriate to offer this alternative label for such innovation. Transformative policing, as understood in this textbook as a refinement or extension of community policing, combines three key elements which constitute the core of this conceptual framework: reactive, proactive, and transactive. Each of these elements is explained in some detail below.

Reactive Elements

There is considerable familiarity with the reactive aspects of policing. This, indeed, was the main focus of police efforts in earlier times, particularly in the context of the professional model of policing in which police administrators were applying available management techniques and technological tools to allow for rapid response to calls for service; e.g., the "911" culture. With police officers assigned to automobiles, randomly patrolling their zones, dispatch units could direct mobile units to crime scenes with considerable speed. Criminal investigators would also react to serious crimes, e.g., through the use of specialized "hold-up" squads, in order to solve them through professional skills, often augmented by the latest developments in forensic sciences.

What is important about this "Transformative Policing" framework is the fact that reactive police services remain important and viable for the delivery of modern police services. In the context of virtually every modern police organization's mandate there remains a significant expectation that the service will include a capacity for rapid emergency response that will deal with a variety of critical situations. One example of this capacity is the provision of police tactical, or emergency response, units. These units are designed to deal with highly volatile emergency situations that often include a hostage-taking, or barricaded person. Such circumstances clearly require an immediate police response with highly trained, specially-equipped officers who have been drilled in such situations. Also within this category would be rapid police response to bomb threats, or the handling of other kinds of explosive devices. While it has been argued that some of these functions may be dealt with through trained military personnel, it remains a convention of urban policing in North America and other jurisdictions, that police organizations should be funded and equipped to deal with such contingencies. Indeed, in democratic societies, there is a profound reluctance to invoke military force when there is an option of using a police service which functions on behalf of the public interest under a civilian governing authority.

As the model above indicates, the reactive element also includes consideration of what is known as a command and control capacity. This refers to the notion that strict, hierarchical discipline continues to hold considerable operational value within a police department. In the reactive mode, there is often little time or opportunity for problem-solving approaches that place a high value on analysis and discussion. It is not difficult to understand the requirement of circumstances in which decisive orders must be able to be given quickly and followed without further deliberation or discussion. The very nature of emergency situations advances this view and requires little argument. Police organizations that have responsibility for certain emergency response functions will always need to retain this capacity, regardless of their innovative approaches to "Transformative Policing."

Proactive Elements

The concept of proactive law enforcement has received considerable attention in the literature on community policing. It grows from the increasing dissatisfaction with an exclusively reactive approach to the problems posed by modern policing. When police organizations were able to justify ever-increasing budgets and were successful in recruiting ever-increasing numbers of officers, it seemed as though the reactive approach would satisfy the public safety needs of the community. However, growing empirical evidence began to show that this type of policing was not only inefficient, it was profoundly ineffective in dealing with crime and disorder. As part of the movement to an innovative approach to policing, the proactive elements are being highlighted. They demonstrate that police organizations can work to anticipate public safety problems in a variety of ways and can plan appropriate responses in a more strategic fashion. The proactive approach also implies that police organizations can undertake internal activities that will make them more efficient and effective in the overall delivery of their services to the public. It should be noted that there are elements of this dimension which are not wholly new to the policing enterprise. For example, crime prevention has been an aspect of policing from the time of the early English constable, and will be discussed in more detail in Chapter 2.

Significantly, the proactive component has focused on what may be termed police-originated initiatives. That is to say, in many progressive police organizations, steps have been taken to improve or alter the department's service to the public without the explicit endorsement of the community. Such improvements may well be the result of detailed and extensive community surveys and questionnaires. However, the specific content of the resulting change derives exclusively from the police department's expertise and its own internally-driven strategies and policies. For example, this can be seen in many North American police departments in which some form of organizational review has been conducted. These organizational initiatives have resulted in the transformation of the entire structure of the department along with many of the service's personnel, procedures, and programs. While such initiatives may be consistent with certain aspects of community policing theory, they do not fundamentally alter the nature of the police/public equation. In spite of the efforts at consultation, the police are the professionals who accept the information, insight, and evidence supplied by members of the pubic and apply their own, internal expertise in order to formulate appropriate responses to that input.

"Transactive" Elements

What is beyond the reactive and proactive dimensions of policing is something that is being termed "transactive." This element is a synthesis, or blending, of these two approaches and provides the central piece of the "Transformative Policing" framework.

Again, when using conventional language, to "transact" is to settle something, often in the context of business. For example, to make a bank transaction is to complete a series of operations that relate to the handling of money, or other negotiable instruments. A transaction implies the interaction of at least two parties in some formal arrangement that is, typically, of some mutual benefit. In the literature on community policing, this approach is frequently referred to as being "co-active" because it envisions the police and the public working in tandem to advance the purposes of public safety. In the framework presented in this textbook, the term "transactive" is being advanced because it allows for broader application. Because, as will be discussed further in Chapter 10, there is rarely one single "community" that the police are in partnership with, it is suggested that this term conveys a wider meaning that anticipates more than simply two "actors." The use of the prefix, "trans-" also helps convey the meaning of "across" or "beyond" which is helpful in attempting to grasp the special nature of what is being accomplished in the most innovative approaches to community policing. In the diagram above, the graphic presentation of the "transactive" component is a revolving circle that is meant to project the image of something that is infinitely perfectible, continuously developing, and capable of drawing into its orbit both the reactive and proactive components. The "transactive" elements are the centrepiece of this framework of "transformative policing" and provide both the ideal of innovative policing, and a focus for the assessment of the other components. While the questions relating to evaluation, measurement, and assessment are dealt with in more detail in Chapter 10, it should be at this level that the most compelling arguments are made for changes to police theory and practice.

The "Transformative Policing" model presented in this textbook will be revisited in Chapter 12. There we will reconsider its application to the challenges faced by community policing and approaches that utilize a problem-solving orientation.

THE SEEDS OF COMMUNITY POLICING IN TRADITIONAL PRACTICE

Much has been written about the radical departure which community policing represents in the context of police practice and how it has come to dominate debate about the future direction of policing (Wilson and Kelling, 1982; Bayley, 1985; Trojanowicz and Bucqueroux, 1990; Chacko and Nancoo, 1993; and Walsh, 1995). It is certainly true that the philosophy and specific applications of community policing have undergone significant advancement in the field of law enforcement. However, it is perhaps more accurate to view its development as a substantial evolution in the process of policing. This evolution draws heavily on certain elements of traditional practice (Bayley, 1985, p.255). By viewing community policing, in all its splendour, as an extension, refinement, or reformulation of positive elements found in earlier modes of police practice, it is possible to better understand the important contributions made by previous generations of police innovators, and to more deeply comprehend the professional commitment that has always resided within the field of policing in Canada and other jurisdictions.

The movement away from the problematic elements of the "political" and "professional" phases of policing characterizes one of the positive contributions of community policing. There was a growing recognition that the professional model (pioneered by law enforcement officials, such as August Vollmer, discussed in more detail below) was no longer succeeding in advancing the enterprise of policing; crime rates were increasing, fear of crime was on the rise, the costs for the delivery of policing services were escalating and, therefore, some new approach had to be developed that would address these, and other challenges facing policy makers at all levels, with regard to public safety and security.

By searching for the seeds of community policing in the context of traditional practice, it is also possible to better understand the character of current experiments being introduced in many police jurisdictions. Innovation does not normally spring fully formed out of nothing, particularly in the world of modern policing that is tied so tightly to its military and hierarchical origins. It is, therefore, helpful to understand the substance of the changes coming about in policing, under the banner of "community policing" in order to see what they are seeking to accomplish and the magnitude of the transformations required. By drawing a somewhat straight line from the theory and practice of policing found in the professional era to the present time, where innovations and experiments that value the elements that were missing from that earlier mode of service delivery, it is possible to see a clear progression. This progression rests on previous good practice and amounts, in many ways, to a synthesis of various approaches applied over time in several jurisdictions. It is critical for a sound understanding of the impact of community policing to view the past as being flawed but not futile in its approach to the problems of crime and public safety.

In their enthusiasm to preach a new gospel according to community policing, many participants in the process have fallen prey to the temptation to discount earlier styles of policing and to offer up their own approach as the only route to effective policing. While major change in policing requires committed leaders who can articulate a clear vision of the future, it is inaccurate and inappropriate to discard the past as being entirely characterized by folly or failure. By looking closely at the historical role of the constable, it will be seen that much of the substance anticipated by those who support community policing was imbedded in the earliest activities of this official. It is fascinating to see the shape of things to come by looking back into the record of the English constable, and some of the fundamentally proactive aspects of this person's duties and responsibilities.

THE ROLE OF THE EARLY ENGLISH CONSTABLE

At the core of policing in Canada, and in other jurisdictions which have inherited British practices in matters of law and law enforcement, is the office of constable. As with other aspects of Canadian policing, the constable represents a basic and persistent link with our English heritage. Without going too deeply into the origins of this office, this section will attempt to point out key elements that make the early constable's role critical to the evolution and growth of community policing. Without the strong roots established by this early officer of the law, without the fundamental focus of this public servant on preventative and community-centred elements of policing, community policing would have been much more difficult to introduce when its historical moment reappeared. These roots also provide some justification for those who argue that they have been engaged in forms of "community policing" all along (Bayley, 1985). This section will reference some early, original documents dealing with the office of constable, and will attempt to sketch a portrait of this official as a genuine precursor to the modern community policing officer.

In Halsbury's Laws of England (3rd ed.), the following treatment of the history of police provides a useful reference:

> The history of the police is the history of the office of the constable and, notwithstanding that present day police forces are the creation of statute and that the police have numerous statutory powers and duties, in essence a police force is neither more nor less than the number of

> individual constables, whose status derives from the common law, organized together in the interests of efficiency.
>
> From the earliest times peace officers under varying titles have been responsible for safeguarding the internal peace of the kingdom and of these the most important for the purpose of analysing the position of the present day constable is the petty constable, or parish constable as he later became styled with the growth of the civil functions of the parish. By the beginning of the seventeenth century it could be said that the parish constable was responsible first, by virtue of his office, for the preservation of the peace within his bailiwick . . . and, secondly, for the execution of the orders and warrants of the justices of the peace. (p. 43)

The historical record is fairly clear in presenting the importance of the constable to life in early England. From the middle ages until the eighteenth century, the constable was an important figure in the life of the community and functioned in many ways as a leader within the village. Price (1971) suggests that the earliest reference to the petty, or village, constable appears in 1252, when townships were required by writ to appoint such an official. However, even in this instance, the requirement appears to merely formalize an already existing practice. The origin of the word "constable" has been a matter of some discussion, although it seems relatively certain that it derives from a combination of the old Saxon words for "king" and "stability," which indicates that it referred to someone who was a key servant directly responsible to the King (Meriton, 1674).

The petty constable during the late sixteenth and early seventeenth centuries in England was a person of considerable importance (Kent, 1986). The constable was a central official within the village or the parish, in spite of the fact that, with a few exceptions, historians of this period have paid that function little sustained attention (Trotter, 1919; Simpson, 1924). The origins of the constable's position and powers are obscured by antiquity. However, most indications would point to the fact that they were indeed chiefs within their communities and were key representatives of those communities to the outside world. Often, these individuals were elected representatives of their communities and became the agents of the community in dealings with the central government. Over time these constables took on responsibilities for the administration of the central government's rules and regulations.

Some have noted that the term "constable" was military in origin and was applied to local officials who were the headmen of the village and, therefore, persons of significant authority. However, their role was never that of a strictly local military authority. The thrust of their responsibilities was always based in the community and derived from community support (Guth, 1994). Among the responsibilities that have been ascribed to these constables were the following:

- Assembling local militia for an inspection of arms;
- Peacekeeping;
- Enforcing provisions against affrays;
- Apprehending thieves and turning them over to sheriffs or gaolers;
- Apprehending and delivering suspicious persons to the sheriff;
- Ensuring the application of standard weights and measures;

- Enforcing statutory provisions relating to servants and labourers;
- Performing duties relating to local taxation; and
- Executing commands of higher officials, e.g., the sheriff, coroner, or justices of the peace.

Even when constables took on greater responsibilities as royal officials, they did not completely relinquish their role as local representatives (Kent, 1986). The source of their authority was two-fold: first and foremost their powers flowed from their original role as leaders within their communities. Some of these powers were recognized in law during the time of the Middle Ages; additional duties were conferred by the state, sometimes explicitly through statute, but also indirectly as a result of the constable's role as an agent of ministers of the Crown. The selection of constables was governed by local custom, which continued to be the practice even after they became royal officials. The constable therefore remained directly accountable to the local community, as well as to the higher authority of the state.

Frequently referred to as "conservators of the peace," these early constables played a particularly important part in the prevention of crime and disorder. It was always intended that their responsibilities include preventative elements along with those that dealt with the suppression and apprehension of offenders. Indeed, there was considerable discretion accorded to these constables in the execution of their duties, particularly when dealing with the more serious felonies and crimes such as murder, assault, or theft (Guth, 1994). Constables were further responsible for setting the watch and ward in order to provide for vigilance throughout the entire day. The watch period was from sunset until sunrise, the ward period being sunrise until sunset. Constables would ensure that villagers were assigned to be alert to suspicious persons, to set up the hue and cry if an offence were committed, and to seize the goods of suspected felons, and hold them for the proper authorities. Village, or petty, constables typically had responsibility for maintaining the local equipment used in the application of justice, such as the stocks, the cage, the pillory, the whipping-post, and the ducking-stool.

It has also been established (Kent, 1986) that constables held certain administrative responsibilities in connection with the repair of local highways and the relief of the poor. They were further expected to assist with the difficult task of "pressing" able-bodied men into service for military purposes. Another duty that fell to these constables, consistent with their leadership role in the community, was related to taxation. Constables were known to lead delegations on behalf of a township to make representations on the community's tax burden. They were also empowered to commit the community to a set level of taxation. The extent of the constables' responsibilities in some jurisdictions highlights how central they were to the overall well-being of the community. Kent (1986) notes that in several villages in Leicestershire, during the early seventeenth century, constables were largely responsible for some of the following duties that benefitted community life:

- Custody of the communal bull;
- Maintenance of village gates, weirs, sheep dams, pinfolds, and hedges;

- Fetching "setts" for planting;
- Trenching and sowing of fields;
- Thatching the herdsman's house;
- Guarding crops from village animals; and
- Eradication of vermin, including foxes and moles.

The historical record is quite clear that the duties of the constable were of significant importance even during the late sixteenth and early seventeenth centuries. Even though there was a growing importance assigned to the newly formed parish officials, such as churchwardens, overseers, and surveyors, the constable is characterized as the "chief link between the central government and local communities" (Kent, 1986, pp. 53–54). This linkage is important to a sound understanding of the development of community in modern times, because it clearly draws upon a tradition of deep engagement of this officer of the law in the immediate life of the community. It also highlights that the constable's source of legitimacy was derived from community. The constable was decidedly not someone who operated on the periphery of village life, he was an official who held a substantial leadership role within the community and carried out a diverse range of responsibilities that contributed to the safety, security, and well-being of the village, town, or parish. As Kent (1986) points out:

> The constableship was a demanding and difficult position due not only to the many duties which it entailed, but also to the conflicting pressures which constables experienced as a result of the very nature of the office, and of their mediating role in linking the village and the state. On many occasions constables were successful in fulfilling the demands that were placed on them by the state, and they often gained the cooperation of other villagers in doing so.(p. 279)

Constables were frequently expected to undertake the following additional tasks:

- Goods and services required by the royal household when the monarch was travelling;
- Requisitions for the saltpeter [i.e., potassium nitrate] men;
- Demands of postmasters for coal, oats, and particularly for the provision of horses;
- Ship-money and military levies; and
- Collecting payments for the victims of plague.

It is possible to discover in the historical record of the English constable an official who had the potential to be highly effective in linking local communities with the state. With the population growth and industrialization that advanced during the eighteenth century, the nature of villages began to change in ways that profoundly altered the nature of community life which the constable was expected to secure. The internally focused nature of the English village with its close ties between villagers was being transformed. The extent of societal change is seen in the limited knowledge held today regarding the nature and role of the early English constable. The dominant image of the London "Bobby," the creature of Sir Robert Peel's nineteenth century innovations, has overshadowed in the popular mind the embodiment of community policing found in the parish constable (Guth, 1994). Furthermore, the fact that these officials derived their authority from ancient local customs and the common law provides the foundation upon which much of the discretion and independence of the modern police officer rests. As noted by Price (1971):

> To understand the nature of the constable's work . . . it must be remembered that, however much he came to be used to serve the purposes of the Crown and its agents, his authority had its origins in ancient local custom, not in statute law: he held it primarily as representing the local community by which to the end he continued to be chosen. (p. xi)

The remarkable role played by the early constable in balancing and mediating the interests of the local community and the central government prefigures much of what is being undertaken by the community policing officer in modern police organizations. It is a common practice in certain approaches to community policing to cast the officer in the role of community "catalyst" in the process of community development. The leadership anticipated from newly empowered police officers is not unlike that demonstrated by petty, or parish, constables from their earliest appearance. What is also useful to emphasize is the strong recognition of the need for community support for the proper functioning of the constable. Police require a mandate from those who are being served in order to operate effectively. The public legitimates the office of constable and provides the basis for their ability to preserve the public peace. This point was made in the work of John Fielding (1758), which builds on the efforts of Henry Fielding, in which we find the following statement:

> . . . it appears that the Public has the greatest share in the Execution of this Police; for without its hearty Concurrence and constant Assistance, little Progress could be made towards the Peace and good Order of this Metropolis: For what Good can the most active Magistrate do, unless he receives Informations? (p. 34)

Not only does this quotation advance the important notion of the "co-production" of public order, whereby the police and the community are linked in the prevention of crime and the maintenance of peace, it also anticipates the insight gained from much recent police research that indicates that crime is more frequently solved on the basis of information provided by the public than through the fruits of police investigations.

Much of the scholarship in this area has shed a fairly poor light on the incumbents of this office. The figure cut by Shakespeare's famous constable, Dogberry, in *Much Ado About Nothing*, presents an image of ignorance and incompetence. This picture is being reversed to some extent (Price, 1971). There is growing evidence to suggest that the quality of those appointed to the position of constable was of great concern. While many of the constables sworn to serve in the late sixteenth and early seventeenth century were from the lower orders, it is certain that many villages were served by constables who were drawn from those who held property and position within their communities (Kent, 1986). The historical record leaves room to reason that the office of constable was often occupied by persons of quality, capacity, and competence. Even contemporary accounts from the seventeenth century place considerable stock in selection criteria that would stand modern police departments in good stead (Meriton, 1674).

> And now as to the choyce of these Constables, you shall understand that the Law requireth that every Constable be 'Idoneus homo', that is, apt and fit for the execution of the said Office, and he is said in Law to be 'Idoneus', who is indued with these three thing[s], honesty, knowledg[e], and ability. (p. 5)

SIR ROBERT PEEL AND TWO DIVERGENT MODELS OF POLICING

Many contemporary writers on the topic of community policing reach back to the work of Sir Robert Peel in order to find a source for much of the thinking that guides this philosophy (Peak and Glensor, 1996; Seagrave, 1997; Bayley, 1989). As can be seen in the discussion above, there is much previous custom and practice in English policing that can be used to inform the theory of community policing. However, Peel's contributions to the evolution of policing are undeniably significant and worthy of careful attention. In addition to formulating his important principles of policing, he was responsible for the appearance of two divergent models of policing that had considerable impact on the growth of policing in Canada and other jurisdictions during the nineteenth and twentieth centuries. We will consider both these models and consider their relationship to community policing in the Canada.

ROYAL IRISH CONSTABULARY

In 1814, Sir Robert Peel was responsible for leading the enactment of the Peace Preservation Act, which created what was to become known as the Royal Irish Constabulary. Initially referred to as the Peace Preservation Force (PPF), this body was essentially a territorial police service assigned to maintain peace within the thirty-five counties of Ireland. It was an experiment in the provision of emergency response in an area of British rule that was open to real, as well as anticipated, disorder. Because this patrol area was largely rural and farm-based, the officers assigned to the PPF were gathered together as a paramilitary unit lead by uniformed officers who were armed and structured under the terms of strict discipline and within a rigid command. These officers lived in barracks and were, in many senses of the word, "detached" from their surrounding community.

METROPOLITAN LONDON POLICE

In 1829, Peel formulated another approach to policing that would again come into being through legislation: Metropolitan Police Improvement Bill (10 George IV, c.44). This bill aimed at the creation of a stable, paid constabulary that would effectively replace the earlier common-law constable (Guth, 1994). The accountability for this police service would rest with the central government (i.e., Cabinet) through the Home Office and the Home Secretary. The Metropolitan London Police would become a model for urban policing that would transform the community-based elements of the earlier tradition of constables drawn from, and responsible to, the local village.

AMERICAN ANTECEDENTS

In mapping the course of community policing in Canada, it is not possible to present a clear picture without some reference to American experience and its impact on the development of Canadian approaches that derive from our neighbours to the south. Historically, a great deal of experience and insight has been imported from the United States relating to the principles and practice of law enforcement. We have been able to imitate many things that have a positive character, and, equally important, we have been able to avoid certain pitfalls and problems faced by several American police departments. Much of our current experience with various forms of community policing has come from an observation of circumstances in the U.S. However, Canada has not merely imitated American experience in its policing philosophy and practices. With a judicious eye, Canadian law enforcement practitioners have gathered information and applied their professional instincts in ways that suit our particular circumstances and systems.

Greene and Mastrofski (1985) divide the history of U.S. policing into three strategic phases; political, reform, and community. In the political phase, there was undue influence held in the hands of the local councillors who made appointments to all levels of the police department based on close party affiliations. It was an approach to policing that was highly prone to corruption and differential police treatment. It was built on a strong political party system not unlike that established in other times by corrupt unions that tightly controlled the work done for various trades and industries. This period is seen to extend from the 1840s until the first signs of the reform phase in the early 1900s. The reform phase, also referred to as the professional period, was introduced by police and other officials discontent with the quality of policing being offered in American cities. These reformers felt that if the effectiveness of policing was going to be maintained, police officers must be distanced from all forms of political influence. In order to achieve professional status, it was felt that police organizations must strive for a higher degree of autonomy. This would free them from the blatant corruption of political machines found in so many American cities. Obviously, police organizations would still be guided by the parameters established by state and federal law; however, political influence would be a sign of failure on the part of the police.

The popular image of the police as the "thin blue line" separating civil society from its criminal and violent elements dates from this era of professionalism. Of course, this independence of the police from political influence also entailed a growing separation of the police from their communities, as patrol officers were assigned to automobiles equipped with radios which kept them in a state of constant motion, awaiting dispatch to the next service call. The engineers of this professional approach valued the distance between the police and the public as being consistent with a professional image. It was an approach that made great use of available technology and allowed for maximum officer mobility and public presence. It further allowed police executives to control their officers more completely and to standardize the services being delivered through deployment strategies and extensively detailed policies and procedures.

However, the reform strategy that is evident in the efforts of August Vollmer and O.W. Wilson in the United States was not able to sustain police organizations into the 1960s and 1970s. Difficulties began to appear in most dramatic relief in American law enforcement, and they were clearly problems that would be of serious concern and interest to their Canadian police counterparts. Fundamentally, police departments, no matter how highly "professionalized" they had become, were not able to maintain public order and stability. The impact of the civil rights movement, large migration of minorities into urban centres, the student protest movement, the aging of the population, all had an impact on the American police department's overall ability to deal with crime and disorder (Kelling and Moore, 1985).

When the policing that falls under the heading of the "political" era is considered, there is much to learn by way of negative example. The rule of law enforcement by local political interests and the use of patronage appointments within the police department was one of the primary causes of the movement to the "professional" model under the leadership of August Vollmer and his disciples (such as O.W. Wilson). Of course, elements of English practice migrated to the U.S. during the colonial period (Trojanowicz and Bucqueroux, 1990). Watch systems were introduced in American cities such as Philadelphia in the early eighteenth century. Here the city was divided into ten patrol zones each assigned a constable who had authority to recruit citizen volunteers to assist him with keeping the watch. The period referred to as the "Spoils Era" (1830s until 1883) was a time when police chiefs were appointed on the basis of their willingness to undertake the wishes of the local politician(s) who gave them their job. The absence of qualified and independent police leaders presented a significant problem for American law enforcement that would persist for several decades.

When it became clear to American law enforcement officials that community policing had considerable potential to assist with the problems of crime and disorder, they rallied behind it in a serious manner that is reflected in Skogan (1995):

> The concept of community spread during the 1980s because of the 'aggressive marketing' of the concept by federal agencies and Washington-based think-tanks. Their target is not just police chiefs, but also the professional managers that run a majority of American cities. The think-tanks have staked their capacity to raise money to the promise that they can effectively promote community policing. (p. 92)

The corruption of policing by politics was condemned by those who were seeking to improve the status and effectiveness of American law enforcement. In 1925, President Calvin Coolidge appointed the first National Crime Commission, a clear sign that there was a real problem in the arena of crime control. This body, known as the Wickersham Commission, made several recommendations on matters before it, including a better approach to the selection of officers, the need for patrol officer testing in order to meet minimum standards, and attention to details relating to police salaries, working conditions, and benefits (Peak, 1993). There was also a focus on the importance of training for officers, not only before they were assigned to duty, but also on an ongoing basis to provide them with updated knowledge, skills, and abilities to meet the changing demands of the job. Of course, these recommendations, and the scrutiny under which the entire policing profession was being put would not have been lost on the police officers and executives operating in Canada. There was a close bond between these two law enforcement communities through such bodies as the International Association of Chiefs of Police that met regularly to discuss issues and concerns of mutual impact.

THE CANADIAN CONTEXT

Canada's unique historical characteristics, and those other attributes that make it a distinct entity, have provided the raw material for the creation of our own form of policing. While a greater degree of detail will be offered in the Chapter 3, it is relevant here to speak generally about the Canadian context as it relates particularly to community policing.

Bayley (1992) has pointed out that Canada has a distinct advantage over the United States with regard to the matter of community policing. He is referring specifically to our engagement in the policing of rural areas. Indeed, there is much to be gained from a close examination of the way in which smaller police organizations in Canada provide their services to a relatively stable, insular community. Bayley refers to these as "... living laboratories in community policing ...," and they have perhaps been significantly overlooked in explorations of this topic. The character of Canadian cities, towns, and villages is not as completely homogeneous as popular presentations would like us to believe. However, there exists a certain cohesive quality across Canada's regions that transcends regional and local differences. It is certainly topical to look for a distinctive Canadian quality in many things. It is clear that we have an acceptance of principles of government and public administration that allows for the growth of community policing. We have a small population that is increasingly urbanized (the importance of this fact will be detailed more fully in Chapter 3). However, it is relevant here to identify features of Canadian life which set our policing goals and objectives apart from those found in the U.S., or any other country's, police departments. The manifestations of community policing in Canadian police services will tend, always, to bear the mark of their origins. Murphy (1985) has noted something that further distinguishes Canadian police departments from their counterparts:

> For a variety of distinctive historical, political, and cultural reasons...Canadian police are publicly popular, well-financed, politically autonomous, organizationally stable, and ideologically powerful. (p. 178)

Once the potential of community policing as an ideology became clear within the policy making and policing community in Canada, it was given serious attention at many levels (Murphy, 1985, p. 179). It is in this context that police organizations and their public overseers began to explore the promise of this approach which had begun to take shape in the United States and other jurisdictions. However, given the differences in Canadian policing, it will become clear that community policing became manifest in ways that were consistent with our law enforcement traditions and public safety expectations.

LEADER IN COMMUNITY POLICING

David Cassels, retired Chief of Police, Winnipeg Police Service

What is Community Policing?

There are two distinctive views of what policing is, or should be. Some of us see the police officer's role as simply a "law enforcement" officer, and the police as the front end of the criminal justice system. Others see the role of a police officer as that of a "peace" officer whose real purpose is to solve community problems in the interest of community well-being and co-existence. However, the skills required to perform as a law enforcement officer are quite different from those required to function as a *peace* officer.

Police agencies, like other institutions, face remarkable change. While the private sector is quick to adjust to the rapid changes of today's environment (or perish), public institutions are slower to react. Most police organizations are no different. Community policing initiatives are surfacing in many departments because most senior managers recognize the need to do things differently. Some realize they must introduce new paradigms, restructure the organization, and develop new ways of doing business.

Two decades have seen a smattering of community relations and crime prevention programs, some innovation in technology and information systems, and a few neighbourhood foot patrol projects. These changes, while somewhat meaningful and productive, have been centred on supporting the patrol function. However, the most pressing, persistent problem for operational personnel is the constant demand to respond to "calls for service."

Traditionally, operations personnel react to calls for service by responding quickly, investigating and recording, contacting the appropriate specialists (i.e., detectives, crime prevention officers, or victim services personnel), and applying a temporary solution within the confines of the criminal justice system. Police agencies tend to focus on major criminal problems, while recurring community or neighbourhood problems are not recognized as being a continual drain on police services. Even if a patrol constable does recognize a recurring problem, the pressure to get back to the car and "in service" for another call dominates thinking, particularly in an urban environment.

While police have become incident driven and reactive, some police executives have recognized the need to move policing from this reactive model to a new model focusing on ownership, problem identification, and problem solving in the interest of community well-being and co-existence. They realize patrol officers must be empowered to use their intellect, energy, and imagination to recruit people within their neighbourhood, and together help shape the local environment to deal with recurring problems. At the same time, the constable must continue to assume responsibility for calls for service so that (s)he is both informed and involved with the community. A sense of ownership will surface. Together, the community and the police become a problem solving group.

David Cassels LEADER IN COMMUNITY POLICING continued

We must position ourselves so we can act as the catalyst to help the community deal with recurring problems, rather than simply responding to individual incidents, and to concentrate on recurring problems, develop a close working relationship with the community, and act as "community problem solvers" rather than simply servants of the criminal justice system. All of this is Community Policing.

Some senior police managers have embraced the Community Policing philosophy and expect front line personnel to do the same. Often, action only includes a short in-service training program or perhaps the creation of special community policing initiatives and, as a result, very little changes. Police departments, deeply entrenched in rules and regulations, continue to select and train new recruits as they have in the past. Managers function in a command and control environment. It is difficult to solve long-standing community problems through creativity and risk-taking in this environment, yet there is an expectation that this will happen.

Police departments are conservative, process oriented, paramilitary organizations that are very resistant to change. Police officers are trained from the beginning of their careers to follow rules and regulations; members are expected to do as ordered. A new policy is developed for every mistake made by someone in the organization. This mindset has dominated policing for decades. Because of the monopoly held by police, managers set the agenda, often without any community input.

Without question, as the traditional method of random patrol, rapid response, and reporting of occurrences diminishes, the constable's role will change. The responsibility for neighbourhood problem solving will become the police officer's primary function. There is also a need to abandon or change much of the traditional classroom instruction that forms part of the recruit and in-service training curriculum. Police officers will always need to learn the history of policing, the major reforms in policing, and the status of contemporary policing in order to understand the nature of policing. Training would need to be expanded to include such subjects as problem identification and problem solving, community dynamics (networking), creative thinking, conflict resolution, peak performance, conducting meetings, management of volunteers, time management, and ownership. Recruits need increased interpersonal skills and human relations training. We must encourage officers to have a different vision about their role.

So it is with police leadership. Strategic plans and setting annual objectives are important, but not enough. Words in a plan do not motivate people. Leaders do. Police executives, middle managers, and supervisors need to liberate talents in people. They need to create vision, guide autonomy, create a climate of trust, improve upward and lateral communication, provide for the freedom to fail and create champions within the organization. Promoting reasonable risk-taking, creativity, participating in decision making, enhancing recognition of accomplishments, and improving organizational processes are just some of the skills required. Blaming, distrust, excessive management of people takes most of our energy

David Cassels LEADER IN COMMUNITY POLICING continued

now and there's very little support in our current culture to provide these new leadership behaviours. These skills need to be identified and promoted within all police organizations. It is said that visionary police leaders have no difficulty abandoning many elements of the traditional police culture.

> If we want the benefits of creativity, we ought to have the wisdom to teach it where it is found wanting, to encourage it where it shows signs of developing, and at the very least, to tolerate it where it threatens to disrupt our comfortable status quo.—R.N. Harder

The present model (reform model) of policing is bureaucratic, impersonal, centralized, highly structured, process oriented, and law-enforcement based. The basic operational components are rapid response, random patrol, and investigation after the fact. The mode that I envisage is much less bureaucratic. It is decentralized (to the community), service oriented, communicative, *consultative,* and problem solving based—not simply law-enforcement based. It focuses on local neighbourhood needs —not police needs. Simply put, it is a new kind of relationship between the police and the public.

I believe the major operational components required today involve decentralization, ownership, increased autonomy to front line officers (they are the ones doing the work!), permanent assignment, neighbourhood consultation, and *problem solving* as the primary "unit of work," not just enforcement. This is not to say that enforcement is not important. Of course it is, but it needs less attention today.

Police should continue to take a tough approach to hardcore, repeat offenders in the interest of public safety. However, there is ample room to consider programs that divert first time offenders for minor offences from the overburdened court system to forms of restorative justice such as mediation, community sentencing circles, and community service work. Victims are much more satisfied and offenders accept responsibility and are remorseful. This changes behaviour.

It is also important to make it very clear that I am not promoting abandonment of the patrol or cruiser car function. There will always be a need to have some police officers in cars responding to urgent calls for service and the types of occurrences that require a police officer to attend the scene to gather evidence, etc. What is important is that managers find a way to free operational personnel from the burden of calls for service so that some of them can begin to focus on long-term problem solving. Those that are assigned to patrol can also begin to experience the personal rewards that come from ownership and problem solving. That is, appreciation and recognition from the people they help.

None of this will happen without organizational restructuring. This restructuring is *not* difficult. In fact, it is exciting and rewarding for people, particularly the senior officers who participate and provide the impetus to make it happen.

New forms of policing broaden the police functions from crime fighting, without any abdication of that role, to responding to the broader range of functions in which the police are engaged. Maintaining order, dealing with quality of life offences, and fixing

David Cassels LEADER IN COMMUNITY POLICING continued

the "broken windows" would now be recognized as much more important than previously thought to be. Understanding the concept of "broken windows" is central to our work. Emphasis placed on ridding the city of graffiti, assisting neighbourhoods in cleaning up the needles and condoms, and generally helping or encouraging people to clean up their neighbourhood has as much to do with reducing the fear of crime and restoring a sense of neighbourhood than providing expensive police visibility.

It is also important that we come clean with people and admit that we cannot be everywhere, respond to all of their calls, and catch all of the bad guys. We need their help to help themselves. The professional model of policing does not work as well as we might think. I believe we have oversold ourselves. We can't handle the backlog of calls in queue on busy days. If we don't act soon we will lose control of our own resources simply because of the volume of telephone calls we accept.

The need to reduce public expectations is every bit as important as the need to broaden the police function. This will make our job more achievable by realigning what we do and do not do; by giving higher priority to some tasks and lower priority to others; by reducing public expectations and levelling with the public about police capacity; by engaging the public to take steps to help themselves and by *connecting* with other agencies.

We need to wean people off the telephone and ask them to come to us to report non-urgent complaints. This is why police need more local neighbourhood police stations. The primary purpose of a local station is to accept walk-in traffic and to take complaints across the counter. To give a *personal* service when it is convenient for people to come to the police. They don't have to wait for police to answer the phone and they don't have to wait for a car to be dispatched. People understand because police will be in a much better position now to respond to the *emergencies* faster. People will also be much more satisfied with the overall service.

As most police officers know, a high percentage of our work comes from a very small number of locations. These are known as "hot spots" and are confined to small geographic areas. By giving ownership of these small areas to individual police officers with some autonomy and freedom to try new approaches *with* neighbourhood residents and other agencies, problem solving becomes easy. This can be called Neighbourhood Foot Patrol. What it is called, or the method of transportation, is really not important. The operational strategy is ownership and problem solving. We should not expect one constable to solve all of the problems or own all of the problems. He or she owns the *area* and gets others, even other police officers, to help.

Ownership is something that is sadly lacking in most police departments. The way we are presently structured and function, few of us have anything we can call our own. In fact, unfortunately, our people often say: "It's not my problem" or, "It's not my area of responsibility, so go see someone else." The specialist approach to policing has complicated this even more. Don't misunderstand me, there will always be a need for some specialists, but it is time for more generalists in policing.

David Cassels LEADER IN COMMUNITY POLICING continued

Back to leadership. A police manager's job is to make this happen. It is not likely to happen if we focus on tight controls and trying to find fault. It will happen if we change some of the bureaucratic processes and become coaches rather than bosses. Think about some of the supervisors you have had over the years. Which ones did you respect the most? My guess is that the ones that found fault with the things you did were not as well respected as the ones that valued your opinion or made you feel like you had a strong sense of worth, and that you were trusted to do the right thing. This created a climate of freedom that you probably found highly motivating. A good leader will provide freedom and respect, but will still keep you focused.

To make this model work, we must allow constables to express themselves and their talent. We must create a climate of trust. We must trust that they will make the right decisions. We must encourage innovation and creativity. We must provide for participation in decision making. We must provide for some freedom to fail without criticizing them. Failure is an opportunity for us to coach in a positive way so they can maintain their self-respect and learn from the mistake. This can be harder for us, but that's what leadership is about. There really isn't a lot of risk for management.

The most precious component of leadership is trust, confidence that the one who leads will act in the best interest of those who follow, the assurance that he or she will serve the group without sacrificing the rights of individuals. I would also like to work in a culture that promotes communication at all levels, whether it is up, down, or sideways—or stated differently, a culture with no walls between sectors, units, groups, or functions within the organization.

We must also reject the demeaning of individuals. We cannot afford to have anyone in a leadership position who give the wrong signals to employees.

> A good leader inspires people to have confidence in himself. A great leader inspires people to have confidence in themselves.—Art Friedman (1992)

I believe that public relations programs are often seen as window-dressing. Yes, improved relations are important, but when citizens ask if Community Policing works, they are not so much interested in knowing if the community likes the police or if the police are getting along with the community. Rather, they really want to know if the Community Policing initiative has had an impact on the problems of concern to them—their fear of using the streets, the abandoned cars in their neighbourhood, the noisy parties, or the gang intimidating them. Local neighbourhood problem solving will have a direct impact on these problems.

Lastly, we should stop believing that we (the police) are the front-end of the criminal justice system. Our purpose is not just to feed the system. Our purpose is to create peace in the community. I believe the system is failing us. It is clogged, adversarial, and does not solve all community disorder problems. It does not provide any reasonable

David Cassels LEADER IN COMMUNITY POLICING continued

form of satisfaction for victims. There are alternatives available to us, alternatives that can provide much more satisfaction for victims, change behaviour of first-time offenders and reduce our workload. These alternatives free the overburdened court system and provide for more court time for serious repeat offenders and those charged with serious offences.

It is also very important for police managers to insist that police officers and support staff treat people with courtesy and respect at all times—*without exception.* Regardless of all of the technology, training, and skills that we are taught, *policing is no more than contact between one human being and another.* Whether you are a complainant, victim, suspect, or witness—you are on the receiving end of anything the police do. A good police officer will always get it right. A true form of community policing will make this happen.

SUMMARY

This chapter has considered various definitions for "community policing" and has cited several examples drawn from academic sources, police services, and government agencies. This background, along with a summary of some of the main assumptions guiding notions of community policing, has been adapted in laying the groundwork for a conceptual framework that offers a vision of something termed "Transformative Policing." Also, this chapter presents the origins of community policing within the context of traditional police practices, especially as they relate to the role and responsibilities of the early English constable. The significance of this role is that it provides a touchstone for understanding the compelling value of crime prevention as a key component of effective policing. The deep attachment of the constable to the community is an early example of the linkage which modern community policing is attempting to redefine and recapture.

QUESTIONS FOR CONSIDERATION

1. What are some of the common elements in the various definitions provided for "community policing"?
2. What distinguishes the "reactive" mode of policing from the "proactive" mode?
3. Discuss some of the major assumptions that have developed around the concept of community policing. How might these assumptions affect the practice of modern policing in Canada or elsewhere?
4. Identify responsibilities and functions carried out by early English constables that are similar to those carried out by police constables in Canada today.
5. What are the key differences between the two models of policing developed by Sir Robert Peel in the nineteenth century?
6. Which police services in Canada today match either the Royal Irish Constabulary model or the Metropolitan London Police model?

7. What are some of the major areas in which American law enforcement has influenced policing in Canada?
8. Discuss key aspects of Canadian society that make its policing needs unique.
9. What distinguishes the three phases of police development identified above: political; reform or professional; and community?

REFERENCES

Bayley, D.H. (1985). *Patterns of Policing: a comparative international analysis*. New Jersey: Rutgers University Press.

Bayley, D.H. (1989). "Community Policing: a report from the devil's advocate." In Green, R.J. and S.D. Mastrofski (eds.) *Community policing: rhetoric or reality?* New York: Praeger.

COPPS: community oriented policing & problem solving. Sacramento, Calif.: Crime Prevention Center, Attorney General's Office, California Department of Justice, 1992

Deakin, Thomas J. (1988). *Police professionalism: the renaissance of American law enforcement.* Springfield, Ill. : Charles C Thomas.

Emsley, Clive (1991). *The English police: a political and social history.* London: St. Martin's Press.

Friedmann, Robert R. (1992). *Community policing: comparative perspectives and prospects.* New York: St. Martin's Press.

Guth, DeLloyd J. (1994). "The traditional common-law constable, 1235–1829: from Bracton to the Fieldings to Canada." In MacLeod, R.C. and D. Schneiderman (eds.). *Police powers in Canada: the evolution and practice of authority.* Toronto: Published in association with the Centre for Constitutional Studies, University of Alberta by the University of Toronto Press.

Kelling, George L. and Mark H. Moore (1985). "From political to reform to community: the evolving strategy of police." In Greene, Jack R. and Stephen D. Mastrofski (eds.), *Community policing: rhetoric or reality.* New York : Praeger.

Kent, Joan R. (1986). *The English village constable 1580–1642: a social and administrative study.* Oxford: Clarendon Press.

Ontario. Ministry of the Solicitor General. Policy Development and Coordination Branch (1991). *Community policing: an introduction to the community policing philosophy and prinicples*. Toronto: The Queen's Printer for Ontario.

Oettmeier, Timothy N. and Mary Ann Wycoff (1997). *Personnel performance evaluations in the community policing context.* Washington, D.C.: Police Executive Research Forum.

Peak, Kenneth J. (1993). *Policing America: methods, issues, challenges.* Englewood Cliffs, N.J.: Regents/Prentice Hall.

Peak, Kenneth J. and Ronald W. Glensor (1996). *Community policing and problem solving: strategies and practices.* Upper Saddle River, N.J.: Prentice Hall.

Radzinowicz, Leon (1956). *A history of English criminal law: and its administration from 1750.* London : Stevens & Sons Ltd. (especially Volume 2: "The enforcement of the law" and Volume 3: "Cross-currents in the movement for the reform of the police")

Richardson, James F. (1974). *Urban policing in the United States.* London: Kennikat Press.

Skogan, Wesley (1995). "Community policing in the United States." In Brodeur, Jean-Paul (ed.), *Comparisons in policing: an international perspective.* Aldershot: Avebury.

Simpson, H.B. (1895). "The office of constable." *English Historical Review,* 10: 625–41.

Strauss, Leo (1953). *Natural right and history.* Chicago: The University of Chicago Press.

Trojanowicz, Robert and David Carter (1988). *The philosophy and role of community policing.* Fort Lansing, Mich.: National Neighborhood Foot Patrol Center, School of Criminal Justice, Michigan State University. (Community Policing Series; No.13).

Trotter, Eleanor (1919). *Seventeenth century life in a country parish.* Cambridge: Cambridge University Press.

Walker, Samuel (1976). "The urban police in America: a review of the literature." *Journal of Police Science and Administration,* 4(3).

Walker, Samuel (1977). *A critical history of police reform.* Lexington, Mass. : Lexington Books.

Walsh, William F. (1995). "Analysis of the police supervisor's role in community policing." In Kratcoski, Peter C. and Duane Dukes (eds.). *Issues in community policing.* Highland Heights, Ky. and Cincinnati, OH: Academy of Criminal Justice Sciences, Northern Kentucky University and Anderson Publishing Co.

TOWARD MORE LEARNING

DEFINITIONS OF COMMUNITY POLICING

Chacko, James and Stephen E. Nancoo (eds.) (1993). *Community policing in Canada.* Toronto: Canadian Scholars' Press Inc.

This publication provides a collection of essays written between 1988 and 1993, drawing from a number of important Canadian sources. Their focus is on the "paradigm shift" which community policing represents and offers a solid background on the beginnings of this approach to policing in Canada. The book includes discussions of the earliest efforts at applying the principles of community policing in Victoria, Halifax, Halton Region, Metropolitan Toronto, Montreal, the Ontario Provincial Police, Edmonton, Vancouver, and the Royal Canadian Mounted Police.

Friedmann, Robert R. (1992). *Community policing: comparative perspectives and prospects.* New York: St. Martin's Press.

This book provides a good overview of relatively recent, and not so recent, literature in the area of community policing. Friedmann begins with a detailed discussion of the various definitions that have been put forward, and consolidates a great deal of information into a fairly comprehensive framework. This work has an extensive bibliography that will be useful to the serious researcher.

Trojanowicz, Robert and David Carter (1988). *The philosophy and role of community policing.* Fort Lansing, Mich.: National Neighborhood Foot Patrol Center, School of Criminal Justice, Michigan State University. (Community Policing Series; No.13).

An important early contribution to the literature of community policing that is clearly and succinctly written. Robert Trojanowicz, in particular, is a key author in this area, and was responsible for the leading work done on the Flint, Michigan Foot Patrol study that fuelled much of the later research and experimentation in community policing. This monograph condenses some of the key points raised by the ongoing inquiries into this idea.

THE OFFICE OF THE EARLY ENGLISH CONSTABLE

Guth, DeLloyd J. (1994). "The traditional common-law constable, 1235-1829: from Bracton to the Fieldings to Canada." In MacLeod, R.C. and D. Schneiderman (eds.). *Police powers in Canada: the evolution and practice of authority.* Toronto: Published in association with the Centre for Constitutional Studies, University of Alberta by the University of Toronto Press.

A valuable summary of the evolution of this office from its earliest times in the middle ages, through the refinements of the English reformers to the appearance of this official in Canada. The importance of the constable in our legal and social history is worth understanding in greater detail. This essay provides the basis for some of that detail.

Kent, Joan R. (1986). *The English village constable 1580-1642: a social and administrative study.* Oxford: Clarendon Press.

This scholarly study offers a thorough and extensively researched review of the role and responsibilities of the English village constable during the late sixteenth and early seventeenth century. It provides a fascinating examination of the importance of this official in the life of the local community and reveals, in some regard, the ideal to which much of today's community policing efforts are directed; namely, the integration of the police with their communities in ways that promote public safety and prevent crime and disorder.

THE AMERICAN CONTEXT

Deakin, Thomas J. (1988). *Police professionalism: the renaissance of American law enforcement.* Springfield, Ill.: Charles C Thomas.

Thomas Deakin was an FBI special agent and editor of the *FBI Law Enforcement Bulletin*. He has written a detailed and complete study of the evolution of policing in the United States, with particular focus on the role of the federal agency to which he belonged and which has a central role in the growth of American law enforcement. It is within this larger context that community policing has come to flourish in policing in the United States.

Peak, Kenneth J. and Ronald W. Glensor (1996). *Community policing & problem solving: strategies and practices.* Upper Saddle River, N.J.: Prentice Hall.

A recent work containing a wealth of insight into the importance of community policing and approaches to problem solving, particularly from the American perspective. Peak and Glensor have collected detail on what they refer to as community-oriented policing and problem solving (COPPS). They look in detail at several U.S. cities that have undertaken recent experiments in this area, including: Austin, Texas; San Diego, California; Reno, Nevada; Hayward, California; and Baltimore, Maryland. They also provide brief summaries of related initiatives in Japan, Singapore, South Australia, Canada, Great Britain, and several Scandinavian countries.

THE CANADIAN CONTEXT

Normandeau, André and Barry Leighton (1990). *A vision of the future of policing in Canada: police-challenge 2000: background document.* Ottawa : Police and Security Branch, Ministry Secretariat, Solicitor General Canada.

An important early contribution to the Canadian debate on matters relevant to the future of policing. This work, along with the brief Discussion Paper that was published simultaneously, were widely circulated at a time when dialogue was beginning to intensify on the value of community policing and its application within the context of Canadian law enforcement.

Seagrave, Jayne (1997). *Introduction to policing in Canada.* Scarborough, Ont.: Prentice Hall Canada.

Chapter 10 of this recent publication looks at the various definitions and interpretations of community policing and cites some of the main sources for further study. Chapter 11 looks at community policing from a Canadian perspective and offers a brief consideration of its development and expression.

CHAPTER TWO

The first thing to understand is that the public peace—the sidewalk and street peace—of cities is not kept primarily by the police, necessary as police are. It is kept primarily by an intricate, almost unconscious, network of voluntary controls and standards among the people themselves.

Jane Jacobs (1961), The death and life of great American cities.

TRACING THE OUTLINES OF COMMUNITY POLICING

LEARNING OBJECTIVES

1. Describe the main elements of crime prevention.
2. Describe the importance of Crime Prevention Through Environmental Design.
3. Outline the importance of Crime Prevention Through Social (Community) Development.
4. Distinguish community services from public relations in policing.Identify several crime prevention programs from various jurisdictions.
5. Describe current applications of the crime prevention model in Canada.

INTRODUCTION

Before the term "community policing" gained currency in discussions about innovative police practice, many police organizations were committed to various forms of activity which fell under the category of crime prevention. The whole notion of crime prevention has, of course, been part of the purpose of policing since its earliest days. The concept of an officer's mere presence being a deterrent to criminal, or unlawful, activity is a strongly and widely held opinion. However, the concentrated effort to develop specific strategies, programs, or activities that focused on crime prevention was a conscious movement in modern policing. Its period of greatest success began in the 1960s and 1970s, especially in the U.S. It coincided with an interest in developing proactive police approaches. More specifically, this meant addressing the causes of criminal activity at their root. This chapter will look at the importance of crime prevention as a precursor to community policing and will consider some of its essential elements. Also, this section will look at how the concept of crime prevention has had a major resurgence, particularly in Canada, largely as a result of a degree of discomfort, if not dissatisfaction, with the "community policing" label.

The most widely known crime prevention program is Neighbourhood Watch, which began in the U.S. in the 1970s and flourished in many communities around the world. The popularity of this type of program is evident in the vast number of Neighbourhood Watch signs seen in many North American communities. However, the effectiveness of such a program is inconclusive at best (Bennett and Lavarakas, 1989). Some observers have noted that they make no substantive contribution to crime prevention (Rosenbaum, 1987). Other programs that grew out of this original interest in crime prevention include: citizen patrols, crime stoppers, and property identification initiatives. While none of these programs has been proven to be highly effective (Griffiths and Verdun-Jones, 1994), they continue to be used in many jurisdictions.

CRIME PREVENTION AS COMMUNITY POLICING

MAIN ELEMENTS OF CRIME PREVENTION

Basically, crime prevention is any activity that tends to reduce violence, disorder, delinquency, or fear of crime by attacking the identified cause(s) of such events. In terms of its academic study, we are typically considering approaches that deal with potential victims or potential offenders. Furthermore, we can consider the physical environment; that is, the spaces where the offender and the victim make contact. In effect, crime prevention initiatives concern themselves, almost exclusively, with:

- *Opportunity reduction*—doing things that make crime more difficult, more detectable, and, therefore, less appealing to the offender. Such initiatives often focus on the potential victim, or on what is called "target hardening."
- *Social development*—doing things that remedy conditions which lead people to commit various forms of criminal activity. This approach naturally focuses attention on the potential offender, particularly the young and the disadvantaged.

In order to be effective, the crime prevention approach typically involves the following components (Home Office, 1991):

- Partnerships with local authorities, along with technical and financial support from the central government;
- Analysis of the causes of crime and the implementation of locally-based solutions to crime problems;
- Introduction of early prevention programs, particularly those aimed at young persons and families;
- Substantial citizen participation; and
- Encouragement of international cooperation.

CRIME PREVENTION THROUGH ENVIRONMENTAL DESIGN (CPTED)

One of the promising initiatives derived from the school of crime prevention is referred to as Crime Prevention Through Environmental Design (commonly referred by the acronym, CPTED). Many police organizations are adding their expertise to this approach that combines several factors that contribute to overall public safety and crime reduction. The popularity of CPTED relates to some of the early work of people such as Jane Jacobs (1961), Jeffrey (1971), and Oscar Newman (1972) who have argued forcefully for the creation and fostering of viable urban communities. Though not new, CPTED has experienced a kind of resurgence in Canada and the U.S. For example, the curriculum of several police academies includes courses dealing extensively with CPTED. This approach to crime prevention derives its insights from several disciplines, including: architecture, urban design, landscape architecture, and planning. Also, valuable insights for CPTED have been drawn from the social sciences (e.g., sociology, criminology, psychology, anthropology, and geography). This overlap of broadly-based knowledge, skills, and abilities is characteristic of the best instances of community policing in that it truly engages members of the public in the "co-production" of public order and safety. The police, while pivotal to the success of CPTED, are not necessarily the leading experts in this area of crime prevention. Police officers, well-versed in CPTED theory and practice, can act as effective catalysts in this regard and can draw together the various subject matter experts require to ensure that proactive environmental design contributes to crime prevention in the community. Obviously, CPTED is most effective at the initial planning stages of a new community, in which it is possible to build a foundation for crime prevention from the ground up. However, it can also contribute significantly to community redesign and rejuvenation. CPTED principles and approaches are based on an understanding of what is known as the "crime triangle" which appears as shown in Figure 2-1, below:

FIGURE 2-1

THE CRIME TRIANGLE

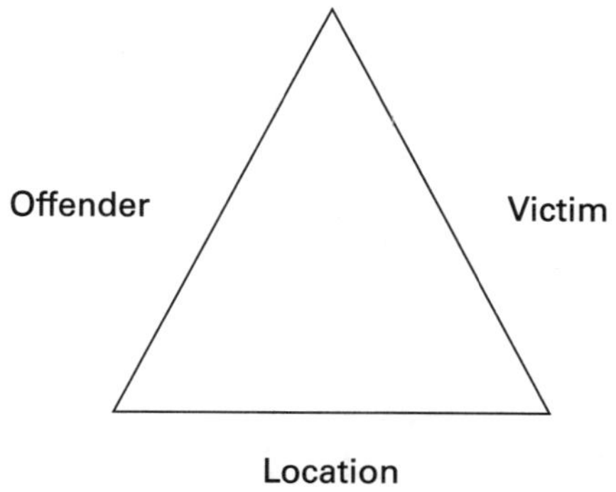

Crimes occur when the three elements noted above come together. Obviously the offender is seeking a vulnerable victim in a location that will maximize the successful conclusion of the offence, as well as providing for an easy means of escape. By removing, or minimizing, one of the elements of the crime triangle, it is possible to reduce not only the occurrence of crime, but also the fear of crime. The proponents of CPTED believe that a better understanding of the crime triangle and a spatial approach to crime prevention and environmental design are keys to improved public safety.

By considering careful design of human spaces, the advocates of CPTED believe that safer, more secure communities can be created. However, taken to its furthest expression, there may be concern raised about the proliferation of gated communities which go far beyond the original precepts of environmentally-designed areas. Frequently, there is also concern that by effectively protecting one location through careful environmental design, crime is merely displaced to another less "hardened" location.

CRIME PREVENTION THROUGH SOCIAL (COMMUNITY) DEVELOPMENT

Another key refinement to modern policing strategy that falls under the general heading of crime prevention is referred to as Crime Prevention Through Social (Community) Development. This approach builds on the premise that crime problems may be addressed by placing considerable resources in building strong communities and ensuring that high-risk groups (e.g., children and young persons) are provided with durable and responsible social networks (John Howard Society of Alberta, 1995). As a result, there will be fewer instances of criminal activity, with proportionate savings in reactive law enforcement, not to mention the other components of the criminal justice system. In 1994, Canada adopted a National Crime Prevention Strategy that placed considerable emphasis on the central importance of social development for any effective approach to crime prevention (Canada. National Crime Prevention Council, 1995a). This view is shared in many other jurisdictions around the world and speaks to the need for governments to invest their human and financial resources in areas in which there can be the most impact (Canada. National Crime Prevention Council, 1995b). Typically, an approach to crime prevention that concentrates on social development will adhere to the following fundamental principles:

- The causes of crime are most effectively addressed when dealing with their antecedents in early childhood and adolescence; and
- Crime prevention programs should be clearly integrated and are best delivered at the local level through engaged stakeholders with overall national support, cooperation, and coordination.

CRIME PREVENTION AS A CORE FUNCTION IN POLICING

Crime prevention is clearly a core function within the typical police organization. Recent amendments to the *Police Services Act* in Ontario have explicitly identified its importance along with other core functions:

> 4. — (1) Every municipality to which this subsection applies shall provide adequate and effective police services in accordance with its needs.

(2) Adequate and effective police services must include, at a minimum, all of the following police services:

1. Crime prevention.
2. Law enforcement.
3. Assistance to victims of crime.
4. Public order maintenance.
5. Emergency response.

However, in the larger context of community policing, crime prevention may be seen as merely one component in an overall organizational strategy. As important as crime prevention is to a comprehensive project of public safety and order, it remains essential that community-based reactive capacities be in place to address the occurrence of crime and disorder. As government agencies tackle the economic realities which confront them, the attraction of effective crime prevention programs is that they attempt to address the root causes of crime and operate to reduce, or eliminate, their appearance.

COMMUNITY RELATIONS VERSUS PUBLIC RELATIONS IN POLICING

In the past, many police organizations experienced a considerable lack of clarity regarding their efforts to engage the community in preventative measures. It was seen as being the right thing to do to provide some formal contact with the public. However, there was little research or practical experience to guide this instinct. When the concept of the community services officer (CSO) was first advanced in many police departments, the individuals assigned to this position had no clear direction regarding their activities. For example, in 1985, the Ontario Provincial Police undertook an analysis of the training needs for their CSOs (OPP. Provincial Police Academy. Systems Approach to Training Unit, 1986). This thorough review (which included interviews, a questionnaire survey, and document search) revealed that members of the Community Services Branch did not have a clear understanding of the purpose or function of their department. Individuals within the branch indicated that they did not have sufficient instruction or guidance with respect to their duties. Most of those interviewed held the view that their function was best fulfilled through classroom presentations to elementary school students. The research pointed to a common confusion, which had also surfaced in U.S. policing in the 1960s, between public relations and community relations. By specifically separating these two sets of activities, police departments could better invest time and effort, not only in crime prevention (OPP. Provincial Police Academy. Systems Approach to Training Unit, 1986):

> . . . but in many social areas, some of which were not normally identified as a police matter but if left unresolved could evolve into police action. (OPP. Provincial Police Academy. Systems Approach to Training Unit, 1986. p. ii)

By defining the role of the community services officer in terms of functions that contribute to community relations and not public relations, by designing and delivering training programs that concentrate on appropriate community relations activities, and by making efforts to evaluate program effectiveness, this area of activity could be substantially improved. The public relations function remains important to every police organization, but its focus is decidedly different from that of community relations. Research also revealed that community services officers (CSOs) frequently feel they are fulfilling a role held in low esteem within the police department, and that their job satisfaction comes from the community rather from within their department.

In preparing the competent CSO, some of the following types of courses were suggested:

INSTRUCTIONAL TECHNIQUES Purpose: to develop the skills and attitudes required to function effectively as a full-time training officer.

Topics:

- Discussion and small groups;
- Evaluation process;
- Lesson content target;
- Lesson planning;
- Motivation;
- Objective setting;
- Programmed learning;
- Teaching hierarchy;
- Testing;
- Theories of learning and instruction; and
- Various learning media.

EFFECTIVE PRESENTATIONS Purpose: to develop the skills and attitudes required to function effectively in a part-time training capacity.

Topics:

- Delivery techniques;
- Lesson planning;
- Principles of instruction;
- Uses of instructional media;
- Lesson types; and
- Objective setting.

CRIME PREVENTION (3 WEEKS) Purpose: to educate crime prevention officers in the attitudes, knowledge, and skills required to research, plan, implement, and evaluate crime prevention programs. Emphasis is placed on community-oriented programs with goals that enhance the effectiveness of the police service by reducing crime risks.

Topics:

- Purpose of crime prevention;
- Problem solving;
- Problem identification;
- Planning programs;
- Evaluation of programs;
- Communications;
- Bank-related crime;
- Rural crime prevention;
- Security surveys;
- Target hardening;
- Effective presentations;
- Use of media;
- Problem-solving meetings; and
- Physical fitness.

The community service officer is the precursor to many forms of community policing, and their expertise in crime prevention theory and practice has proven valuable on many fronts.

LEADER IN COMMUNITY POLICING

Robert Lunney, Consultant & former Chief of Police (Edmonton & Peel Region)

What is your understanding of "community policing" and its importance to the evolution of policing in Canada?

Community Policing is as much a philosophy as a method of operation. A police service committee to community policing undertakes to listen actively and attentively to the citizens, conduct intelligent, rational reviews of the conditions of crime and disorder, and respond in ways that effectively satisfy the public need. Prevention is a major strategy of community policing, but its practice does not mean the abandonment of tactics to suppress crime. On the contrary, suppressive policing methods are more successful because they are based on better information and involve community partners.

While problem solving and partnerships are the primary elements, community policing only achieves its potential when facilitated by organizational change. Front line officers are given the responsibility and authority to initiate problem solving activities, and teams of officers are permanently assigned to specific geographic areas to promote their sense of ownership. The successful implementation of community policing requires a major

Robert Lunney LEADER IN COMMUNITY POLICING continued

commitment to change, a capacity to persevere with longer-term goals, and above all, a sincere and constant commitment to public safety.

Canadians enjoy a unique sense of community and a commitment to the constitutional values of peace, order, and good government. The practice of community policing has permitted Canadian policing to renew its bonds with the people in a true reflection of democratic principle.

Outline your own contribution to the development of "community policing" in Canada.

I had the good fortune to be appointed Chief of Police at Edmonton in 1974 when the early theories and practices of community policing were emerging from research and the first experimental projects. I discovered a rich vein of talent and creativity within the organization, and we partnered with citizens, buoyed along by a wave of community volunteerism. Together we experimented widely, succeeding brilliantly in some areas, failing utterly in others, but always learning. As theories became principles, I had the opportunity to influence other community-oriented policing systems at Winnipeg as a City Commissioner responsible for the police, and as Chief of Peel Regional Police. Laying the foundations for community policing in the Region of Peel was my best technical work. Perhaps my major contribution was not so much in the promotion of any single strategy or system, but in fostering an internal environment where change and challenge are welcomed with enthusiasm, and empirical learning is a major part of the organizational culture.

What is your view of the future of "community policing" in Canada?

Community policing has become the "new orthodoxy" of policing style in Canada, but there are huge opportunities for improvement in the practice of problem solving. In many instances cooperation and coordination are limiting conditions of partnerships, rather than the achievement of true collaboration. There remains a tendency to leap from preliminary scanning to solutions, and those solutions are too often law enforcement without proper consideration of prevention or other options, including community involvement. Program evaluation is the weakest element of all. There remains some vestige of the earliest days of community policing, when public relations were emphasized over public safety. Now that the pioneering phase is over, future progress depends on improving performance, and above all, commitment and perseverance.

Business-oriented theories applied to public service at the insistence of some governments threaten to destabilize some community policing programs. Perhaps the greatest challenge today is the marketing of community-oriented policing through an emphasis on results.

THEORETICAL BACKGROUND

Crime prevention is based on a theoretical position that links three elements of a crime event: victim, offender, and situation (see the Crime Triangle presented above). Anything that operates to avoid the convergence of these three elements can be seen an effort in crime prevention. A great deal of thought has been given to this relationship and has resulted in much of the debate over a movement to community policing which was discussed in Chapter One. Hastings (1993) has outlined a crime prevention typology that places various levels of intervention (i.e., primary, secondary, and tertiary) into the context of the elements of a crime event (i.e., offender, situation, and victim). Beginning with the primary level of intervention, Hastings places significant emphasis on community initiatives aimed at preventing crime, and includes a focus on the social, political, and economic aspects typically associated with crime and disorder in Canadian society. Indeed, a great deal of effective crime prevention activity can take place without any police or law enforcement engagement. One of the distinguishing characteristics of crime prevention theory is its virtual absence of attention to the organizational dimension of policing. The focus of crime prevention theory is to understand, articulate, and evaluate those approaches to crime, disorder, and delinquency which have a preventative impact. Crime prevention strategies rarely address the organizational design issues relevant to police departments wishing to transform themselves into innovative enterprises. In the extreme, crime prevention aims at making police, particularly in their reactive mode, unnecessary.

PRACTICAL APPLICATIONS

What follows are a few of the many initiatives that have been introduced in several jurisdictions to advance the purpose of crime prevention. Many of these programs have been developed with a great deal of care to ensure that their effectiveness can be properly evaluated.

CANADA

National Crime Prevention Council (NCPC)

In 1994, the National Crime Prevention Council was established to pursue Phase I of a National Strategy for Community Safety and Crime Prevention. Resulting from a careful study of issues undertaken by the federal government, this strategy was announced by the Solicitor General and the Minister of Justice along with funding to support an array of initiatives and activities designed to meet the following overall objectives:

- To foster partnerships and collaboration between local communities, provincial and territorial governments, the federal government and non-governmental organizations;
- To empower community-based and community-driven crime prevention solutions;
- To foster a comprehensive approach with both short and long-term goals; and
- To recognize that the national strategy will develop reflecting regional differences and realities.

With the creation of the National Crime Prevention Council, the federal government made a substantial commitment to a mechanism that would ensure that trends and effective strategies for dealing with crime prevention become a shared resource for a range of stakeholders. During its tenure the National Crime Prevention Council has accomplished a great deal in meeting its mandate and delivering on the following key elements of the strategy:

- A national focus;
- Partnerships and coordination;
- Involving citizens and communities;
- Enhancing knowledge;
- Incorporating crime prevention in legislation and official mandates; and
- More efficient use of existing funding.

Of particular importance to the various groups, organizations, and associations committed to the objectives of crime prevention were the following databases supported by the National Crime Prevention Council in support of prevention initiatives, ideas, and innovations:

- *Canadian Crime Prevention Practices Database*—designed to assist community groups to analyze local concerns and seek possible solutions. Also intended to assist in the identification of skills and resources needed to implement solutions;
- Federal/Provincial/Territorial Working Group on Community Safety and Crime Prevention Models of Practice—full text descriptions of effective programs;
- *National Crime Prevention Council Program Surveys*—designed to direct users to specific areas of program interest;
- *PREVENT Database*—a library and database of materials dealing with crime prevention, criminal justice, and social development maintained by the NCPC. Users may submit queries for bibliographic information relevant to specific subject areas;
- *Federal Inventory on Crime Prevention*—a comprehensive record of various crime prevention and social development activities carried out under the auspices of the federal government;
- *International Centre for the Prevention of Crime Resource Bureau*—a database maintained by the International Centre for the Prevention of Crime (ICPC) that includes model practices gathered on a world-wide basis; and
- Access to Justice Network Databases—this is considered to be a definitive source of information on criminal justice and public legal education information. Maintained by the Access to Justice Network, it represents a series of interrelated databases.

Phase II of the National Strategy on Community Safety and Crime Prevention came into effect in June 1998, with an announcement by the Minister of Justice and Solicitor General of Canada. This new investment in Canadian crime prevention is dealt with in more detail in Chapter 13 ("Future Prospects for Community Policing"). It is evident, however, that crime prevention continues to be seen as an integral part of the federal government's strategy for building safer communities. It is also a clear sign that investments made in addressing the causes of crime and social disorder are seen to be effective measures for social order.

Federation of Canadian Municipalities (FCM)

In 1985, the Federation of Canadian Municipalities (FCM), recognizing the importance of inter-agency cooperation in the prevention of crime, formed a Task Force on Crime Prevention. This group prepared a "National Philosophy for the Creation of a Stronger, Safer Community Life in Canada." Following this process, the FCM collaborated with the U.S. Conference of Mayors, the European Forum on Urban Safety, and the Montreal Urban Community to organize the first international conference on crime prevention in 1989. With more than 900 delegates from 31 countries, this conference brought together elected municipal officials, police executives, and crime experts in order to focus on the issues surrounding crime prevention. As a result of this conference, the *Agenda for Safer Cities* was adopted.

The *Agenda for Safer Cities* created a sense of urgency within the community of local government officials in that it identified fear of crime, particularly among women and the elderly, as a pressing problem. Beyond the need for policies and programs to deal with crime and drugs, the *Agenda* highlighted a parallel need for approaches to the following issues:

- Employment;
- Housing;
- Education;
- Opportunities for young people;
- Support for families;
- Equal opportunities with respect to race and gender;
- Improvements in the physical environment; and
- Cooperation between community agencies and the criminal justice system.

With the federal department of Justice, the FCM held a national symposium on community safety and crime prevention in Toronto in 1993. Again, organizations and individuals were brought together with a view to better understanding ways in which to improve and advance public safety. These activities led to FCM participation in a Mayors' Summit on Safer Cities, held in Vancouver, B.C. on March 30 and 31, 1996. This gathering was attended by mayors and councillors from three continents and fifteen cities, and resulted in the development of a "Final Declaration" which contained recommendations for the advancement of crime prevention. There was particular emphasis placed on social and economic development, stakeholder involvement, and attention to specific actions that would reduce and prevent crime and confront root causes of crime and delinquency.

In keeping with its action orientation, the FCM advocates a range of legislative and regulatory measures that would contribute to crime prevention. As a result, the FCM has formulated policy statements pertaining to the *Young Offenders Act* and the *Criminal Code of Canada* which it believes will contribute to community safety and crime prevention. Furthermore, the FCM has adopted a series of recommendations to support and advance crime prevention in key areas, including:

- National leadership;
- International support;
- Youth;
- Family violence and sexual assault;
- The elderly, visible minorities, and other vulnerable groups;
- Housing; and
- Policing.

In more detail, the following policy statement on recommendations relating to policing offers a good example of the FCM's ongoing commitment to crime prevention:

- Foster police support for citizen and community agency crime prevention efforts;
- Encourage police agencies to work more in the community to prevent crime and reduce fear, particularly through problem-oriented policing approaches;
- Ensure that adequate programs and inter-agency approaches are available for those already convicted of crime;
- Provide better information, access to remedies, and reparation for victims of crime;
- Encourage municipal leaders to ensure greater use of mediation and conciliation as a response to criminal conflict;
- Encourage municipal leaders to provide their staff, police, and community leaders with the appropriate training on inter-agency approaches to crime prevention; and
- Ensure that police personnel receive training to assist them in dealing sensitively with racial and ethnic minorities.

In order to consolidate its wide-ranging commitment, the FCM has a Standing Committee on Community Safety and Crime Prevention which has representatives from across Canada.

Rural Crime Prevention Strategy

In May 1997, the Ontario Solicitor General and Minister of Correctional Services announced the commencement of a high enforcement initiative that would target break-and-enters in rural communities. The unit responsible for this initiative will be part of an OPP detachment in eastern Ontario. The Rural Crime Prevention Strategy includes the establishment of a provincial Cottage Watch program that is aimed at crime prevention. Cottage Watch will be developed in cooperation with the Federation of Ontario Cottagers' Association Inc. and its corporate sponsors. It will include local cottage associations and local OPP detachments.

This Rural Crime Prevention Strategy is part of the Ontario government's commitment to community safety, which includes a number of initiatives funded through $150 million in new budget allocations. Other ventures include the Community Policing Partnership program, which will assist municipalities in hiring up to one thousand new front-line officers, a new Cadet program for the OPP, and increased funding for the Anti-Biker Gang Squad.

Prince Edward Island—Justice Resource Service

Since 1982, the notion of community involvement with issues relating to criminal justice has been fostered through the Justice Resource Service. It was identified that victims of crime, vandalism, and domestic violence were areas of particular concern within the community, requiring involvement and public education. With support from the federal Solicitor General, the service developed its programs in these areas.

In 1985, a Community Legal Information Association was formed to promote public legal information. A Victims Services program was introduced in 1989, along with the Victims of Crime Act. A number of groups dealing with family violence issues were extended or strengthened during this period, and a provincial Family Violence Prevention Coordinator position was created in 1991.

Using a social development model, personnel within the justice system have come to increasingly understand the interdependence of the various elements of the criminal justice system. The Justice Resource Service is a province-wide entity that seeks to promote a wide range of activities including:

- Crime prevention and community safety efforts, including a long-term provincial strategy;
- Continuing efforts to deal with impaired driving, including a long-term provincial strategy;
- Assisting Justice and related services in addressing violence and abuse issues, particularly in the context of the criminal justice response to family violence; and
- Continuing prevention, community development, and interagency approaches in supporting traditional Justice services as well as involvement of community organizations.

The service has had several accomplishments that would indicate that it is having a positive impact on community safety and crime prevention in the province of Prince Edward Island.

Nova Scotia—Family Service Of Support Organization

This program offers its services to families in which child abuse and/or neglect has occurred previously and there is a risk of some reoccurrence. The Family Service of Support (SOS) program pursues the following objectives:

- Early identification of, and intervention in, families deemed to be at risk in order to prevent abuse or neglect; and
- The preservation of the family through the use of skilled intervention in areas known to cause major family crises.

This program was begun in 1978 and builds on the understanding that parents, when provided with appropriate support and understanding, can develop effective parenting techniques and avoid conditions in which children are abused or neglected. The program often relies on self-referrals by parents. However, other agencies may be in a position to identify high-risk parents and refer them to this program.

Family SOS provides training for parents in infant care and attempts to offer an extended family to young parents who are without such supports. The program places an emphasis on giving emotional support to parents and providing them with practical skills necessary for successful parenting, including: homemaking, budgeting, life-skills, communication skills; and anger management.

British Columbia—Surrey Place Prevention Program

This program was developed to address a growing problem with auto theft and other crimes occurring in a regional shopping centre. Following consultation with a number of community groups and police officials, this initiative was set in place with the following objectives:

- To demonstrate a commitment to the community by responding to a growing fear of crime at the Surrey Place Mall; and
- To make Surrey Place Mall a safe and socially-responsible shopping experience.

As a result of the consultation, a program was developed to meet the objectives noted above. This initiative had three main components:

- Assignment of a Volunteer Patrol within the mall. These volunteers would be highly visible and would patrol parking areas looking for any suspicious activity. Outfitted with official identification, orange reflector jackets, binoculars, vests, hats, and a radio, the volunteers ranged from students to senior citizens. All volunteers were reviewed by the RCMP, trained, and operate in cooperation with existing mall security staff;
- A public information campaign, "Lock Out Auto Crime" presented in partnership with the Insurance Company of British Columbia (ICBC). This program concentrates on ways to avoid becoming the victim of car theft; and
- Sponsorship of a candidate in the RCMP's summer constable program. The Surrey Place Mall sponsors an individual who is attached to the RCMP and liaises on a regular basis with mall security staff.

In evaluating this program, it was found that while auto theft in the City of Surrey had increased 35% (by 1996), the Surrey Place Mall had experienced an 80% decrease in car thefts (comparing May 1995 to May 1996). There is some expectation that public opinion about the mall as a safe place to shop and visit will improve accordingly. Media coverage of the Surrey Place Crime Prevention program has been favourable, and shoppers, merchants, and community members have an improved sense of safety within the mall. Markborough Properties, the owner of the mall, are adopting this model for application to their other property sites and are working with the RCMP to that end.

Quebec—Opération Surveillance Anjou (OSA)

This program was inaugurated to deal with a problem of community and industrial crime in the Ville d'Anjou, Quebec. The Opération Surveillance Anjou (OSA) is a non-profit organization operating in the area of crime prevention since 1986. The Ville d'Anjou has a population of approximately 38 000 and the OSA seeks to increase the participation of citizens and businesses in local crime prevention activities, particularly in relation to robberies, break-and-enters, and thefts relating to vehicles.

The program seeks to apply an original approach to crime prevention and has used theatre, puppet shows, and videos to present its messages. The OSA design team consults with a variety of multidisciplinary resources in the preparation of materials and relies on consultation committees, conferences, and symposia for new ideas. Included in the OSA repertoire of initiatives are the following:

- An evaluation of the security of residences;
- A monthly prevention bulletin;
- Guest speakers on the topic of security;
- "Operation Jeans" (a program for 12- to 17-year-olds);
- Training program for parents on youth crime prevention; and
- An industrial security project.

While no formal evaluation has been undertaken with regard to OSA, the program was the recipient of a Solicitor General of Canada Crime Prevention Award for excellence in crime prevention in 1989. Statistically, it would appear that favourable results can be traced to the program, with break-and-enters at 128 in 1996, compared to 215 in 1995.

Ontario—Safe Neighbourhood Initiative (SNI)

Under the auspices of the (then) Metro Toronto Housing Authority (MTHA), this program was introduced to provide a comprehensive and effective strategy for dealing with crime and disorder problems in a number of neighbourhoods providing housing to the economically disadvantaged. By focusing on identified threats to security, it was the objective of this program to improve security and reduce crime.

Launched in March 1991, this Safe Neighbourhoods Initiative (SNI) was applied by the MTHA in eleven communities, located in both downtown and suburban locations, and included high-rise, low-rise, and townhouse buildings. The program adhered to a "Strategic Security Framework" which included elements relating to planning, development, management, and evaluation. This was a community-driven initiative that involved specific threat assessments. Residents and staff in the individual communities were responsible for the implementation of a response to the threats. SNI encouraged an approach that would build on existing initiatives, organizations, and resources available within the community.

Response measures used by MTHA residents and staff included physical security (e.g., cameras, locks), direct security (e.g., building and ground foot patrols), and community development (e.g., newsletters, community development workers). From an evaluation perspective, a joint review conducted by the MTHA and Canada Mortgage and Housing Corporation indicates that the strategies applied effectively increased security in both public and private housing communities. It was also seen that the Strategic Security

Framework contained general principles that were suitable to other communities. Also, residents and staff have noted that with the introduction of SNI, there has been an increase in the sense of security felt in these communities, as well as an increase in what can be referred to as "neighbourliness" among the residents. Police has reported a decrease in calls for service at SNI sites and better rapport between the residents and the officers assigned to those locations.

Alberta

In 1986, the Edmonton Police Service began by evaluating 153 000 calls for service within the city. As a result, they were able to identify 21 areas that could be considered "hot spots," and that fully 81% of calls were coming from repeat addresses. The Edmonton Police were thus able to realize that they had to address the problem of repeat users in order to improve police service effectiveness.

In 1988, the Edmonton Police Service began to apply what is referred to as a problem-oriented strategy in dealing with crime and disorder issues. While constables would continue to respond to calls for service, they would also operate in a decidedly more proactive manner to engage citizens in the identification of local problems and enlist the support of the public in preventing crime. To achieve the goals of a problem-oriented approach, the Edmonton Police Service was restructured to put more officers on front-line service delivery. Officers were also given more authority to engage in street-level decision making. In 1987, a Neighbourhood Foot Patrol program was put in place in those twenty-one areas identified with high repeat calls for service. Each foot patrol area had a neighbourhood constable assigned to that location. To promote joint community/police problem-solving, Neighbourhood Advisory Committees were created. Finally, twelve community police stations were established, each staffed by volunteers and one officer, where people were encouraged to report non-emergency problems, thereby relieving the emergency dispatch system.

As a result of this initiative, a 41% drop in the total number of criminal offences was documented, the largest reduction in any major Canadian city. By way of comparison, there were 121 278 incidents in 1991 and 71 595 incidents in 1995. Also, there was a 31% decrease in violent crimes between 1991 and 1995. It is noted that these reductions have not been confirmed by independent scientific evaluation. As an exercise in community-based policing, this experiment probably allowed the Edmonton Police Service to avoid the creation of additional police officer positions, as well as a matching increase in operating and capital expenditures for the police department.

More recently, the Edmonton Police Service has been actively engaged in the development of a registered crime prevention program known as: "The Heads Up Fraud Prevention Association." Spearheaded by Economic Crime Section within the Edmonton Police Service, this initiative grew out of a brainstorming session in November 1995 aimed at dealing with high workload and limited personnel. The Edmonton Police Service was deeply aware of the value inherent in the broader community, and that many individuals, groups, and organizations could be mobilized to assist with a community-driven initiative aimed at education in the reduction of fraud. A steering group was established to drive this project forward and subdivided the thirty-three educational initiatives into the following key categories:

- Seniors;
- Written publications;
- Communications;
- Financial services; and
- Legislation.

In the area of education, the "Heads Up" group has produced a series of pamphlets dealing with the following topics:

- Phony bank inspector fraud;
- Automated teller/bank card fraud;
- Guide for parents, children, and computers: the hidden danger;
- Counterfeit currency;
- Home renovation fraud;
- Credit card fraud prevention checklist; and
- Counterfeit credit cards.

UNITED STATES

National Crime Prevention Council (NCPC)

During the 1970s, many people in the U.S. felt powerless in the face of growing urban crime. There was a strong view that nothing could be done by members of the public to control or prevent crime and disorder. In an attempt to counteract this view and present a more positive attitude to the challenge of crime, a diverse group of individuals formed the National Citizens' Prevention Campaign. This campaign, which featured McGruff, the U.S. symbol of crime prevention, was developed by government policy makers, police representatives, as well as members of the business and labour community.

In 1977, the senior members of the FBI approached the Advertising Council with a proposal for a crime prevention public service advertising campaign. Discussions around this campaign led to interest from the AFL-CIO and resulted in contact being made within the U.S. Department of Justice's Law Enforcement Assistance Administration (LEAA). Later that year, the LEAA, the FBI, and the AFL-CIO joined with the National Sheriffs' Association and the International Association of Chiefs of Police to form the core of a Crime Prevention Coalition, which has now grown to include a membership of 136 organizations.

Around 1982, with funding from the federal government, the National Crime Prevention Council was formed as a non-profit organization to continue the McGruff Campaign and coordinate the various activities of the Crime Prevention Coalition. Over time the NCPC has developed a vast array of services, publications, and programs aimed at crime prevention in the U.S. It has formulated and implemented widely-respected youth development programs, as well as demonstration projects such as the Community Responses to Drug Abuse and the Texas City Action Plan To Prevent Crime. Significant program assistance can be acquired through the NCPC in the following areas:

- Public education;
- Community mobilization;
- Youth mobilization;
- Police-community partnerships;
- Safe public places;
- Violence prevention;
- Bias crime; and
- Property crime.

Over its lifetime, the NCPC has grown into a significant agency which concentrates its efforts and activities in ways that have made it a valuable resource for crime prevention in the U.S. It has been estimated that more than half a million young persons have been impacted by the NCPC's youth programs (NCPC, 1998).

Texas—City Action to Prevent Youth Crime and Victimization

In 1993, the city of Houston (population, approximately 1.6 million) created a Coalition involving citizens, youth and childcare service providers, police, judicial, and health care representatives and sixteen local youths. The problem that the Coalition was attempting to address was the cycle of arrest and re-arrest experienced by many young offenders. There was a sense that these young offenders were unable to access the educational and recreational activities that might reduce their recidivism. The following preventative actions were set out in order to reduce delinquency and victimization:

- *Education action plan*—programs in diversity and cultural sensitivity were developed along with learning approaches for youth who were not in attendance at school. Prevention programs aimed at a range of children from pre-school onward;
- *Employment action plan*—developing links between employers and providers of youth employment services, offering job-readiness programs and promoting greater public awareness of youth employment needs;
- *Health/environment action plan*—taking action to prevent street gang activity, referring at-risk young people to appropriate services, and taking action to reduce portrayals of violence in the media;
- *Recreation action plan*—development of a database on youth organizations and services, focusing on youth recreation activities, and providing assistance in preparing young persons with communications and problem-solving skills.

While these interrelated programs have not been validated by independent scientific evaluation, it appears that there has been a 14% reduction in overall crime per 100 000 population. It was found (Saenz, 1993) that the successful implementation of a comprehensive program of crime prevention requires the following:

- Strong and influential leadership from local leaders (both government and police);
- Planning groups that are representative of the community and can enlist local expertise;
- Public awareness and support; and
- Stable staff support and appropriate training for planning groups.

This initiative was carried out within the limits of existing municipal funds. However, some technical and administrative assistance was provided by the National Crime Prevention Council (through a U.S. Department of Justice, Bureau of Justice Assistance grant).

GREAT BRITAIN

Crime prevention has been pursued in Great Britain since the early 1960s. In the early 1980s, however, a considerable increase in activity took place in order to strengthen the pursuit of crime prevention initiatives. This was justified according to the following considerations:

- Community decay and rising levels of property crime, violence, and fear of crime;
- Limited ability of prisons to prevent recidivism; and
- Impact of research showing that situational crime prevention measures could be highly effective.

There has been a Standing Committee on Crime Prevention in existence in Great Britain since 1966, and it has been successful in bringing together representatives from trade, industry, and insurance companies, as well as trade unions and chief constables. In 1983, the Standing Committee was fortified through the creation of the Home Office's Crime Prevention Unit. This was followed, in 1986, with the appearance of the Crime Prevention Training College through an Inter-departmental ministerial working Group on Crime Prevention. Finally, in 1995, the Crime Prevention Agency was introduced in order to provide improved coordination and promotion for crime prevention initiatives and strategies for local implementation. As with many other such organizations, the Crime Prevention Agency focuses on best practices, and aims to:

- Initiate, develop, and promote innovative ideas to help prevent crime and reduce fear of crime;
- Work with the police, local authorities, business, industry, volunteers, and the community to target specific types of crime;
- Ensure all sections of the community have an opportunity to be involved in crime prevention; and
- Advise ministers on the major strategic and policy implications concerning issues and other strategic approaches to crime prevention.

In 1988, the Home Office Crime Prevention Division introduced the "Safer Cities Programme" which was designed to reduce crime, fear of crime, and contribute to the improvement of the quality of life in cities. Over time, this program has assisted in the establishment of more than 4400 crime prevention projects in 30 inner city and urban areas. Evaluation of the various initiatives has resulted in the following findings:

- An evaluation of domestic burglary initiatives discovered that they have reduced incidents of burglary by a minimum of 10%;
- Some negative effects, such as crime displacement, are beginning to emerge; and
- Situational techniques and crime opportunities reductions can be effective in reducing crimes. This includes the use of closed-circuit television in certain high-crime areas.

In terms of funding, the Safer Cities program received more than $50 million U.S. in 1996. The support for crime prevention is clearly significant and the Home Office provides substantial support for the evaluation of model projects.

Northumbria—City Action to Prevent Crime and Reduce the Fear of Crime

In this jurisdiction, the Northumbria Community Safety Strategy (NCSS) was begun in 1993 to prevent crime and reduce the fear of crime among local citizens. In order to raise public safety, the NCSS focused on drug use, car crime, violent crime, burglary, young offenders, and those most vulnerable to involvement in such crimes. The strategy attempted to bring about change through initiatives that would:

- Create a safer environment through improved street lighting, and addressing vandalism and graffiti;
- Improve both residential and commercial security by the installation of sound locks and alarm systems, as well as the use of closed-circuit television in public areas;
- Divert young people from crime through the involvement of local employers in collaborative projects; and
- Create community projects and assist victims of crime, utilizing volunteers and organizational supports.

These initiatives are assisted by the Northumbria Police and include both governmental and non-governmental organizations. The NCSS is financially supported by funding from the Home Office's Single Regeneration Budget.

Merseyside—Coordinated Action to Tackle Repeat Victimization

In order to address a serious problem with repeat victimizations, various local authorities collaborated with the police on a system of mapping the location of criminal activity. This information was used to alert schools, housing authorities, and social service agencies in Merseyside. As a result, it was hoped that the physical opportunities for crime would be reduced or eliminated. While the outcomes of the overall program have not been formally evaluated (Farrell and Pease, 1993), it was learned that the Youth Action Programme, that provides positive sports and recreational opportunities for the youth in high crime areas, resulted in a 20% reduction in calls for police assistance in certain areas.

Newcastle—Closed-circuit Television Surveillance of Public Places

Within the centre of the city of Newcastle, it was found that several problems were developing that impacted on public order (e.g., vandalism), personal safety (e.g., assaults and robberies), and property crime (break-ins and theft). In order to address these problems, the city has been utilizing closed circuit television cameras (CCTV) since 1992. The use of this technology is an attempt to deal with the situational factors that contribute to crime. It was found that the following benefits attach to CCTV:

- Discrete surveillance of high crime areas;
- Coordination of rapid, or targeted, response through direct radio communication between civilian monitors and front-line officers; and
- Capacity to produce photographs and retain videotape records for use in court preparation and prosecutions.

Using crime data from two years prior to the installation of fully operational CCTV, in March 1993, it was determined that average monthly crime reductions were as follows:

- Burglary 56%;
- Thefts from motor vehicles 50%;
- Thefts of motor vehicles 47%;
- Incidents of criminal damage 34%; and
- Overall crime reduced by 19.2% (with little evidence of displacement of crime to neighbouring areas).

The use of CCTV has been adopted by many other city councils throughout England and Wales. However, the commentators for this project warn that changes to the physical environment must be undertaken with care so as to create clear sight lines for the cameras without disturbing the quality of life within the city centre (Edwards and Tilley, 1994; Brown, 1995).

OTHER JURISDICTIONS

International Centre for the Prevention of Crime (ICPC)

The International Centre for the Prevention of Crime (ICPC) is an international, non-governmental organization founded in order to assist cities and countries with the reduction of delinquency, violent crime, and insecurity. The ICPC collaborates directly with local, regional, national, and special purpose organizations to collect experience in the area of prevention from around the world. In order to assist with this purpose, the ICPC provides the following services:

- *Best practices bureau*—created to identify, collate, and disseminate information on successful crime prevention initiatives gathered from contacts with cities, government agencies, prevention experts, and research institutes throughout the world, as well as from the governments of Canada, Belgium, France, the United Kingdom, and the Netherlands. By 1997, the bureau provided reference to nearly one hundred promising programs with details on their methodology and results;
- *Technicial assistance*—staff of the ICPC will assist urban areas in developing public safety diagnoses and in formulating crime prevention strategies. They will also undertake needs assessments and develop work plans applying their international expertise. Also, the ICPC will support entities in the development of crime prevention strategies.
- *Exchange of expertise*—the ICPC supports such an exchange through cross-national meetings designed to extend the implementation of effective crime prevention programs and to broaden the exposure of practitioners in this area of activity. The ICPC promotes the development of a network of experts in order to facilitate the exchange of information and experience on an ongoing basis. As an example, in 1997 the ICPC coordinated a seminar in Burkina Faso for city and police leaders from six African countries dealing with the theme of "public security and crime prevention: roles and responsibilities of the police and local authorities in the context of democratization."
- *Tools to raise awareness*—the ICPC is developing tools that focus on leadership in crime prevention efforts, demonstrated by several categories of institutional and community partners, and the cost/benefits of crime prevention programs. Materials are also produced that show the contributions to be made to sustainable social and community development through crime prevention initiatives. In 1997, the ICPC launched a newsletter, Action, that presents current issues in crime prevention and cites examples of "best practices." The audience for this newsletter is intended to include decision makers in governments (at all levels), municipal managers, police executives, and a range of practitioners in areas related to crime prevention.

The ICPC is governed by a board of directors, bringing together the expertise of prevention specialists, private sector institutes, and other resources. The ICPC receives support from various governments, including Canada, France, the Netherlands, the United Kingdom, and Quebec. It publishes relevant materials and reports, and is managed on a day-to-day basis by a small staff of experienced personnel.

United Nations Interregional Crime and Justice Research Institute (UNICRI)

This organization was established to undertake research, training, and field activities that would advance the collection and dissemination of information, and the development and implementation of policies aimed at the improvement of crime prevention and control. Presently based in Rome, Italy, UNICRI maintains a website that provides details about its ongoing projects, conferences, and meetings. Information is also available on UNICRI publications, documentation services, and links to other relevant sites.

LILLE, FRANCE

Recognizing that in many industrialized nations citizens have become largely removed from several aspects of their social life (e.g., employment, housing, education, and culture), and that this form of alienation has contributed to higher levels of delinquency, officials have begun to look for more integration of citizens and society.

In 1984, a Conseil Communal de Prévention de la Délinquance (Community Crime Prevention Council) was established in the region of Lille, France to address problems associated with delinquency and insecurity, as well as to promote social integration of residents. Recognizing that police, justice officials, and social service professionals could not arrive at long-term solutions while they were operating in isolation, the Community Crime Prevention Council represented an attempt at drawing representatives of these groups together. Once combined, they could pursue an integrated strategy that would address the following priorities:

- Dividing the city into ten districts with local government councils that would encourage greater citizen participation in local decision-making, as well as the development of a stronger sense of community and civic pride;
- Situational measures that would result in better public lighting, refurbishment of rental housing estates, computerization to permit the central police to engage in crime mapping and diagnosis;
- Social development initiatives to prevent violence and drug-related activity among the youth; and
- Targeted measures to counteract the social exclusion of the youth in the city, and to assist them in finding solutions to their problems by linking them with social service, housing, and employment agencies.

ANTWERP, BELGIUM

In order to address what were seen as unacceptable crime rates and insecurity among its citizens, the city of Antwerp (population of approximately 463 000) has been successful in entering into "security and society contracts" with the central government. These contracts link financial assistance to targeted programs in crime prevention. Antwerp's concentration has been on initiatives aimed at improving quality of life through effective policing and social development. Among the specific measures introduced are the following:

- Efforts at preventive policing have been increased through the creation of an auxiliary police force of 90 officers (60 permanent and 30 contract). These officers are intended to increase informal social control by patrolling parking spaces, using public transportation to increase their visibility and surveillance capabilities;
- Social development initiatives aimed at improving the management and security of neighbourhoods with integrated inter-agency approaches. This includes a mediation and guidance program ("City-Coaching") for high-risk urban youth, as well as drug prevention, treatment, and reintegration programs for youth; and

- Improving security by engaging the public in neighbourhood observation to increase social control, and using of street workers to deal with problems stemming from drug abuse and prostitution.

The federal government had allocated approximately $4.7 million (U.S.) to the city of Antwerp for these initiatives in 1993. This amount was increased to nearly $6.3 million (U.S.) in 1996, with nearly 75% of that funding going to support police initiatives and to reinforce the number of police officers.

The Netherlands

In 1995, a Major Cities Memorandum was prepared to assist in the promotion of public safety and improved quality of life in the cities of Amsterdam (population, approximately 724 000), Rotterdam (population 599 000), the Hague (population 445 000), and Utrecht (population 234 000). In specific terms, the memorandum establishes the following targets for action in order to address crime and anti-social behaviour:

- Reductions in long-term unemployment, especially in depressed areas and among disadvantaged groups;
- Improvements in educational achievement levels, and reducing rates of secondary school drop-out by 35% over a four-year period;
- Reductions in robberies, interpersonal violence, commercial and residential break-ins, and vandalism through specific prevention measures; and
- Renewal of residential environments with increased public participation and improved racial harmony contributing to a better quality of life.

This project has not been formally evaluated. However, it represents a substantial financial commitment for the central government with approximately $1.5 billion (U.S.) being earmarked for various initiatives to be undertaken between 1996 and 1999 (Netherlands. Ministry of the Interior, 1995).

New Zealand

In order to address the issues relating to crime prevention, New Zealand established a Crime Prevention Unit (CPU) within the Prime Minister's office, which represents the highest level within this political structure. This unit, created in 1993, was in response to the recommendations of the Roper Committee on Violence (1987) and a body known as the Crime Prevention Action Group (1992). Leading a national strategy for crime prevention, the CPU seeks to reduce the overall number of offences, increase community involvement in crime prevention, and improve levels of public security. Specifically, the CPU has three important roles:

- To act as a key advisor to government on crime prevention strategies;
- To undertake planning, coordination, monitoring, and advisory functions relating to the implementation of the crime prevention strategy; and

- To ensure coordination and cooperation among all levels of government and representative community groups.

As a result of its efforts, the CPU has pinpointed seven priority areas for action:

- Improving effectiveness of support for "at risk" families;
- Reducing the incidence of family violence;
- Targeting preventative programs for young persons who are at risk of offending;
- Minimizing the formal involvement of casual offenders with the criminal justice system;
- Developing a coordinated and strategic national approach to manage programs that deal with the use and abuse of drugs and alcohol;
- Generating a strategy to deal with the problem of white collar crime; and
- Addressing the concerns of victims of crime and potential victims.

The CPU has been instrumental in helping to establish more than 58 "Safer Community Councils" throughout New Zealand, and is involved in several important prevention initiatives. Evaluation of these programs is considered critical and should be completed by the Year 2000.

Australia

In 1997, the government of Australia established a National Campaign Against Violence and Crime. This was intended as a high profile project aimed at reducing crime and violence within the community. The campaign is part of the Attorney General's department and works closely with the Lead Ministers' National Anti-crime Strategy. It is part of a strategic approach to crime prevention that places a priority on practical methods for dealing with related issues. The overall purpose of the campaign is as follows:

- To build stronger and more viable communities;
- To reduce levels of fear and crime;
- To protect vulnerable groups; and
- To change attitudes that are prone to tolerate crime.

By placing a priority on specific areas of crime prevention, the campaign hopes to meet its mandate in the near term. It is also planning to address issues relating to crime and violence against the country's indigenous and ethnic communities.

South Africa

As a country plagued with serious crime problems, compounded by significant political upheaval, South Africa is attempting to address its public safety issues. With the highest murder rate in the world, and increasing levels of property crime, it is important that the South African government respond to these demands. In 1995, the South African National Crime Prevention Strategy was developed as a broad approach to move from crime control to crime prevention. In undertaking this transformation, there was recognition of the following obstacles:

- Overcoming the unreliability of existing crime data;
- Influence of the media in shaping public perceptions about crime;
- Complications associated with political transformation and the negative inheritance of the former political regime, characterized by policies of apartheid;
- An established culture of violence;
- Significant poverty and lack of development;
- Easy availability of weapons and firearms; and
- Issues pertaining to gender inequality and substantial ethnic tensions.

With these challenges in mind, the National Crime Prevention Strategy (NCPS) planned action along four fronts:

- Streamlining the criminal justice system and its related processes;
- Applying the theory and practice of crime prevention through environmental design;
- Improving public values relating to crime and undertaking widespread public education; and
- Addressing transnational criminal activity.

With leadership from the Ministry of Safety and Security, the NCPS works with key agencies and departments to pursue the actions noted above. The business community has been actively involved in the development of this strategy, and has even recommended the establishment of a Council of Public Safety and Crime Prevention within the President's office.

SUMMARY

From its beginnings as a minor component of a police department's efforts at developing good relations within the community, crime prevention has steadily grown into something that, in many jurisdictions, constitutes the leading edge of public safety initiative. When approached as an integrated strategy for the removal of causes of crime, it is an area that can operate almost entirely outside of conventional police channels. Many of the instances cited above hinge more significantly on community mobilization and social development than on active police interaction or intervention. This is not to imply that police are not critical stakeholders in the crime prevention enterprise. On the contrary, their involvement and expertise is central to many successful initiatives. However, in order to distinguish crime prevention from other modes or forms of community policing, it would appear that crime prevention is something that can be pursued with increasingly less direct police engagement. This, of course, makes it very attractive for police organizations who want to concentrate their proactive and reactive activities in areas that most fully demand the knowledge, skills, abilities, and aptitudes of fully-sworn officers. With the increasing need to rationalize the funds invested in law enforcement, it makes eminent good sense to place an emphasis on crime prevention, not only because it affords programs that are effective in reducing crime and improving public safety, but also because it can contribute to the best focus for highly trained, highly specialized police officers.

QUESTIONS FOR CONSIDERATION

1. How does crime prevention differ from community policing?
2. What are some possible ways in which crime prevention initiatives could be measured?
3. Discuss the differences between community relations and public relations for the police.
4. How could crime prevention initiatives be tailored to meet the needs of the following groups: young persons, the elderly, low income neighbourhoods, ethnic groups, business people/merchants?
5. What are some possible sources of friction between the various groups that would make up a community-based crime prevention council?
6. What topics would be relevant to the training, education, and development of community services officers?
7. Discuss various crime prevention initiatives and their application within your own community.
8. How can the application of the principles of CPTED improve safety and security in a variety of public settings, for example, parks, housing developments, and downtown areas?

REFERENCES

Anderson, D., S. Chenery, and K. Pease (1995). *Biting back: tackling repeat burglary and car crime*. London : Home Office Police Department. (Police Research Group, Crime Detection and Prevention Series paper; no. 58).

Bennett, S.F. and P.J. Lavarakas (1989). "Community-based crime prevention: an assessment of the Eisenhower foundations neighborhood program," *Crime and Deliquency*, 25(2).

Berry, G. and M. Carter (1992). *Assessing crime prevention initiative: the first steps*. London: Home Office (Crime prevention unit paper; no. 31).

Bodlakova, I, and J. Nemec (1994). *Project of criminality prevention in the town of Vsetin*. Czech Republic: Board of Criminality Prevention.

Bridgeman, C. and A. Sampson (1994). *Wise after the event: tackling repeat victimisation: a report of the National Board for Crime Prevention*. London: Home Office.

Brown, B. (1995). *CCTV in town centres: three case studies*. London: Home Office Police Department.

Buendia, H.G. (ed.) (1989). *Urban crime: global trends and policies*. Tokyo: The United Nations University.

Canada. Ministry of the Solicitor General (1984). *Working together to prevent crime: a practitioner's handbook*. Ottawa: Minister of Supply and Services.

Canada. National Crime Prevention Council (1995a). *First annual report*. Ottawa: National Crime Prevention Council.

Canada. National Crime Prevention Council (1995b). *Dollars and sense: analyzing the cost-benefit of crime prevention through social development*. Ottawa: National Crime Prevention Council.

Canada. Standing Committee on Justice and the Solicitor General (1993). *Crime prevention in Canada: toward a national strategy*. Ottawa: House of Commons.

Cirel, P. et al. (1977). *An exemplary project: community crime prevention*. Seattle, Washington: National Criminal Justice Reference Service.

Clarke, R.V. (ed.) (1982). *Situational crime prevention: successful case studies*. New York: Harrow and Heston.

Crime Concern (1993). *A practical guide to crime prevention for local partnerships*. London: Home Office.

Crime Concern (1996). *Reducing crime and criminality in high crime neighbourhoods*. Swindon, U.K.: Crime Concern. (Briefing paper; no. 3)

Crime Prevention Unit (1994). *A guide to setting up a safer community council*. Wellington, New Zealand: Crime Prevention Unit, Department of the Prime Minister and Cabinet.

Crowe, T.D. (1991). *Crime prevention through environmental design*. Stoneham, Mass.: Butterworth-Heinemann.

De Liège, M.P. (1993). "Social development and the prevention of crime in France: a challenge for local parties and central government," In Heidensohn, F. and M. Farrell (eds.) *Crime in Europe*. New York: Routledge, pp. 121–132.

De Waard, J. (1995). "Special issue: crime prevention in the Netherlands: policies, research and practice, part 1." *Security Journal*, Vol. 6, No. 1.

De Waard, J. (1996). "Special issue: crime prevention in the Netherlands: policies, research and practice, part 2." *Security Journal*, Vol. 7, No. 2.

Developmental Research and Programs (1994). *Communities that care: risk-focused prevention using the social development strategy*. Seattle, Washington: Development Research and Programs.

Eck, John E. and David Weisburd (eds.) (1995). *Crime and place*. Monsey, N.Y.: Criminal Justice Press and the Police Executive Research Forum (Crime prevention studies; vol. 4).

Eckblom, P. (1992). "Preventing post office robberies in London: effects and side effects," In Clarke, R.V. (ed.), *Situational crime prevention: successful case studies*. New York: Harrow and Heston, pp. 66–74.

Eckblom, P. and K. Pease (1995). "Evaluating crime prevention," In Tonry, M.H. and M.H.Farrington (eds.) *Building a safer society: strategic approaches to crime prevention*. Chicago: University of Chicago Press.

Edwards, P. and N. Tilley (1994). *CCTV: looking out for you*. London: Home Office.

Erikson, I., M. Hakansson, and M. Olsson (1992). *Some observations on crime prevention in Sweden*. Sweden: The Swedish Institute.

European Forum for Urban Safety, Federation of Canadian Municipalities, and the United States Conference of Mayors (1989). *Agenda for safer cities: final declaration: European and North American conference on urban safety and crime prevention, October 10–13, 1989, Montreal*.

Farrell, G. and K. Pease (1993). *Once bitten, twice bitten: repeat victimization and its implications for crime prevention*. London : Home Office Police Research Group. (Monograph; no. 46).

Farrington, D.P. (1993). "Success stories in the prevention of adolescent aggression and youth violence," paper presented at the second world conference on criminal injury, Atlanta, Georgia, May 1993.

Farrington, D.P. (1996). *Understanding and preventing youth crime*. New York : Joseph Rowntree Foundation.

Felson, M. (1992). "Routine activities and crime prevention: armchair concepts and practical action." *Studies on Crime and Crime Prevention*, Vol. 1, No. 1, pp. 30–34.

Fenley, M.A. et al. (1993). *The prevention of youth violence: a framework for community action*. Atlanta, Georgia: Centers for Disease Control and Prevention.

Fisher, B. (1993). "What works: block watch meetings or crime prevention seminars?" *Journal of Crime and Justice*, Vol. 16, No. 1.

Fleissner, Dan and Fred Heinzelman (1996). *Crime prevention through environmental design and community policing*. Washington, D.C.: U.S. Department of Justice, National Institute of Justice.

Forrester, D. et al. (1990). *The Kirkholt burglary prevention project: phase II*. London: Home Office. (Crime Prevention Unit Paper; no. 23).

Grabosky, P. and M. James (eds.) (1995). *The promise of crime prevention: leading crime prevention programs*. Canberra: Australian Institute of Criminology.

Graham, J. and T. Bennett (1995). *Crime prevention strategies in Europe and North America, 2nd ed*. Helsinki, Finland: Helsinki Institute for Crime Prevention and Control.

Greenwood, P.W. et al. (1996). *Diverting children from a life of crime: measuring costs and benefits*. Santa Monica, Calif. and Washington, D.C.: RAND

Griffiths, Curt T. and Simon N. Verdun-Jones (1994). *Canadian criminal justice*. Toronto: Harcourt Brace.

Hahn, A. and C. Lattimore (1995). *Quantum opportunities program: a brief on the QOP pilot program*. Waltham, Mass.: Brandeis University, Center for Human Resources.

Harvey, L., P.Grimshaw, and K. Pease (1989). "Crime prevention delivery: the work of crime prevention officers," In Morgan, R. and D. Smith (eds.) *Coming to terms with policing*. London: Routledge.

Hastings, Ross (1993). "Workshop on crime prevention and community policing: presentation for the OPP Organizational Review Project, December 2, 1993" (unpublished).

Hauber, A. et al. (1996). "Some new forms of functional social control in the Netherlands and their effects." *British Journal of Criminology*, Vol. 36, No. 2 (Spring), pp. 199–219.

Home Office (1991). *Safer communities: the local delivery of crime prevention through the partnership approach*. London : HMSO.

Hunter, R.D. and C.R. Jeffrey (1992). "Preventing convenience store robbery through environmental design," In Clarke, R.V. (ed.) *Situational crime prevention: successful case studies*. New York: Harrow and Heston, pp. 194–204.

Inter-Departmental Strategy Team (1996). *National crime prevention strategy*. Pretoria, South Africa.

Jacobs, Jane (1961). *The death and life of great American cities*. New York: Random House.

Jeffrey, C.R. (1971). *Crime prevention through environmental design*. Beverly Hills, Calif.: Sage Publications.

John Howard Society of Alberta (1995). *Crime prevention through social development: a literature review*. Edmonton: The Society.

Johnson, V., J.Shapland, and P. Wiles (1993). *Developing police crime prevention management and organizational change*. London: Home Office (Crime prevention unit paper; no. 41).

Lavrakas, P.J. and J.W. Kushmuk (1986). "Evaluating crime prevention through environmental design: the Portland commercial demonstration project," In Rosenbaum, D.P. (ed.) *Community crime prevention: does it work?* Beverly Hills, Calif.: Sage Publications, pp. 202–227.

Linden, R. (1996). *Building a safer Canada: a community-based crime prevention manual*. Ottawa: Justice Canada and Supply and Services Canada.

Loree, Donald and R.W. Walker (1991). *Community crime prevention: shaping the future*. Ottawa: RCMP.

Lurgio, A.J. and D.P. Rosenbaum (1986). "Evaluation research in community crime prevention: a critical look at the field," In Rosenbaum, D.P. (ed.) *Community crime prevention: does it work?* Beverly Hills, Calif.: Sage. (Sage Criminal Justice System Annual; Vol. 22).

MacLeod, L. (1995). *Mobilizing political will and community responsibility to prevent youth crime*. Ottawa: National Crime Prevention Council.

Mayhall, Pamela D. (1985). *Police-community relations and the administration of justice*. New York: John Wiley & Sons.

Mayor's Task Force on Safer Cities (1992). *Toward a safer Edmonton for all: final report*. Edmonton, Alta.: Safer Cities Initiatives Office.

Mayor's Task Force on Safer Cities (1996). *Achieving a safer city together: making Edmonton a safer city*. Edmonton, Alta. : Safer Cities Initiatives Office.

McLaughlin, K.A. and K.J. Brilliant (1997). *Healing the hate: a national crime prevention curriculum for middle schools*. Newton, Mass.: Educational Development Center, Inc.

Mendel, R.A. (1995). *Prevention or pork? a hard-headed look at youth-oriented anti-crime programs*. Washington, D.C.: American Youth Policy Forum.

Milton S. Eisenhower Foundation (1990). *Youth investment and community reconstruction: street lessons on drugs and crime for the nineties*. Washington, D.C.: Milton S. Eisenhower Foundation.

Ministry of Justice (1997). *Our collective responsibility: a national programme for crime prevention*. Stockholm, Sweden: National Council for Crime Prevention.

Morris, P. and K. Heal (1981). *Crime control and the police: a review of the research*. London: HMSO (Home Office research study; no. 67).

Mulvey, E.P., M.W. Arthur, and N.D. Reppucci (1993). "The prevention and treatment of juvenile delinquency: a review of the research." *Clinical Psychology Review*, Vol. 13, pp. 133–167.

National Crime Prevention Council (1994). *Taking the offensive to prevent crime: how seven cities did it*. Washington, D.C.: National Crime Prevention Council.

National Crime Prevention Council (1995). *350 tested strategies to prevent crime: a resource for municipal agencies and community groups*. Washington, D.C.: National Crime Prevention Council.

National Crime Prevention Council (1997). *Designing safer communities: a crime prevention through environmental design handbook*. Washington, D.C.: National Crime Prevention Council.

National Crime Prevention Council (1998). National Crime Prevention Council on-line resource center. Website—http://www.ncpc.org/

National Crime Prevention Institute (1986). *Understanding crime prevention*. Stoneham, Mass.: Butterworth Publishers.

National League of Cities (1994). *Rethinking public safety: the 1994 futures report*. Washington, D.C.: National League of Cities.

Netherlands. Ministry of the Interior (1995). *Covenant: voluntary agreement between the government and the four major cities*. The Hague: Ministry of the Interior, The Netherlands.

Newman, Oscar (1972). *Defensible space: crime prevention through urban design*. New York: Macmillan.

Norris, F.H. and K. Kaniasty (1992). "A longitudinal study of the effects of various crime prevention strategies on criminal victimization, fear of crime, and psychological distress." *American Journal of Community Psychology*, Vol. 20, No. 5.

O'Keefe, G.J. et al. (1996). *Taking a bite out of crime: the impact of the national citizens' crime prevention media campaign*. Thousand Oaks, Calif.: Sage Publishing.

Ontario Provincial Police. Provincial Police Academy. Systems Approach to Training Unit. (1986). *Community services officers: training needs analysis phase report*. Brampton, Ont.: OPP, Provincial Police Academy.

Osborne, S. and H. Shaftoe (1995). *Safer neighbourhoods? successes and failures in crime prevention*. London: Safe Neighbourhoods Unit.

Poyner, B. (1992). "Situational crime prevention in tow parking facilities," In Clarke, R.V. (ed.) *Situational crime prevention: successful case studies*. New York: Harrow and Heston, pp. 174–184.

Roehl, J. (1984). *Evaluation of the urban crime prevention program*. Washington, D.C.: U.S. National Institute of Justice, Institute for Social Analysis.

Rosenbaum, D.P. (1986). *Community crime prevention? Does it work?* Beverly Hills, Calif.: Sage Publications.

Saenz, G.G. (1993). *Texas City action plan to prevent crime*. Houston: Coalition of the Texas City action plan to prevent crime.

Safe Neighbourhoods Unit (1993). *Crime prevention on council estates*. London: Her Majesty's Stationery Office.

Schinke, S.P., M.A. Orlandi, M.A., and K.C. Cole (1992). "Boys and girls clubs in public housing developments: prevention services for youth at risk." *Journal of Community Psychology*, Office of Substance Abuse Prevention Special Issue: 118–128.

Scottish Crime Prevention Council (1994). *Turning the tide: conference report*.

Sherman, L. et al. (1997). *Preventing crime: what works, what doesn't, what's promising: report to the United States Congress*. Washington, D.C.: National Institute of Justice.

Solicitor General of Canada (1983). *Working together to prevent crime: a practitioner's handbook*. Ottawa : Solicitor General of Canada.

South Australia, Government of (1989). *Confronting crime: the South Australian crime prevention strategy*. South Australia : Attorney-General's Department.

Task Force on Crime Prevention (1993). *Partners in crime prevention: for a safer Québec*. Quebec: Ministry of Public Security.

Tilley, Nick (1995). *Thinking about crime prevention performance indicators*. London : Home Office Police Department, Police Research Group (Crime detection and prevention series; paper no. 57).

United Nations (1995). "Guidelines for cooperation and technical assistance in the field of urban crime prevention." *Trends: UNCJIN Crime and Justice Newsletter*, Vol 2, No. 3, pp. 11–12.

United Nations. Economic and Social Council (1993). *Report of the Commission on Crime Prevention and Criminal Justice: guidelines for the prevention of urban crime*.

United Nations. General Assembly (1990). *Prevention of urban crime: report of the 8th U.N. congress on the prevention of crime and treatment of offenders*.

United Nations. General Assembly (1990). *United Nations guidelines for the prevention of juvenile delinquency (the Riyadh guidelines): report of the 8th U.N. congress on the prevention of crime and the treatment of offenders*.

United Nations. General Assembly (1991). *Conclusions and recommendations: the need for technical cooperation in the field of crime prevention and criminal justice: ministerial meeting on the creation of an effective United Nations crime prevention and criminal justice program, November 21–23, Paris*.

United Nations Interregional Crime and Justice Research Institute and Research Institute, Ministry of the Interior of the Russian Federation (1994). *Crime and crime prevention in Moscow*. Rome: UNICRI.

U.S. Department of Justice. Bureau of Justice Assistance (1993). *The social impact of the national citizens' crime prevention campaign: focus on what works*. Washington, D.C.: Bureau of Justice Assistance.

Van Andel, H. (1989). "Crime prevention that works: the care of public transport in the Netherlands." *British Journal of Criminology*, Vol. 29, No. 1, pp. 47–56.

Van Limbergen, K., S. Walgrave, and I. Dekegel (1996). *Promising practices: crime prevention in Belgium, 10 examples*. Brussels, Belgium: Permanent Secretariat for Crime Prevention Policy.

Van Soomeren, P. and T. Wilendorp (1996). "Secured by design in the Netherlands." *Security Journal*, Vol. 7, No. 2.

Waller, Irvin (1982). "What reduces residential burglary: action and research in Seattle and Toronto," In Schneider, H.J. (ed.) *Victim in international perspective*. New York: de Gruyter.

Waller, Irvin (1989). *Current trends in European crime prevention: implications for Canada*. Ottawa: Department of Justice Canada.

Waller, Irvin (1991). *Putting crime prevention on the map: introductory report to the International Conference on Urban Safety, Drugs and Crime Prevention*. Paris : European Forum for Local Authorities on Urban Safety.

Waller, Irvin (1992). "Strategic planning and policing," In *Managing diversity and change: a training strategy, May 25-19, 1992*. Regina, Sask.: Police Educators Conference Board of Canada, pp. 32–41.

Wekerle, G.R. and C. Whitzman (1995). *Safe cities: guidelines for planning, design and management*. New York: Van Nostrand Reinhold.

Willemse, H.M. (1994). "Developments in Dutch crime prevention," In Clarke, R.V. (ed.) *Crime Prevention Studies*, Vol. 2, pp. 33–47. Monsey, N.Y.: Criminal Justice Press.

Wise-Harris, D. (1991). *A safer city: the second stage report of the safe city committee*. Toronto: Planning and Development Department.

Yoshikawa, H. (1994). "Prevention as cumulative protection: effects of early family support and education on chronic delinquency and its risks." *Psychological Bulletin*, Vol. 115, No. 1, pp. 1–27.

Zigler, E., C. Taussig, and K. Black (1992). "Early childhood intervention: a promising preventative for juvenile delinquency." *American Psychologist*, Vol 47, No. 8, pp. 997–1006.

TOWARD MORE LEARNING

CRIME PREVENTION THROUGH ENVIRONMENTAL DESIGN

Eck, John E. and David Weisburd (eds.) (1995). *Crime and place*. Monsey, N.Y.: Criminal Justice Press and the Police Executive Research Forum (Crime prevention studies; vol. 4).

Jeffrey, C.R. (1971). *Crime prevention through environmental design*. Beverly Hills, Calif.: Sage Publications.

This is considered to be a central work on the use of CPTED and continues to provide useful insight into this particular approach to crime prevention.

National Crime Prevention Council (1997). *Designing safer communities: a crime prevention through environmental design handbook*. Washington, D.C.: National Crime Prevention Council.

A recent publication that offers a good overview of the history, theory, and practice of CPTED. It deals with establishing a neighbourhood CPTED team and provides several case studies to demonstrate the effectiveness of CPTED in dealing with public safety problems. The booklet contains some useful references and provides "neighbourhood survey" and "neighbourhood inventory" forms for local application.

PRACTICAL APPLICATIONS

Canada

National Crime Prevention Centre

This organization, which is part of the federal government, has moved into Phase II of its National Strategy on Community Safety and Crime Prevention. It is complemented by a range of other government programs and initiatives, particularly focused on children and youth. It can be contacted as follows:

National Crime Prevention Centre
Department of Justice
St. Andrew's Tower
284 Wellington Street
Ottawa, Ontario
K1A 0H8
Toll-free: 1-877-302-NCPC
Tel.: (613) 941-9306
Fax.: (613) 952-3515
Email: ncpc@web.net
Website: http://www.crime-prevention.org

United States

National Crime Prevention Council

This is an invaluable on-line resource with a considerable amount of constantly updated information on American crime prevention activities, resources, publications, and initiatives. It includes details about the history and development of the NCPC, news about upcoming events, and details about the National Citizens' Crime Prevention Campaign.

The NCPC can also be reached at:

National Crime Prevention Council
1700 K Street, NW, Second Floor
Washington, D.C.
20006-3817
Tel.: 202-466-6272
Fax.: 202-296-1356
Website: http://www.ncpc.org

Great Britain

Within the Home Office website there is a series of pages devoted to crime prevention. Included are useful guidelines for securing your home and vehicle, as well as information on Neighbourhood Watch and Home Watch inititatives. The site also provides data relevant to businesses looking for information on the prevention of fraud, burglary, and theft.

Of particular importance is the link to the Home Office's Crime Prevention Agency, which was established to ". . . stimulate, foster and develop strategies and tactics for the prevention of crime." Many of this agency's publications are available over the internet and provide detailed background on initiatives and programs that may be of interest to other jurisdictions.

Crime Prevention Agency
50 Queen Anne's Gate
London SW1H 9AT
Website for Home Office Crime Prevention:
http://www.homeoffice.gov.uk/cpmain.htm
Website for the Crime Prevention Agency:
http://www.homeoffice.gov.uk/cpa.htm

Other Jurisdictions

International Centre for the Prevention of Crime

The International Centre for the Prevention of Crime is based in Montreal, Canada and maintains an extremely detailed website with many examples of effective crime prevention programs throughout the world. It can be reached as follows:

507, Place d'Armes, bureau 2100
Montreal, Quebec
H2Y 2W8

Fax.: 1-514-288-8763
Email: cipc@crime-prevention-intl.org
Website: http://www.crime-prevention-intl.org

United Nations Interregional Crime and Justice Research Institute

The United Nations Interregional Crime and Justice Research Institute maintains a useful website that offers global coverage for many interesting projects relating to crime prevention. It reveals that such efforts are being undertaken everywhere from Albania to Zimbabwe. It can be reached as follows:

Via Giulia 52
00186 Rome, ITALY
Tel.: (+39-6) 6877437
Fax.: (+39-6) 6892638
Email: unicri@unicri.it
Website: http://www.unicri.it/

CHAPTER THREE

To exist as a North American is an amazing and enthralling fate. As in every historical condition, some not only have to live their fate, but also to let it come to be thought. What we have built and become in so short a time calls forth amazement in the face of its novelty, an amazement which leads to that thinking. Yet the very dynamism of the novelty enthralls us to inhibit that thinking.

George Grant (1969). "In defence of North America."

UNDERSTANDING THE CANADIAN CHARACTER

LEARNING OBJECTIVES

1. Identify key demographic trends in Canada that might influence the development of policing.
2. Explain the impact of key social trends on policing in Canada.
3. List five political trends that could affect the development or direction of policing in Canada.
4. Describe the legislative trends that may influence Canadian policing.
5. Explain the influence of technological change on police organizations in Canada.
6. Describe several crime trends and patterns currently affecting police organizations in Canada.
7. Identify several global trends that could influence policing in Canada.
8. Identify relevant sources of information dealing with social, political, legislative, technological, crime, and global trends.

INTRODUCTION

This chapter is devoted to an examination of the Canadian character, particularly as it pertains to policing. In order to attempt an understanding of this Canadian character, it is useful to consider a number of pertinent topics, including:

- Demographic trends;
- Social trends;
- Political trends;
- Legislative trends;
- Technological trends;
- Crime trends and patterns; and
- Global trends.

Immediately it should be stated that the treatment provided in this chapter is by no means comprehensive. However, by closely tracking key directions in these areas, it is hoped that an impression of the Canadian character, and its environmental setting, will emerge. That impression will contribute to a better understanding of the nature and challenge of community policing. Furthermore, by seeking out the background allowing us to construct this picture of the Canadian character, it will be possible to pursue further lines of inquiry in order to provide more detail and definition. It is important to learn where to look for insight into the Canadian character and its changing elements. For policing, it is critical because police organizations are called upon to serve the needs of specific and distinct communities in matters of the utmost importance: their safety, security, and well-being. To know the Canadian character at some considerable level of detail is essential to effective policing. If community policing is rooted in a true understanding of the "community" in which police function, the Canadian community constitutes a major reference point for this understanding.

Of course, a deep appreciation of the history of Canada would also be significant in comprehending the nature of the Canadian public. However, the Canadian landscape is changing rapidly and the forward glance is immediately more pertinent for understanding our policing enterprise.

DEMOGRAPHIC TRENDS

The word "demographics" is a combination of the Greek words "demos" meaning the people, and "graphein"—to write. The task of making determinations about the possible future by reckoning on people patterns is an extremely complex and challenging one (Adams, 1997). However, it is essential to the proper planning of all sorts of services, programs, and functions within the country. The statistics gathered with regard to population figures give us a picture of the country and allow for making reasonable projections into the future.

In planning police services, accurate population figures need to be available. Policing levels have typically been tied to police-to-population ratios, and decisions about policing expenditures frequently rest on comparative costs per capita. Certainly, beyond the raw numbers, demographics allows police decision makers to formulate community policing judgments relevant to a host of statistical categories, including the following:

- Aboriginal population;
- Citizenship;
- Ethnic origin;
- Household income;
- Household size;
- Immigrant population;
- Labour force activity;
- Language in the home;
- Place of birth;
- Population density;

- Schooling; and
- Visible minorities.

On several levels, it is vital that police organizations have a firm knowledge with respect to these types of demographic indicators. It constitutes part of the environmental scanning process that is central to the strategic management of any modern police department.

Canada has a well-established body that provides these figures. Statistics Canada (the Dominion Bureau of Statistics, created in 1918, became known as Statistics Canada in 1971) is a federal organization that takes responsibility for gathering, storing, analyzing, and disseminating statistics in three broad subject areas: demographic and social, socio-economic, and economic. Of considerable value to police organizations are those statistics that offer breakdowns of population figures according to race, age, ethnic origin, linguistic background, and other aspects. These provide enormous assistance that relates directly to community policing. Only through careful, methodical counting can police organizations know the precise makeup of their communities. With this kind of knowledge, they can begin to make determinations about engaging the members of those communities in meaningful programs aimed at effective policing services. By tracking the changes in population patterns over time, it is possible to make certain predictions about the size, age, and diversity of the population you will be serving. In specific terms, it is meaningful to consider some of the following demographic realities and their impact on policing in Canada:

- Aging population;
- Immigration patterns; and
- Youth culture.

Aging Population

The fact that our population is aging is obvious. However, the impact of that aging population is significant, especially for police organizations. Because fear of crime is an important aspect of policing, and because the elderly are typically more prone to such fears, it places a special onus on police services to ensure that they are addressing these needs. The figures below show the projections for population growth in Canada in various age groups:

Population Estimates for 1996 and Projections[1] for the Year 2001 (Thousands)

	1996[2]			2001		
	Both sexes	*Male*	*Female*	*Both sexes*	*Male*	*Female*
All ages	29 969.2	14 847.3	15 122.0	31 877.3	15 781.2	16 096.1
0–4	1 951.3	1 000.2	951.1	1 924.3	988.2	936.2
5–9	2 016.7	1 031.8	984.9	2 082.2	1 069.0	1 013.1
10–14	2 020.4	1 032.3	988.1	2 124.8	1 089.2	1 035.6
15–19	2 003.7	1 026.7	977.0	2 124.5	1 088.0	1 036.4
20–24	2 037.4	1 033.9	1 003.5	2 115.2	1 080.1	1 035.0

25–29	2 225.4	1 122.3	1 103.1	2 177.7	1 103.0	1 074.7
30–34	2 633.3	1 335.0	1 298.3	2 366.4	1 192.9	1 173.4
35–39	2 668.2	1 323.3	1 323.3	2 723.4	1 376.1	1 347.3
40–44	2 388.8	1 196.4	1 196.4	2 716.3	1 363.9	1 352.3
45–49	2 159.9	1 085.2	1 074.7	2 399.6	1 193.9	1 205.7
50–54	1 672.6	838.5	834.1	2 140.1	1 069.8	1 070.3
55–59	1 333.1	662.4	670.8	1 651.4	820.1	831.3
60–64	1 214.6	597.1	617.5	1 300.9	636.8	664.0
65–69	1 130.3	536.9	593.4	1 154.0	554.2	599.7
70–74	981.4	433.9	547.5	1 027.1	470.6	556.6
75–79	704.9	289.3	415.6	831.9	345.9	486.1
80–84	467.6	174.8	292.8	541.8	201.3	340.5
85–89	239.5	77.6	161.8	308.5	98.3	210.2
90 & over	120.0	31.9	88.1	167.4	39.8	127.6

Source: Statistics Canada, CANSIM, Matrices 6367 (estimates) and 6900 (projections) Extracted May, 1999. Reproduced with permission.

[1] Figures represent a medium-growth projection and are based on 1993 population estimates.

[2] Post-censal estimate of 1996 population.

The aging population should be an important concern to the police, and significant effort should be placed on understanding that community. Part of such an understanding may come from the careful analysis of this category of people. While they will not always constitute a true cohesive group, it is useful to consider detailed demographic information for the purposes of planning. It may also assist police organizations in developing strategies for effectively engaging the elderly in coherent problem-solving approaches.

Statistics Canada has produced a valuable report; *A portrait of seniors in Canada. 2nd ed.* (1997) that offers demographic insights that may impact on the delivery of police services. In long-range terms we learn that:

- By the year 2041, approximately 23% of the Canadian population will be over the age of 65.

Also, significant numbers of seniors take part in formal, unpaid volunteer activities, something that may relate directly to community policing initiatives on a number of important levels. The willingness of seniors to become involved in community policing, crime prevention, and other proactive approaches to the problem of crime and disorder will be a major benefit to many police agencies in Canada. The Statistics Canada study also examines criminal victimization and fear of crime among the elderly in Canada. Although seniors are less likely to be victims of a crime than younger people, they are more likely to feel unsafe when walking alone in their neighbourhoods, particularly after dark. This sense of fear is greater in senior women than senior men, and Statistics Canada has found that, in 1993, over half (i.e., 57%) of women aged 65 and over reported feeling unsafe alone in their neighbourhood after dark, as compared with only 19% of senior men. This is the reverse of the actual crime statistics as determined by Statistics Canada, with senior men being more likely the victims of personal or property crimes. However, with a rapidly aging population, the majority of whom will be senior women, it is appropriate for police organizations to be aware of, and sensitive to, these statistical realities. It has also been determined that a large majority of seniors live in private homes. For example, in 1992, 92% of all people 65 years of age and over, lived in a private household. Furthermore, 38% of all senior women, and 53% of those women aged 85 and over, lived alone. These statistics are particularly relevant when considering the impact of the fear of crime and victimization statistics quoted above.

As Normandeau and Leighton (1990) indicate, the size of the older population can be attributed to both low fertility rates and increased life expectancy. These elderly Canadians will increasingly be women, and will concentrate themselves in Canada's largest centres. Police organizations in those larger cities will need to be attuned to such demographic realities and prepare themselves accordingly.

IMMIGRATION PATTERNS

Another important area that pertains to the planning of community-based police services is the size and nature of immigration to Canada. By examining trends in immigration, it is possible to determine the diversity of cultural, ethnic, linguistic, and related services that will be expected of your police department.

Statistics Canada is, again, an essential source of information on this front and allows police planners to make some reasonable predictions to assist in addressing important service requirements. While some of the practical challenges facing census takers in the U.S. regarding accurate reporting of minorities (Ogawa, 1998) are shared in Canada, it is helpful to use available figures as a guide for planning purposes. Table 3-1, below, represents the top ten places of birth for total immigrants. Table 3-2 shows immigrants coming before 1961 and, finally, Table 3-3, details recent immigrants to Canada, taken from the 1996 census:

TABLE 3-1

TOTAL IMMIGRANTS[3]

	Number	%
Total	4 971 070	100.0
United Kingdom	655 540	13.2
Italy	332 110	6.7
United States	244 695	4.9
Hong Kong	241 095	4.8
India	235 930	4.7
People's Republic of China	231 055	4.6
Poland	193 375	3.9
Philippines	184 550	3.7
Germany	181 650	3.7
Portugal	158 820	3.2

[3] Non-permanent residents are not included in this table.

Adapted from 1996 census data. Reproduced with permission.

TABLE 3-2

IMMIGRATED BEFORE 1961

	Number	%
Total	1 054 930	100.0
United Kingdom	265 575	25.2
Italy	161 730	15.3
Germany	107 270	10.2
Netherlands	88 810	8.4
Poland	57 820	5.5
United States	45 050	4.3
Hungary	33 215	3.1
Ukraine	27 640	2.6
Greece	21 555	2.0
People's Republic of China	17 545	1.7

TABLE 3-3

RECENT IMMIGRANTS[4]

	Number	%
Total	1 038 995	100.0
Hong Kong	108 915	10.5
People's Republic of China	87 875	8.5
India	71 335	6.9
Philippines	71 325	6.9
Sri Lanka	44 235	4.3
Poland	36 965	3.6
Taiwan	32 140	3.1
Viet Nam	32 060	3.1
United States	29 020	2.8
United Kingdom	25 425	2.4

* This table contains data selected from Catalogue No. 93F0023XDB96003 in the Nation series.

[4] Recent immigrants are those who immigrated between 1991 and the first four months of 1996.

The shifts in population have been profoundly effected by global events and this is certainly true in the case of immigration to Canada. The period prior to 1961 reveals substantial emigration from those countries impacted by the Second World War, such as Italy, Poland, Hungary, and the Ukraine. More recent immigration patterns have shifted to the Asian and Asia Pacific countries, like Hong Kong, the People's Republic of China, the Philippines, and Viet Nam.

In 1996, Canada was home to about five million immigrants. This constitutes a 14.5% increase since 1991, and represents a little more than three times the growth rate (i.e., 4%) of the Canadian-born population. These immigrants have had the largest impact on the big cities in Canada. For example, 46% of Toronto's population in 1996 was composed of immigrants, which represents the largest immigrant population in all 25 census metropolitan areas (CMAs) in Canada. Vancouver had the second largest immigrant population, with immigrants accounting for more than one-third of its population. The source of Canada's immigrant population has shifted from the United Kingdom, the United States, and Europe to Eastern Asia (e.g., Hong Kong and the People's Republic of China) and Southern Asia (e.g., India and Sri Lanka). The third largest group originates from Central and South America, followed by immigrants from Africa and the Caribbean.

Tables 3-4 to 3-6 provide an instructive comparison of the recent immigrants to the major cities of Toronto, Vancouver, and Montreal, respectively:

TABLE 3-4

RECENT IMMIGRANTS[5] TO TORONTO

	Number	%
Total	441 035	100.0
Hong Kong	48 535	11.0
Sri Lanka	36 735	8.3
People's Republic of China	35 330	8.0
Philippines	33 210	7.5
India	33 185	7.5
Poland	18 605	4.2
Jamaica	16 780	3.8
Guyana	13 195	3.0
Viet Nam	12 290	2.8
Trinidad and Tobago	11 375	2.6

[5] Recent immigrants are those who immigrated between 1991 and the first four months of 1996.
Adapted from 1996 census data. Reproduced with permission.

TABLE 3-5

RECENT IMMIGRANTS TO VANCOUVER

	Number	%
Total	189 660	100.0
Hong Kong	44 715	23.6
People's Republic of China	27 005	14.2
Taiwan	22 315	11.8
India	16 185	8.5
Philippines	13 610	7.2
South Korea	6 335	3.3
Iran	4 640	2.4
United Kingdom	4 040	2.1
Viet Nam	3 855	2.0
United States	3 640	1.9

Adapted from "Immigrant population by place of birth and sex, showing period of immigration, for Canada, provinces, territories and census metropolitan areas, 1996 census," Catalogue No. 93F0023XDB96003. Reproduced with permission.

TABLE 3-6
RECENT IMMIGRANTS TO MONTREAL

	Number	%
Total	134 535	100.0
Haiti	9 995	7.4
Lebanon	9 610	7.1
France	7 540	5.6
People's Republic of China	6 650	4.9
Romania	5 225	3.9
Sri Lanka	4 675	3.5
Philippines	4 640	3.4
India	4 380	3.3
Viet Nam	4 135	3.1
Morocco	3 820	2.8

Adapted from "Immigrant population by place of birth and sex, showing period of immigration, for Canada, provinces, territories and census metropolitan areas, 1996 census," Catalogue No. 93F0023XDB96003. Reproduced with permission.

There is considerable insight embedded in these figures because they impact on the kind of police services required to meet the needs of these ethnic, cultural, or linguistic communities. This is true on several levels, including the variations in the nature of potential crime victims, and even in the types of crimes committed within these communities. The background of new community members coming from places like Iran, Haiti, Viet Nam, or India will be dramatically different, and will summon forth unique approaches from a sensitive police organization seeking to provide appropriate services.

Increasingly, Canadian police organizations are being called upon to deal with some of the following specialized crime areas:

- Asian gangs;
- Religious conflicts (e.g., Muslims against Sikhs);
- Drug crimes; and
- Ethnic organized crime.

The extension of ethnic conflicts within a Canadian setting places a real burden on police when dealing with the various combatants. This is true in several areas, including the high incidence of home invasions amongst people from the Asian community. It has also taken shape in Vancouver, British Columbia with the ongoing controversy relating to leadership in the Sikh community.

The process of dealing with ethnic groups must be informed by the special characteristics of those groups. It is important for operational reasons to have this type of insight (Cryderman and Fleras, 1992). The process of arriving at this insight begins with a clear understanding of the demographic nature of your community.

Youth Culture

Dealing with youth presents a particular problem for police agencies in Canada and calls for expertise and knowledge based on demographic trends. The importance of detailed knowledge about young people, especially young offenders, is crucial to law enforcement professionals. Making decisions about the nature and concentration of youth crime is essential to effective policing. Knowledge about the numbers of young people in your jurisdiction is central to a good understanding of this category. Police deployment, assignment, and training decisions should be informed by these demographics. Youth gangs are growing in many urban centres in Canada and they have become a significant flashpoint in places like Toronto.

SOCIAL TRENDS

Making determinations about social trends in Canada is not a simple matter and is not prone to the kinds of actuarial precision available to statisticians (Adams, 1997). Of course, there is a relationship between the demographic makeup of Canada and its social character. However, it becomes very difficult to offer predictions about our social institutions when the pace of change is taken into account. In spite of this difficulty, it is inevitable that predictions will be made by those who study the attitudes and values of the Canadian people. For example, Michael Adams (1997) notes that:

> According to the UN at least, we've succeeded in achieving the highest quality of life in the world. Our geography, size and history of accommodation of different groups have resulted in a more affluent and egalitarian society. Ideologically, we are a serendipitous mixture of toryism, liberalism and socialism. Personally, I believe that our success as a nation is also due, in part, to the declining influence of traditional demographic characteristics in defining this country and its people. We are creating a new post-modern Canadian culture. (pp. 20–21)

In any event, the Canadian character does have a number of social aspects that can be pointed to and studied. The first point is to clarify what we mean by "social trends." Essentially, the social dimension refers to the formal and informal arrangements within society. This includes considerations about the family, marital arrangements, group interactions, and any number of topics that describe how people within society deal with one another, individually and/or collectively. Clearly this is an area of immense importance to Canadian police organizations, as the police are profoundly part of what may be called the social network that maintains the fabric of our society.

An examination of Canada's social context will offer insights into the way our country has developed and should assist in making assessments of how we should be policed. The social dimension is part of an extended context that includes the political, economic, technological, and other trends cited in this chapter. Indeed, it is frequently difficult to differentiate between a social, political, or technological trend because they are often so tightly connected in their origins, aspects, and impacts. When we consider an individual social phenomenon, such as the appearance of hate literature on the Internet, we are dealing with something that penetrates the social, legal, political, religious, and technological realms. All these areas are influenced by this type of phenomenon and have an impact on the meaning of this issue. Therefore, the "sociology" of policing should include a careful study of the society in which policing takes place and the characteristics, qualities, and patterns established by individuals and groups within that society.

As noted by Giddens (1982):

> We live in an age of massive social transformation. In the space of only something like two centuries a sweeping set of social changes, which have hastened rather than lessened their pace today, have occurred. These changes emanating originally from Western Europe, are now global in their impact. They have all but totally dissolved the forms of social organisation in which humankind had lived for thousands of years of its previous history. (pp.4–5).

Among the topics to be dealt with when reviewing social trends and the sociological context of community policing are the following:

- Impact of the industrial revolution;
- Influence of capitalism;
- Class division and social transformation;
- Role of the modern state;
- Aspects of urban life;
- Family, gender, marriage, sexuality; and
- Morality, values, and ethics.

Sociology (i.e., the study of society) essentially begins with the Enlightenment and is tied to the idea of progress and human perfectibility. In the nineteenth century, through social theorists like Auguste Comte, Herbert Spencer, and Karl Marx, sociology began to take on a particular shape that was focused on capitalist society. With the turn of the century, Emile Durkheim and Max Weber gave sociology its definitive form and methodology and provided the groundwork for an extended analysis of urban, industrial capitalist society.

Because the police serve to provide stability within the social order, it is a matter of some importance that the social trends be studied carefully for their possible impact on the delivery of police services. To understand the social elements of Canadian life, it is necessary to cast widely for insight.

POLITICAL TRENDS

Again, as noted above, there is a great deal of cross-pollination between social and political trends. However, it may be possible to more easily identify the main features of the political realm that impact on policing. Politics, within a democratic society, has some of the following components:

- Open and accountable democratic institutions;
- Free elections;
- Constitutional protections;
- Legislative, executive, and judiciary branches of government;
- Ministerial responsibility;

- Judicial independence; and
- Universal suffrage.

Political trends are those events or developments occurring in the areas noted above at the federal, provincial, and municipal levels that may impact on police organizations directly. Because the federal jurisdiction bears upon police services across Canada, the trends at the national level have the potential to impact strongly. This is certainly true in the following areas:

- Drug legislation;
- Gun control (including restricted and prohibited weapons);
- Treatment of young offenders;
- Search and seizure;
- Private security and investigation;
- Identification of criminals and high-risk offenders;
- Creation of a national DNA databank;
- Other areas of a national scope (including the deployment of the RCMP, as a national police service).

It is also important to realize that the specifics of these so-called "political" issues, as they take legislative or regulatory form, have a **direct** impact on the limits imposed on police and their local constitutents in the context of community policing. The more prescriptive legislation is at the federal level, the more restrictive the police must be in terms of neighbourhood policing when local solutions to crime and security problems are being explored.

Changes in political jurisdictions have immediate and direct implications for policing and can be easily discerned in the following areas:

- Quebec sovereignty and separation;
- First Nations self-government and land claims negotiations;
- Western Canada separation;
- Creation of new political entities (e.g., Nunavut);
- Municipal amalgamations and restructuring.

These political trends reveal why the major stakeholders in policing and public safety have national organizations to speak to their members' professional interests. The efforts of these organizations are discussed in more detail in the section "Legislative Trends," below.

Civil rights trends may be considered a distinct subset within the political arena and should have an important place in any discussion about the Canadian character. The *Canadian Charter of Rights and Freedoms* is central to the practice of law enforcement in Canada and has, over its life span, had considerable impact on the daily working life of a police officer. Canadian respect for the rule of law and the protection of individual rights and liberties is embedded in the Charter and draws upon earlier commitments found in the *Canadian Bill of Rights*. Related to this area is the increased attention to civilian governance and oversight of police activities. There have been substantial developments in this field over the last two decades that have a bearing on police organizations. This topic is dealt with in more detail in Chapter Four.

The general trend to less government has also had a direct impact on Canadian police organizations. When federal, provincial, and municipal/regional levels of government consider outsourcing and privatization, this has clear implications for the delivery of police services. It means that police organizations will be questioned as to their specific requirements and whether or not they can do more with less. It may also press police organizations to do things differently. Of particular interest in some jurisdictions is the question of integrating emergency services (i.e., police, fire, and ambulance). Public officials are often seeking out ways in which they can reduce or redistribute the costs of the services they offer, and the subject of combining apparently similar services is frequently raised for discussion and debate. This certainly impacts on the very labour-intensive programs provided by police departments.

Political ideologies and the existence of new parties on the Canadian political landscape have a bearing on police organizations. There is also the issue of political activity of police officers themselves. In the political realm, the ideological agenda of any given party will often inform the overall direction of policy with respect to policing and law enforcement issues. Knowing a particular party's platform on relevant issues around law and public order, crime prevention, protection of victims of crime, and the punishment of offenders will obviously influence the overall direction of law enforcement and policing in a given jurisdiction.

CRIME TRENDS AND PATTERNS

When considering various trends in crime occurring in Canada, it is, once again, worthwhile looking to Statistics Canada, and particularly at the work of the Canadian Centre for Justice Statistics. Some of the following topics are indicative of the current trends in crime, and have been examined on a statistical basis:

- Child pornography;
- Terrorism;
- Abortion clinic shootings;
- Animal rights extremists;
- Environmental militants;
- Gang violence;
- International crime networks; and
- Youth violence.

In the 1980s, a careful study of crime trends was undertaken by William Tafoya, using something called the Delphi technique. Tafoya's 1986 doctoral dissertation constituted a forecast of future issues facing police agencies. Among his findings were the following concerns:

- Drug-related crimes of violence involving high-powered weapons would significantly increase;
- Computer-related crime would undergo a dramatic increase; and
- Terrorist acts would occur with much greater frequency in the United States.

The Delphi technique was first developed by the RAND Corporation (Helmer, 1967) in the late 1960s as a tool for the purpose of forecasting. Subsequently, the U.S. government refined the technique as a group decision-making tool. This approach allowed a group of experts to arrive at some form of consensus. It is particularly valuable when the critical factors involved are subjective and not specifically knowledge-based. It has been suggested that the Delphi technique functions best in areas in which decisions are taking place within a political or emotional context, or, in which there are strong factions involved with opposing preferences. It is useful in arriving at a prioritization of topics through a series of assessments and evaluations that do not require the participants to debate or defend their individual preferences. Through a cycle of rankings, the listing will stabilize at a point where opinions have become fixed.

Beyond the use of the Delphi technique, there are many ways to capture crime trends and take them into account when planning for both proactive and reactive police services. Typically, police departments in Canada prepare detailed crime statistics that are used internally for such purposes. These statistics must be submitted to Statistics Canada for reporting purposes, but their immediate relevance is at the local level. The vast wealth of information maintained by Statistics Canada, allows for trends to be determined through a comparative analysis of various statistical measures. This is where the Canadian Centre for Justice Statistics offers an incredibly valuable service.

The use of statistical analysis can be extremely useful in the context of crime prevention or community policing in that it becomes possible to concentrate efforts in areas where crime is high, both in terms of types of crime and the specific locations where particular crimes are occurring. Because police budgets are under the same severe pressures as other public service organizations, police departments must be able to target areas where they can be most effective and concentrate prevention, problem-solving, and enforcement efforts where they are most needed. This prioritization of police activity is, of course, not the exclusive domain of the police themselves. It then becomes important for the public to understand the nature of crime trends and patterns in order that appropriate priorities and objectives can be set for the police.

LEGISLATIVE TRENDS

Closely related to the broader political trends, discussed above, are specific legislative trends that have the potential to alter or redirect law enforcement practices. The degree to which police organizations are able to engage in proactive, community-based policing is substantially influenced by the legislative framework in place at the federal level and within individual provinces. Again, it is valuable to consider the impact of federal legislation (e.g., the *Criminal Code*, the *Narcotic Control Act*, the *Identification of Criminals Act*, the *Canadian Charter of Rights and Freedoms*) as well as provincial legislation (e.g., *Highway Traffic Act*, *Provincial Offences Act*, *Police Services Act*, among others) on how police go about doing their job. In fact, the nature of the legislative agenda over the long term will have a profound influence over the delivery of police services in Canada.

Tied closely to the political agenda of the government in power, the legislative trends are, however, not formulated in complete isolation. The impact of professional police organizations and public interest groups often have a compelling influence in this area of government activity (Marquis, 1993). Some of the following groups have historically been highly influential in the context of federal and provincial legislative trends impacting on police matters:

- Canadian Association of Police Services Boards;
- Canadian Association of Chiefs of Police;
- Canadian Police Association;
- Canadian Bar Association;
- Canadian Civil Liberties Association; and
- Provincial and local counterparts of these organizations.

This is only a small sampling of the many organizations in place that attempt to guide and inform the law-makers with regard to law enforcement and public safety legislation. Beyond the crime trends identified above, which clearly call for legislative action, the following additional areas of concentration are relevant, at either the federal or provincial levels:

- Firearms legislation;
- Search and seizure (including strip searches);
- High-speed pursuits;
- Corrections and conditional release;
- Victims' rights
- High-risk offenders
- Young offenders;
- Immigration and deportation enforcement;
- Police adequacy and effectiveness;
- Municipal restructuring;
- Use of force (especially lethal force);
- Drug legislation (including decriminalization);
- Prostitution (including child prostitution);
- Hate crimes; and
- Crimes over the Internet.

Governments alone cannot tackle the complexities that pertain to these areas of legislation. It is essential, therefore, that the federal Parliament and provincial legislatures rely on the insight, instruction, and support of policing and public safety experts in areas of legislative reform. Beyond this level of critical input, the government must also satisfy public expectations in the realm of law enforcement and public order. The process of legislative change is, in and of itself, complex and requires a significant bureaucracy to manage the transition from policy direction to final legislative enactment. There are several useful ways of tracking the legislation in this context. Many of the tools available allow access to pending legislation in pertinent areas of law enforcement, criminal law, and constitutional law.

TECHNOLOGICAL TRENDS

This is an area that is exceedingly difficult to cover adequately. Technology is not only changing at a rapid rate, it has the ability to reach into many areas that have a possible impact on the delivery of police services. Several advances in technology also have the potential to be adapted for criminal purposes, the Internet and cellular telephones being but two obvious examples. Modern technology may be viewed as the refinement of technique, and combines the Greek words for *art* ("techne") and *systematic study* ("logos"). Technique has been defined by Jacques Ellul (1962) as:

> The totality of methods, rationally arrived at, and having absolute efficiency (for a given stage of development) in every field of human activity.

Technology is the application of wisdom in ways that operate directly and indirectly on both the human and material world. Therefore, technology is *the* guiding presence in the modern world.

Some of the areas of technology that directly impact policing include advances in communications, information technology, and database management. A police officer from the 1960s would not be able to function, without substantial remedial training, in a patrol capacity in the late 1990s. So much has changed due to interventions and inventions of modern technology. Significant in this area is the work of the Canadian Police Research Centre (CPRC). This is a partnership involving the Royal Canadian Mounted Police (RCMP), the National Research Council of Canada (NRC), and the Canadian Association of Chiefs of Police (CACP). The objectives of the CPRC have been summarized as follows:

- To develop the best tools (equipment and information sources) for the police community;
- To strive to keep necessary technology affordable; and
- To forge partnerships with Canadian industry and the national and international research community.

The work of the CPRC is extensive, and it has recently undertaken a variety of innovative projects under the following main categories:

- *Health and safety—protecting the police in hazardous situations*: bomb disposal suit testing, soft body armour, miniature emergency response vehicle (MERV), evaluation of "smart gun" technology;
- *Operational effectiveness—fighting crime, gathering information, intelligence and evidence*: detection and visualization of fingerprints on human skin, forensic

entomology, geographic profiling, shift scheduling software, evaluation of use of force training simulators; and
- *Protecting the public—traffic, custody, crime prevention*: evaluation of spiked barrier strips, less-than-lethal technologies.

While it is appropriate to pursue technological innovations that will assist the police in their law enforcement efforts, it is also critical that the focus on technological advancements does not lead to the dilemma faced during what is referred to as the "professional" era of policing. In this context it became increasingly apparent that officers were distanced from their communities by being assigned to police vehicles engaged in motorized patrols. This dilemma is addressed in more detail in Chapter Eight. Also of profound importance to police organizations will be the impact of the Year 2000 on all manner of computerized systems. Commonly referred to as the Y2K problem, it remains largely unknown what the repercussions will be. However, police organizations must be prepared to ensure that their own operating systems have measures in place to deal with possible malfunctions and system failures, as well as to protect the public in the event of possible civil disruption should there be widespread system breakdowns.

GLOBAL TRENDS

This is one of the most difficult areas of concentration because it delves into the uncertainties of global political, economic, social, and technological development. The impact of events and occurrences in other countries has both direct and indirect repercussions on policing in Canada and influences in subtle ways the delivery of community policing. The following examples may suffice:

- Cyber-terrorism (i.e., the use of high technology to assist in the destruction of people or property through the deployment of computer viruses, logic bombs, and "trojan horses");
- Ethnic violence;
- Religious strife (e.g., Vancouver and the Sikh temple controversy);
- Breakdown of the Soviet Union;
- Right-wing extremism;
- Asian gangs;
- Disruption in Eastern Europe (e.g., Bosnia, Kosovo, mass emigration of ethnic minorities);
- Unrest in the Middle East, Iraq;
- Drug trafficking
- Indonesia and the APEC crisis; and
- International terrorism.

The police in Canada cannot operate in a vacuum and expect to deliver enlightened, community-oriented services. The wide world impinges upon the "ordinary" functions of a police organization in unpredictable ways. However, there is a degree of foresight that can be applied in order to become more aware of the various potential areas of impact. There are several sources of information that can be relied on by police organizations in gaining that kind of insight:

- Environmental scans;
- Current events magazines and publications (e.g., *Economist*, *Atlantic Monthly*, etc.);
- Internet sites (including the World Wide Web); and
- Annual reports of organizations that study the global trends (e.g, Canadian Security Intelligence Agency (CSIS), Criminal Intelligence Service Canada (CISC), among many others).

In this context, it is worthwhile studying the efforts of organizations that have made the study of the future their focus. This includes the Society of Police Futurists International (PFI), an organization formed in 1991 as a result of the International Symposium on the Future of Law Enforcement held at the FBI Academy in Quantico, Virginia. PFI has the following goals:

- To encourage partnerships between law enforcement, the academic community, and private industry;
- To facilitate information transfer concerning police futures research to its members;
- To develop, analyze, and interpret long range forecasts;
- To provide and promote education and training in the application of long range forecasting;
- To promote the use of technology;
- To serve as a clearing house for the communication of creative, innovative, and proactive policing strategies; and
- To advance ethical behaviour in law enforcement.

The society produces a newsletter, *Police Futurist*, which provides valuable insight into global trends and forecasts that relate to law enforcement. It is also linked with the World Futurist Society that is extremely active in the areas of futures research and forecasting. This latter organization publishes two periodicals that frequently contain insight of direct relevance to policing: *The Futurist* and *Futures Research Quarterly*.

From a Canadian perspective, the Canadian Security Intelligence Service (CSIS) prepares and disseminates a series of papers under the collective title *Commentary* that speak to global trends that may impact on domestic law enforcement agencies and could influence the approach to community-based policing services. Also, the public report of CSIS is a valuable source of data for matters pertaining to national security that could influence public safety and law enforcement at the local level. For example, the *Canadian Security Intelligence Service 1997 public report* includes a summary of the global security environment, counter-terrorism, counter-intelligence, and other national security programs.

LEADER IN COMMUNITY POLICING

Connie Snow, Inspector, Royal Newfoundland Constabulary

What is your understanding of "community policing" and its importance to the evolution of policing in Canada?

The police are representative of the general populace. We must be a part of the community and not apart from it. This approach to policing enables us to proactively solve problems with citizens before they become crime and disorder issues. People look to us for a safe and secure environment. Therefore, nationally the police should be forerunners in helping to provide a better quality of life for all citizens. Globally, our policing approach is held up as an example, and a source of envy.

Outline your own contribution to the development of "community policing" in Canada.

I am a staunch supporter of the philosophy and spread the "word" at every opportunity. I co-developed and co-facilitated a seven-day training program—"Community Policing: Train the Trainers Course" during October 1998 (I worked with Staff Inspector Bill Stephens of the Windsor Police Service, and we know of no other course like it in existence).

What is your view of the future of "community policing" in Canada?

I hope we never revert back to that "total" law enforcement mode—I don't think the public will let us. People like the involvement from us and they like being involved. The public demands more from police officers and police departments, and we are all being held accountable in a whole new way.

Ordinary citizens *must* have the right to articulate their views on issues that affect them. We have to continue down the Community Policing path.

What other comments would you like to make on this topic?

The key word in the philosophy is "Community." We are already members of our diverse and varied communities . . . we already know how we like to be treated. Everyone deserves a voice, an opportunity, and respect. We must replace cynicism and skepticism with acceptance and tolerance. When people take ownership and responsibility, healthier individuals result. "Healthy" people make healthy communities.

SUMMARY

This chapter has set out some of the elements that can be used to construct an understanding of the Canadian character. It points to the significance of the following areas in arriving at that understanding:

- Demographic trends;
- Social trends;
- Political trends;
- Legislative trends;
- Technological trends;
- Crime trends and patterns; and
- Global trends.

While it has not been possible to cover in detail all of the aspects of these areas, it should be valuable in leading the reader to sources of information and insight that will offer analysis and data relevant to such topics. In trying to discover the nature of the Canadian character, particularly as it relates to policing services, there is much to be gained from a very broad perspective that draws from many different sources. This chapter also tries to provide some idea of the scope and range of these various topics and points to several sources for further, ongoing study.

QUESTIONS FOR CONSIDERATION

1. In what ways could social trends affect policing in Canada?
2. What are some of the key demographic trends that will impact Canadian police organizations?
3. Are advances in technology a positive influence for community policing in Canada?
4. Select a significant global trend currently being discussed in the media. Does this trend have any potential impact on policing in Canada?
5. How do trends in victimization, crime reporting, or attitudes toward the police affect the way in which police organizations "do business"?
6. What are the best sources of information for background on the social, political, legislative, technological, and global trends that relate to policing in Canada?

REFERENCES

Adams, Michael (1997). *Sex in the snow: Canadian social values at the end of the millenium.* Toronto: Viking.

Anti-Defamation League (1996). *Danger: extremism: the major vehicles and voices on America's far-right fringe.* New York: Anti-Defamation League.

Ashley, David and David Michael Orenstein (1998). *Sociological theory: classical statements.* New York: Allyn and Bacon.

Bellah, Robert N. (ed.) (1973). *On morality and society: selected writings*. Chicago: University of Chicago Press.

Canadian Police Research Centre (1998). *Canadian Police Research Centre annual report 1997–1998*. Ottawa : Minister of Supply and Services Canada.

Castel, Jacqueline R. and Omeela K. Latchman (1996). *The practical guide to Canadian legal research*. Toronto: Carswell.

Cryderman, Brian K. and Augie Fleras (1992). *Police, race and ethnicity: a guide for police services. 2nd ed.* Toronto: Butterworths.

Dalkey, N.C. and O. Helmer (1963). "An experimental application of the Delphi method to the use of experts." *Management Science*, Vol. 9, No. 3, pp. 458–467.

Delbecq, A.L., A.H. Van deVen, and D.H. Gustafson (1975). *Group techniques for program planning: a guide to nominal group and Delphi processes*. Glenview, Ill.: Scott, Foresman & Co.

Downey, T.J. and R.J. Williams (1998). "Provincial agendas, local responses: the *common sense* restructuring of Ontario's municipal governments." *Canadian Public Administration*, (Summer 1998), Vol. 41, no. 2, pp. 210–238.

Eck, John E. and Nancy G. LaVigne (1994). *Using research: a primer for law enforcement managers. 2nd ed.* Washington, D.C.: Police Executive Research Forum.

Ellul, Jacques, translated from the French by John Wilkinson (1964). *The technological society*. New York: Vintage Books.

Giddens, Anthony (1982). *Sociology: a brief but critical introduction*. San Diego : Harcourt Brace Jovanovich Publishers.

Graham, K.A. and S.D. Phillips (1998). "*Who does what* in Ontario: the process of provincial-municipal disentanglement." *Canadian Public Administration*, (Summer 1998), Vol. 41, no. 2, pp. 175–209.

Grant, George (1969). *Technology and empire: perspectives on North America*. Toronto: House of Anansi.

Grant, George (1986). "Thinking about technology." In *Technology and justice*. Toronto: House of Anansi.

Hammond, Allen (1998). *Which world?: scenarios for the 21st century, global destinies, regional choices.* New York: Island Press.

Helmer, O. (1967). *Analysis of the future: the Delphi method*. Santa Monica, Calif.: RAND Corporation.

Holden, Richard N. and Richard D. Sluder (1995). "American policing: prospect for change in the twenty-first century." *Futures Research Quarterly*, (Fall), Vol. 11, no. 3

Hum, Derek and Wayne Simpson (1998). *Wage opportunities for visible minorities in Canada*. Ottawa: Statistics Canada (Income and Labour Dynamics Working Paper Series).

Linstone, H.A. and Turoff, M. (1975). *The Delphi method: techniques and applications*. Reading, Mass.: Addison-Wesley.

McIntosh, Ian (ed.) (1997). *Classical sociological theory: a reader*. Washington Square, N.Y.: New York University Press.

McKenna, Paul F. (1998). *Foundations of policing in Canada*. Scarborough: Prentice Hall Canada.

Marquis, Greg (1993). *Policing Canada's century: a history of the Canadian Association of Chiefs of Police*. Toronto: The Osgoode Society and University of Toronto Press.

Mommsen, Wolfgang J. (1989). *The political and social theory of Max Weber: collected essays*. Chicago: University of Chicago Press.

Moore, Jr. Richter H. (1995). "Twenty-first century law to meet the challenge of twenty-first century organized crime." *Futures Research Quarterly*, (Spring), Vol. 11, no. 1.

Normandeau, André and Barry Leighton (1990). *A vision of the future of policing in Canada: police-challenge 2000: background document*. Ottawa: Solicitor General Canada, Police and Security Branch.

Peak, Kenneth J. and Ronald W. Glensor (1996). *Community policing and problem solving: strategies and practices*. Upper Saddle River, N.J.: Prentice Hall.

Pearson, Ian (ed.) (1998). *The Macmillan atlas of the future*. New York: Macmillan Reference

Ringer, Fritz K. (1997). *Max Weber's methodology: the unification of the cultural and social sciences*. Cambridge, Mass.: Harvard University Press.

Robertson, Douglas S. (1998). *The new renaissance: computers and the next level of civilization*. London: Oxford University Press.

Smith, G. Davidson (1998). "Single issue terrorism: commentary no. 74, a Canadian Security Intelligence Service publication." Ottawa: CSIS.

Statistics Canada. Housing, Family and Social Statistics Division (1997). *A portrait of seniors in Canada. 2nd ed.* Ottawa: Minister of Industry.

Swol, Karen (1998). "Private security and public policing in Canada." *Juristat*, Vol. 18, no. 13. Ottawa: Statistics Canada, Canadian Centre for Justice Statistics.

Traugott, Mark (ed.) (1978). *Emile Durkheim on institutional analysis*. Chicago: University of Chicago Press.

Vojnovic, I. (1998). "Municipal consolidation in the 1990s: an analysis of British Columbia, New Brunswick, and Nova Scotia." *Canadian Public Administration*, (Summer 1998), Vol. 41, no. 2, pp. 239–283.

Volti, Rudi (1995). *Society and technological change. 3rd ed.* New York: St. Martin's Press.

Westby, David (1991). *The growth of sociological theory: nature, knowledge and social change*. New York: Prentice Hall.

TOWARD MORE LEARNING

DEMOGRAPHIC TRENDS

Statistics Canada has a portion of its website devoted to providing statistical profiles of individual Canadian communities at http://ww2.statcan.ca/english/profil/. This site contains information from the 1996 census of population undertaken by Statistics Canada. A statistical profile is provided for all Canadian communities (including cities, towns, villages, Indian Reserves and settlements). The profiles highlight information on education, income and work, families and dwellings. It also has information on the general population.

SOCIAL TRENDS

Of particular interest to those looking for information relevant to the social dimension in Canada, Statistics Canada publishes an important quarterly that is available by subscription. *Canadian social trends* is a publication that discusses social, economic, and demographic changes that affect the lives of Canadians. It frequently offers insights that are directly relevant to policing and contains the latest figures from major social indicators.

The Environics Research Group, started by Michael Adams, the author of *Sex in the snow: Canadian social values at the end of the millennium*, provides a sample questionnaire based on its 3SC Social Values Monitor. The 3SC is a survey that Environics conducts in Canada and the United States. It is an international planning tool that maps changing consumer values, attitudes, and behaviour. The survey has been used in strategic planning by business in over 20 countries and has been applied in Canada since 1983. This sample will allow the reader to see where they sit on this map. The questionnaire can be found through the Environics website: http://www.environics.ca

POLITICAL TRENDS

Again, the Environics Research Group provides some valuable insights with respect to political trends in Canada through two series of surveys:

- *FOCUS CANADA Report*—this is Canada's most comprehensive survey of attitudes toward public policy issues, as well as political, economic, and social trends. It is a quarterly product that began in 1976. It consists of in-home interviews with 2000 adult Canadians. Subscribers to this service have access to the extensive database that has been developed over the existence of the survey, and may commission confidential add-on questions for their own use.
- *FOCUS ONTARIO Report*—this is another quarterly survey service provided by Environics. It offers a survey of Ontarians' attitudes toward major issues on the provincial agenda. It also monitors public satisfaction with the provincial government's performance and gauges public attitudes toward public police issues, as well as political, economic, and social trends. This survey has been conducted since 1985 and consists of telephone interviews with 1000 adult Ontarians. FOCUS ONTARIO also provides, on a biannual basis, a survey of 1000 adult residents of the Greater Toronto Area (GTA) which focuses on citizen satisfaction with public services and addresses public attitudes toward local issues and problems. Subscribers to this service have access to an extensive database on GTA opinion going back to 1988 (when the survey was initially introduced as the "Metropoll").

LEGISLATIVE TRENDS

There are several tools available for tracking the changes in legislation that may directly impact on policing services in Canada. Of particular relevance are those that provide constant updating on federal statutes and regulations that pertain to policing, law enforcement, and public safety. It should be remembered that the changes in statutes and regulations are only one component of the broader legislative context, since case law may also influence and redirect police activity in certain areas. Both case law and statute amendments are key sources of insight in those areas relevant to law enforcement.

Constitutional Law

Law Reports

Canadian Rights Reporter

Charter of Rights Decisions

Looseleaf Services

Hogg, *Constitutional law of Canada*. 3rd ed. (Toronto: Carswell).

Laskin et al. (eds.), *Canadian Charter of Rights Annotated*. (Aurora: Canada Law Book).

Criminal Law

Law Reports

Canadian Criminal Cases

Criminal Reports

Weekly Criminal Bulletin

Annotated Statutes

Greenspan, *Martin's Annotated Criminal Code*. (Aurora: Canada Law Book).

Hamilton and Shilton, *The 1998 Annotated Ontario Police Services Act*. (Toronto: Carswell).

Rodrigues (ed.), *Crankshaw's Criminal Code of Canada, R.S.C. 1985*. (Toronto: Carswell).

Watt and Fuerst, *The Annotated 1998 Tremeear's Criminal Code*. (Toronto: Carswell).

Looseleaf Services

Gibson, *Canadian Criminal Code offenses*. (Toronto: Carswell).

Charter

Altrens, *The Charter and criminal procedure*. (Toronto: Butterworths).

Hutchison and Morton, *Search and seizure law in Canada*. (Toronto: Carswell).

McLeod et al. *The Canadian Charter of Rights*. (Toronto: Carswell).

Driving Offences

Gold, *Drinking & driving law*. (Toronto: Carswell).

Segal, *Manual of motor vehicle law*. (Toronto: Carswell).

Drug Offences

MacFarlane, *Drug offences in Canada*. 3rd ed. (Aurora: Canada Law Book).

Evidence and Procedure

Altrens, Burns, and Taylor, *Criminal procedure: Canadian law and practice*. (Toronto: Butterworths).

Gibson, *Criminal law evidence, practice and procedure*. (Toronto: Carswell).

McWilliams, *Canadian criminal evidence. 3rd ed.* (Aurora: Canada Law Book).

Salhany, *Canadian criminal procedure. 6th ed.* (Aurora: Canada Law Book).

Police

Ceyssens, *Legal aspects of policing*. (Toronto: Carswell)

Search and Seizure

Hutchison and Bury, *Search and seizure law in Canada*. (Toronto: Carswell).

Weapons

Harris, *Weapons offences manual*. (Aurora: Canada Law Book).

Young Offenders

Harris, *Young Offenders Act manual*. (Aurora: Canada Law Book).

Platt, *Bala and Liles Young offenders service*. (Toronto: Butterworths).

The Canadian Police Association, headquartered in Ottawa, has spent a great deal of time and money preparing submissions relevant to the legislative agenda of the federal government.

TECHNOLOGICAL TRENDS

The work of the Canadian Police Research Centre is extremely important in this regard, and their annual report lists a number of completed and ongoing projects. The CPRC provides secure Internet services to law enforcement agencies, as well as to certain community-based public safety organizations. It also offers a Computer Investigators discussion list, an international Crime Stoppers web board, and a variety of communications areas for police research, including a platform for a police futures research group. The CPRC can be contacted as follows:

Canadian Police Research Centre
P.O. Box 8885
Ottawa, Ontario
K1G 3M8
Telephone: (613) 998-6343
Facsimile: (613) 952-0156
Website: http://www.cprc.org

CRIME TRENDS AND PATTERNS

In the area of justice, Statistics Canada offers a publication series that is of great significance. *Juristat* is a periodical produced on an irregular schedule that serves a vital need. It provides a detailed insight into various categories relevant to law enforcement, public safety, and the Canadian justice system. This series includes detailed statistics and analysis on topics and issues, such as the following:

- *Arson*—including property violations by location of incident, and young offenders, rates and trends;
- *Assaults*—including rates of wife assaults, extent of injuries suffered, psychological consequences, reasons for not reporting to police, residential incidents

and occupancy status, young offenders. A special issue (Vol. 12, no. 15) examined victims, including seniors, relationship of the accused to the victim, and level of injury sustained;

- *Break and Enters*—including rates, definitions, international comparisons, national trends, provincial and municipal comparisons, characteristics, and accused profiles (see Vol. 12, no.1);
- *Crime statistics*—including public perceptions, level of neighbourhood crime, changing levels of crime, personal risk, trends 1962-1990 (see Vol. 12, no. 7), fear of crime and public perceptions (see Vol. 15, no. 9), and crime in major metropolitan areas (see Vol. 17, no. 4);
- *Drugs*;
- *Firearms*;
- *Fraud*;
- *Homicide*—including rates 1961–1990 (see Vol. 11, no. 5);
- *Impaired Driving*;
- *Police*—including public perception of police performance (see Vol. 18, no. 6);
- *Prostitution*—including street prostitution in Canada (see Vol. 17, no. 2);
- *Robbery*—including rates, 1962–1989 (see Vol. 10, no. 15);
- *Sexual Assaults*—including rates, 1962–1989 (see Vol. 10, no. 5);
- *Victims*—including rates, 1962–1989 (see Vol. 10, no. 5), and criminal victimization, an international perspective (see Vol. 18, no. 6);
- *Violence*—including violence committed by strangers (see Vol. 18, no. 9);
- *Weapons*—including weapons and violent crime (see Vol. 17, no. 7);
- *Women*—including the findings of a survey (see Vol. 14, no. 9); and
- *Young Offenders*.

GLOBAL TRENDS

The study of global trends is possible through several means, including a careful study of a variety of respectable daily newspapers or online news sources. For a long-range perspective it is valuable to consider the depth of insight offered in the public documents released by agencies that study international and global issues, including CSIS, and the U.S. government through its various intelligence agencies. The following websites are useful points of entry for this type of inquiry:

Canadian Security Intelligence Service (CSIS):
http://www.csis-scrs.gc.ca/eng

U.S. State Department:
http://www.state.gov/www/global/terrorism/1996report/

CHAPTER FOUR

Most leaders instigating change are like gardeners standing over their plants, imploring them: "Grow! Try harder! You can do it!" But if a seedling has no room to grow, no soil, and no water, it will never become a tree. Similarly, if organizations don't foresee the obstacles that arise naturally wherever growth and learning take place, their change initiatives will fail. All effective organizations take part in this "dance of change," a balance between growth and the limits of growth, whose impact can be anticipated and mastered.

Peter Senge

MODERN ORGANIZATIONS—THE GREAT RECONSTRUCTION

LEARNING OBJECTIVES

1. Identify key components of change in modern organizations.
2. Identify major elements of change within the Public Service of Canada and within the civil service of various provinces.
3. Explain key elements of organizational review and reform in private and public corporations in Canada.
4. List the essential elements of effective governance.
5. Identify the twelve "attributes of effectiveness" identified by the Canadian Comprehensive Auditing Foundation.
6. Explain the meaning of alternative service delivery as applied in a Canadian context.
7. Identify the benefits of continuous improvement and constant change in organizational design theory.
8. Explain the basic elements of postmodern theory.

INTRODUCTION

This chapter looks at some of the major changes in modern organizations that have taken place over the last several decades. While it should not be seen as a complete treatment of this topic, it is important to consider some of the fundamental aspects that may be considered part of the great reconstruction of the modern organization. This review is particularly helpful in light of the enormous impact these changes promise to have on police services in Canada. Whether actual, or anticipated, virtually every police organization in Canada will be influenced by these fundamental changes in the larger corporate environment.

It is appropriate to highlight the fact that this transformation of the modern organization is not reserved for private corporations. Governments, at all levels, are being altered in radical ways that directly overhaul the agencies and services found within those levels of government. Because policing in Canada emanates from the federal, provincial, and municipal/regional levels, it is not surprising that the waves of change are washing over police services in an unprecedented manner. This chapter focuses on the larger context of corporate change, and Chapter 5 will offer a concentrated analysis of the impact of the change culture in Canadian police organizations.

Additionally, the whole concept of governance has undergone a revision, mirroring the major changes in organizational design and function. This, too, has implications for Canadian policing, due to our longstanding commitment to civilian governance of police agencies and to the notion that "the public are the police and the police are the public," to reiterate Sir Robert Peel's famous *dictum*.

The concept of alternative service delivery (ASD) will also be considered, as well as the importance of continuous improvement and constant change. These ideas have influenced the ways in which organizations do business and have altered the expectations placed on individuals and the growth of teams within those organizations. Finally, this chapter will touch on postmodern theory, a theory that can be seen to inform much of the transformation occurring in our world today. A general understanding of this theory will make it easier to trace its appearance in the changing practice of organizations, and in society at large.

RECONSTRUCTING THE MODERN ORGANIZATION

An enormous amount has been written about transformation in modern organizations. No longer is it feasible to adhere to the strict hierarchy that characterized corporations growing out of the industrial revolution. Today, all sorts of public and private corporations are changing in ways that will allow them to survive, perhaps even thrive, into the future. The intensity of change within the modern organization has given rise to a great outpouring of theories, approaches, strategies, and blueprints for dealing with this challenge. Much of this information is extremely helpful in providing guidance. However, the sheer volume of competing approaches may produce a degree of bewilderment for many new organizational designers.

Some of the major components of this transformation of the modern organization are as follows:

- Fewer layers of management;
- Greater application of automation and information technology;
- More emphasis on teams, task forces, and special projects;
- Greater cooperation between union and management;
- Higher levels of service orientation;
- Increased use of knowledge-based approaches;

- Increased focus on experts or specialists directing their own work performance;
- Movement toward the "virtual" office and remote management; and
- Increased focus on processes rather than products.

Inevitably, modern organizations have had to come to grips with new technology, increased global competitiveness, changing customer expectations and loyalties. In doing so, they frequently discover that they have not structured in a way that allows them to easily adjust to rapidly changing circumstances. In essence, their organization is too rigid to permit change, without considerable effort and disruption. The days of simply churning out your product, or offering a routine service are over. There is a significant push toward flexibility, customization, and innovation. The example of the Swiss watch industry failing to recognize the value of the new quartz crystal is highly instructive. This missed opportunity caused them to fall behind in the face of an important "paradigm shift" that immediately and profoundly advantaged the Japanese innovators who refined the quartz technology with fabulous results. Figure 4–1, below, captures the various elements that are required to work in harmony for effective organization.

FIGURE 4–1

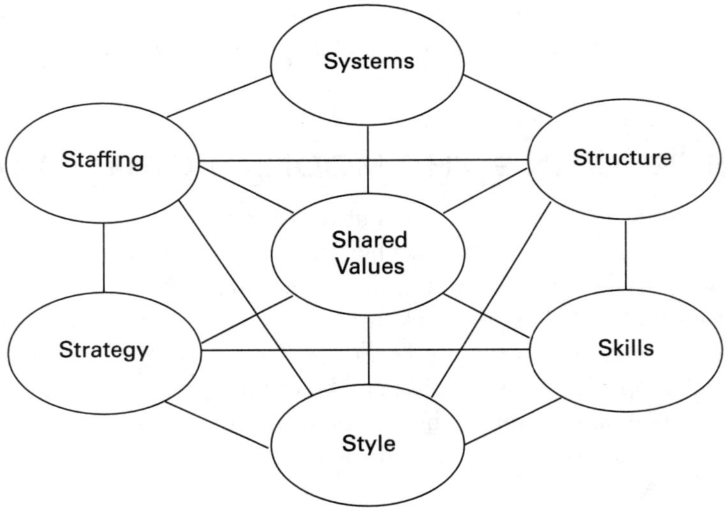

Adapted with the permission of Simon and Schuster, Inc., from *The art of Japanese management* by Richard Tanner Pascale and Anthony Athos. Copyright © Richard Tanner Pascale and Anthony Athos.

REENGINEERING IN THE PUBLIC SERVICE

One of the areas most keenly affected by the great reconstruction in Canada is the public service. Much has been made of the downsizing of the civil service at all levels of government (Osborne and Gabler, 1992). The reengineering that has already been undertaken has had an impact on policing, which may be seen as the quintessential public service. This section will consider the transformation of the civil service at the federal level, as well as in several provinces. It is important to remember that the process of change in the public service is ongoing and, therefore, cannot be considered complete. It is, appropriately, a "work in progress."

Federal Public Service

The beginning of major change in the Canadian federal public service occurred around 1989 with the commencement of the Public Service 2000 (PS2000) project. Its aim was to prepare the civil service for the next millennium. As Dewar (1994) notes:

> PS2000 has been helpful in defining the task; it has refocused our attention on service to the public as our external purpose and has suggested how we can go about being more efficient, responsive and creative. These ideas, plus the idea of continuous feedback and improvement that is gaining currency, provide us with the means to respond effectively to the challenges we face. (p. 8)

Dewar goes on to identify the dominant values he found in the public service:

- Serving the public;
- Political neutrality;
- Loyalty to the elected government;
- Anonymity;
- Integrity;
- Professionalism; and
- Fairness and equality for all citizens.

To which he adds new areas of public service concentration:

- Values of efficiency and prudence in using resources;
- Value of representativeness; and
- Idea of respect for others as a result of new social and political values and management concepts.

In looking at the latest developments affecting public administration in Canada, Manion (1994) makes the following observation:

> The role of government is changing from one of élite accommodation to one of public participation and power-sharing; from one of commanding to one of informing, persuading and leading; from one of doing things for people to one of getting people to do for themselves; from one of setting the rules to one of setting the example and providing the vision; from one of regulations to one of voluntary standards and self-regulation.

PS2000 began in December 1989 as a comprehensive process of public service renewal and reform. It was led by the Clerk of the Privy Council (the person in charge of Canada's federal bureaucracy) and included eleven task forces headed up by appropriate Deputy Ministers. PS2000 culminated in the publication of a federal White Paper on December

1990. This project led to the passage of the *Public Service Amendment Act* which received Royal Assent in December 1992.

As Ingstrup (1992) notes:

> PS2000 and the broader renewal process have legitimized new kinds of "conversation" in Canadian federal organizations, a conversation about service, about excellence, about people, about leadership, and about accountability and integrity, a conversation quite different in character than would have been possible ten or fifteen years ago. This is what I call public service renewal, and is why I call it a success! (p. 12)

The impact of PS2000 has been to fuel a more deeply rooted concern for, and consciousness of, innovation within the civil service. The complacency that was often associated with the public service was, to some degree, dissipated by the PS2000 example.

As Thomas (1994) observes:

> No organization is ever completely at rest, but public sector organizations were seen to change reluctantly and only incrementally. Today this negative stereotype no longer applies. Many praiseworthy innovations involving the governance and management of public sector departments and agencies are occurring at all levels of government throughout Canada.

Clark (1994) cites the eight principles that have been taken from the private sector and which have served as a guide for public service renewal within the context of PS2000:

- Focus on customers;
- Review the relevance of programs and processes;
- Eliminate duplication;
- Assess value added;
- Focus on the bottom line;
- Increase performance measurement;
- Empower employees; and
- Implement boldly.

These principles are outlined in more detail by Paul Tellier (1994). However, there is a need to recognize certain limitations on the extent and degree of possible change within the public service. As Clark (1994) indicates:

> It is crucial that advocates of reform acknowledge the traditional constraints on public servants and the need to fully engage governments in the process of change, particularly when such reform will have consequences for the direction of policy and programs. By focusing on liberating and empowering entrepreneurial managers, the New Public Management approach overlooks the legitimate constraints presented by the institutional environment in which public servants work—namely, the statutory responsibilities and conventions that comprise the body of pertinent public law, and the legitimate interests and responsibilities of elected representatives in public management. (p. 20)

Therefore, while innovation and change may be positive influences within the public service, there are certain limits to the kind of experimentation that can take place. With high expectations around accountability and acceptability, there are significant constraints on the entrepreneurial possibilities available to public *servants*.

In his recent examination of the overall public service renewal agenda, however, Marson (1997) discovered the following key pressures and issues:

FIGURE 4-2

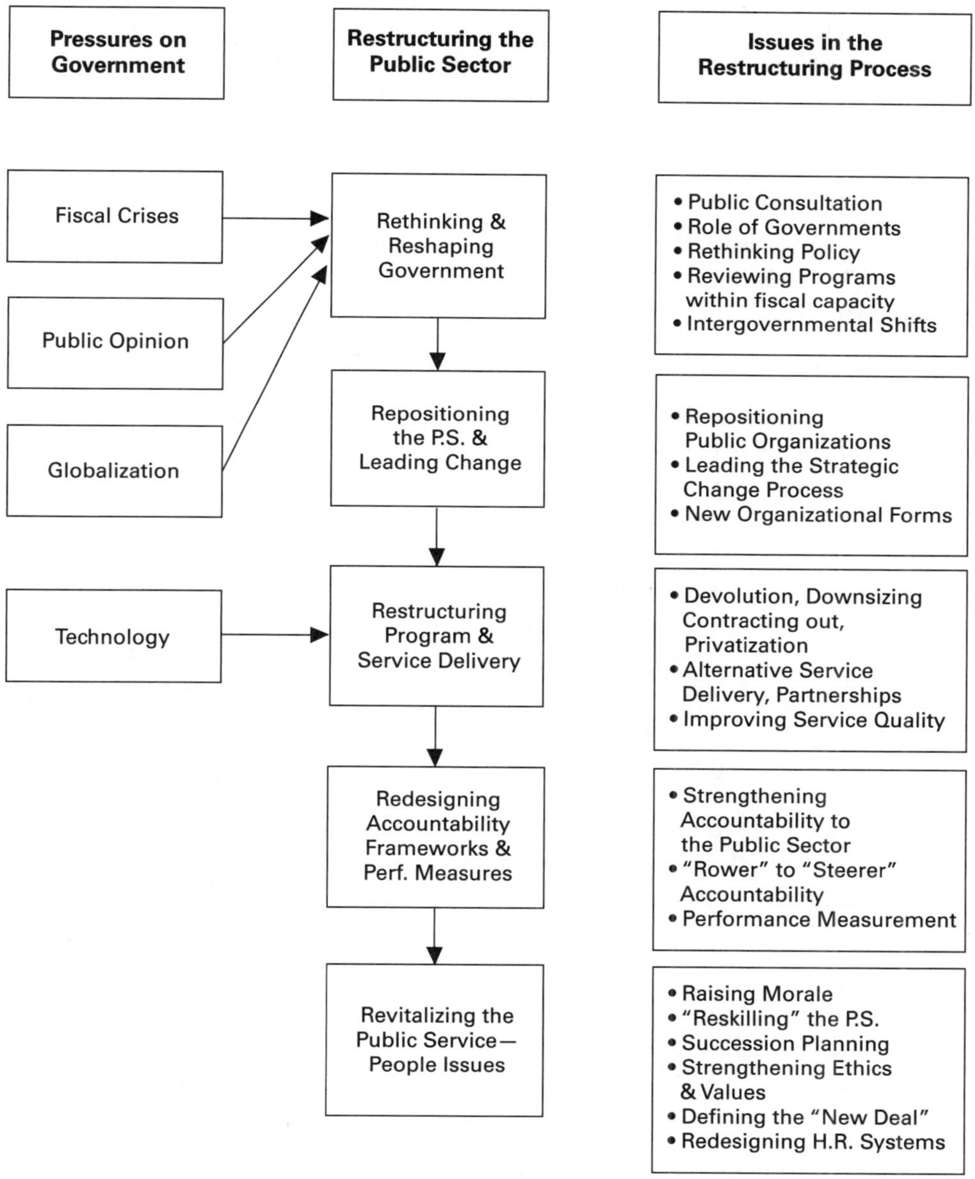

Source: Marson, Brian (1996). "Report on The IPAC 1996-97 Deputy Ministers' Issues Survey: Public Service Renewal Agenda." *Public Sector Management*, Vol. 7, no. 4, p. 13. Reproduced with permission of the Institute of Public Administration of Canada.

Roberts (1996) outlines some of the impediments to meaningful change introduced by what he refers to as the "control lobby" with respect to the PS2000 enterprise:

> . . . the control lobby had three important effects on the PS2000 exercise. In some cases it slowed or cut back reform efforts, particularly when the implementation of reforms required legislative action. The lobby also undercut the central agencies' attempt at cultural change, by criticizing public servants who broke rules and behaved unconventionally or "entrepreneurially." Finally, the public discussion of PS2000, which was largely shaped by the control lobby, may have diminished the prospects for future reform exercises by reinforcing popular beliefs about the trustworthiness of public servants and the wisdom of deregulating the public service. (p. 491)

Holmes (1996) points to the analysis of the PS2000 exercise undertaken by the Office of the Auditor General in the following terms:

> PS2000 proposes a new management approach—a new management philosophy— that places increased emphasis on being open and consultative, and on:
>
> Results (rather than process); Flexibility (rather than conformity); Judgment (rather than rules); Innovation (rather than risk-aversion); Accountability (rather than detailed, central controls); Public servants as assets to be developed (rather than as resources to be used).

In spite of the criticisms, reservations, and shortcomings of the PS2000 enterprise, meaningful change has been undertaken in a serious and systematic manner that will have repercussions for years to come within the public service, at all levels across Canada.

ALBERTA

Changes in the government in Alberta led to some considerable efforts at reform in that province's public service. The *Measuring Up* exercise was directed at improvements in public service performance throughout the province. In several areas the civil service rose to the challenge, as evidenced by a considerable number of innovative projects. The so-called "Alberta Revolution" began in 1993 and entailed a complete transformation of the provincial government. The approach taken in Alberta was inspired, in part, by the State of Oregon's "Oregon Benchmarks" project (Warrian, 1995), and involved the use of performance indicators to define the "businesses" the Alberta government was involved in pursuing. As a result, the Government of Alberta determined that it would concentrate on the following five "core businesses":

- Investing in people and ideas;
- Building a strong, sustainable, and prosperous province;
- Providing essential services for the health and well-being of Albertans;
- Maintaining a quality system of roads and highways, telecommunications and utilities; and
- Providing law, order, and good government.

On this basis, it was then possible to establish specific performance measures that could be applied to gauge government effectiveness. These performance measures were defined as follows (Alberta, 1994):

Literacy/numeracy	Crime rate	Taxation load
Employment rate	Public debt per capita	Disposable income
Air quality	Land quality	Resource wealth index
Quality of life	Attractive working conditions	Health status
Educational attainment	Level of community development	Skill development
New jobs created	Overall family income	Gross domestic product
Water quality	Life expectancy	Economic effect of research & development environment

Many of the areas identified above have the potential to impact substantially on public safety and law enforcement. This helps to clarify that the police operate within an environment that is tied very closely to the entire social fabric for which governments are, in large part, responsible.

ONTARIO

In Ontario, a determined effort to transform the public service, introduced by the NDP government under Premier Bob Rae, received a redirected thrust under the auspices of the "Common Sense Revolution" of the Progressive Conservative Party which won the 1995 election. This "Common Sense Revolution" had a number of key components that were partially inspired by the successes of the Klein government in Alberta. It led to a continued interest in business process reengineering. Several ministries undertook the challenge of transforming themselves according to the principles of business reengineering, including:

- Ministry of Revenue;
- Ministry of Natural Resources;
- Ministry of Commercial and Consumer Affairs; and
- Ministry of the Solicitor General and Correctional Services.

Within many other government departments, business reengineering was applied in the hopes of accomplishing significant change. Essentially, Hammer and Stanton (1995) define business reengineering as follows:

> The fundamental rethinking and radical redesign of business processes to bring about dramatic improvements in performance. (p.3) [emphasis in original]

In this context, it is also helpful to consider the following key elements of any reengineering exercise:

- *Redesign*—critically examining new ways of applying business processes;
- *Retool*—engaging information technology in the change process; and
- *Reorchestrate*—modifying the human resource dimension in several ways, including the job design, reward systems, performance measurements, and organizational design.

Basically, the promise of reengineering lies in its focus on a complete understanding of the business processes under review. If the entire sequence of steps taken, and the time spent engaged in those steps, is fully understood, it becomes possible to remove, modify or automate specific elements and arrive at streamlined processes that require less time to complete and fewer "hand-offs" from one worker to another. Overall, the work that is done will be faster, better, and cheaper. As we will see in Chapter 5, this approach has proven to be beneficial for many police organizations in Canada.

The general approach to business reengineering includes the following eight steps:

1. Identification of redesign opportunities;
2. Deconstruction of functions;
3. Development of process maps and descriptors;
4. Development of targets and redefined processes;
5. Determination of redesign impacts;
6. Determination of redesign completeness;
7. Development of requirements for retooling; and
8. Development of requirements for reorchestration.

In Ontario, the Management Board Secretariat provided its expertise in assisting various ministries, departments, and agencies with the campaign of business reengineering (Ontario. Management Board Secretariat, 1993). The overall purpose is to:

- Streamline processes;
- Introduce information technology as an enabler of change; and
- Provide improved services to the public.

Of course, governments have been attempting to enhance public service effectiveness for many decades. However, business reengineering represented something of a high watermark in the most recent efforts at applying, in a somewhat systematic manner, the new public management theories. Because it entails such a radical approach to organizational design, it has both advocates and adversaries. However, many government agencies have benefitted from the fundamental re-examination of their purpose, processes, procedures, and principles. This is true even if results fall short of the dramatic ideals established by business reengineering theorists.

Over several decades, the public service has undertaken efforts to improve its efficiency. For example, in 1985 Price Waterhouse Associates and the Canada Consulting Group produced, *A study of management and accountability in the Government of Ontario* that made recommendations to improve:

- The government's accountability structure, by clarifying the responsibilities of and relationships between ministries and central agencies;
- Management policies, by determining how administrative rules and management practices can further encourage prudence, probity, and efficiency; and

- The attitudes and motivations of managers in the civil service and the related development of human resources, by promoting a working climate that fosters adherence to the rules, value for money, and innovation.

Clearly, one of the messages implied in all of these studies is that change is a constant in the public sector. The principle of continuous improvement, discussed in more detail below, has been accepted and applied with increasing rigor and broader scope.

ORGANIZATIONAL REVIEW AND REFORM

Along with the specific application of business process reengineering in both public and private corporations, a more general trend in the modern organization has been to large-scale organizational review and reform. These types of undertakings are intended to transform the organization into something more effective, efficient, and economical. It is often done as a result of the development of some compelling "case for action" which typically involves an assessment of the organization's strengths, weaknesses, opportunities, and threats.

Much of the new management literature is devoted to the assessment of, or assistance with, this type of fundamental corporate transformative process. Major review and reform has been undertaken in many Canadian organizations in order to permit them to remain viable in the future. A substantial amount of this work is inspired by management thinking coming out of the U.S., Japan, and other countries grappling with some of the following pressures:

- Fiscal constraints;
- Export/import markets;
- Changes in technology;
- Education and training patterns;
- Productivity;
- Government deficits and debts;
- Political leadership;
- Global stability; and
- Environmental challenges.

One of the most valuable guides to organizational renewal can be found in the twelve "attributes of effectiveness" developed by the Canadian Comprehensive Auditing Foundation (CCAF). This approach was developed after considerable study and examines the following areas, and relevant questions, in detail:

1. *Management direction*—does everyone understand what they are meant to be doing?
2. *Relevance*—do our activities continue to make sense in terms of the conditions, needs, or problems to which they are intended to respond?
3. *Appropriateness*—are levels of effort and selected methods of pursuing objectives sensible and sufficient?

4. *Achievement of intended results*—how challenging are our goals, and have they been accomplished?
5. *Acceptance*—are clients and other key stakeholders satisfied with the organization and its products or services?
6. *Secondary impacts*—what are the unintended effects of our activities, be they positive or negative?
7. *Costs and productivity*—are the relationships between costs, inputs and outputs favourable?
8. *Responsiveness*—how well are we anticipating and responding to change?
9. *Financial results*—how good are the financial results in terms of matching costs with revenues and appropriations, and financial assets with obligations?
10. *Working environment*—is it a happy, healthy, and constructive working environment in which staff are motivated to work together, adapt to change, and develop?
11. *Protection of assets*—how well do we protect against surprise events or losses of key personnel, critical occupations, client information, facilities, equipment, inventories, processes, or agreements?
12. *Monitoring and reporting*—does management have the information it needs to support its decision-making and its own accountability, and is the information used appropriately?

By examining the organization in the context of these attributes of effectiveness, it is possible to arrive at new approaches that would help redesign the corporation (public or private) accordingly. There is a wealth of information and support around the entire process of organizational review and reform that has provided support and guidance to all sorts of organizations. In Chapter 5, the specific task of organizational review and reform in policing will be considered in detail.

There is no shortage of information and instruction on approaches to organizational review and renewal. Academics and practitioners have contributed to a burgeoning literature that offers advice on a wide range of approaches. The best of these publications rely on lessons learned in a variety of applications and offer a path through the immediate turbulence of organizational change.

LEADER IN COMMUNITY POLICING

Christine E. Silverberg, Chief of Police, Calgary Police Service

Christine Silverberg has served for 26 years in policing. She was appointed Chief of Police of the 1700-member Calgary Police Service in 1995 after serving in progressively responsible positions within the Peel Regional Police Service, the Ministry of the Solicitor General (Ontario), and the Hamilton-Wentworth Regional Police. She holds an M.A. (Criminology) from the University of Toronto; APR (Professional Accreditation in Public Relations); Executive Development, Queen's University School of Business, Kingston, Ontario; and numerous certifications from both Ontario and Canadian Police Colleges. At the time of this writing, she is Chair of the Criminal Intelligence Services Alberta (CISA); serves on the Board of Directors and Law Amendments Committee of the Canadian Association of Chiefs of Police (CACP); is Prairie Region Chair of the National Coordinating Committee on Organized Crime; and serves on the Board of Directors of several charitable and community organizations. She is a member of IACP, IAWP, PERF, and the Major Cities Chiefs Association.

One of Chief Silverberg's major contributions to Canadian policing is the building of *organizational capacity* for the advancement of *community policing*; in essence, the development of the police organization from the inside out, to increase the *potential for collaborative problem-solving* in relation to crime prevention and law enforcement in communities served by the organization. She has worked to apply this notion in Calgary on the assumption that it is one very logical and practical way of evolving proactively with rapidly changing times in major cities which, at the end of the twentieth century, are witnessing shifting demographic patterns, rising youth unemployment, real or perceived erosion of family values, increasing fear of violence, rising technological innovation and access to information, increasing diversity of ideology, and societal expectations of immediate satisfaction. Parochial, hierarchical management models of old, founded on rigid planning and interpersonal control, simply cannot cope effectively or flexibly with such demands; police can no longer *tell* communities what they need, in reaction to sporadic statistics, but must rather *listen* to dynamic social processes in order to be organizationally responsive and prepared for the future.

In a sense, this requires a different kind of police officer, who requires different kinds of organizational systems to support him or her in that endeavour. There is need for the organization to become actively involved in external partnerships and dialogue, and in the creation of an internal emphasis on values, learning, and creative problem-solving. In short, there is a need to become *people-oriented* and people-focused, inside the organization as well as outside of it.

How is the Calgary Police Service creating such organizational capacity? Under the leadership of Chief Silverberg, it is working to decentralize delivery of policing services,

Christine E. Silverberg LEADER IN COMMUNITY POLICING continued

empower front-line officers in problem-solving, collaborate with diverse publics and service agencies, and share resources and responsibilities with local organizations. It is sharing the vision and values of the organization at all levels; focusing on the development and wellness of employees; emphasizing 'leadership' rather than 'supervision'; encouraging continual individual and organizational improvements and evaluation; promoting quality, cost-effectiveness, and efficiency in all policing services; enhancing the use of technology to strengthen business processes and customer service; and very importantly, it is committed to promoting a *learning culture,* in the belief that it is the learning and transformation of individual members that will ultimately produce the collective shift in organizational learning which intervenes, educates, and builds preventative capabilities in the community.

NEW MODELS OF GOVERNANCE

An important feature of much of this great transformation in modern organizations is a renewed interest in the importance of effective governance. What is meant by governance is relevant to both private and public corporations. It takes its meaning from the Greek word for "steering." This is extremely relevant to organizational leadership. As Paquet (1994) notes:

> Governance is about guiding. It is the process through which a socio-economy or any other complex organization is steered. (p. 29)

With the increasing complexity of the modern world, the corporation has undergone changes which make the earlier modes of governance unworkable, unrealistic, and unproductive. When organizations were of a size and simplicity that allowed them to be guided by a single strong leader, governance could be controlled by one individual. With growth came complexity, with complexity came the need for more elaborate organizational systems. Stable environments fostered the growth of complex bureaucracies. However, the current climate of uncertainty has deeply unsettled that environment, bringing about the present re-examination of governance and the search for effective new models. Again, Paquet (1994) is helpful on this point:

> As goals become ambiguous and means-end relationships less reliable, the governance system can no longer work if it is engineered top-down by the leader. The organization can only govern itself by becoming capable of learning both its goals and the means to reach them as it proceeds. This sort of governance system is built, fundamentally, on intelligence and innovation: a capacity to tap knowledge and information held by active citizens and to get them to invent ways out of the predicaments they are in. A leader in this new context must be able to ensure that the institutional setting is capable of promoting and ensuring the requisite amount of social learning. (p. 35).

The concept of quality governance expresses itself in the effectiveness of a group of people normally referred to as a board of directors. In Chapter 5, the role of police services boards, or police commissions, will be viewed in a context similar to that of a corporation's board of directors.

On one important level, good governance relies on the supply of adequate information. The Canadian Comprehensive Auditing Foundation (CCAF) has identified the qualities of good information as follows:

- Explains options;
- Compares intentions with results;
- Forward-looking;
- Illuminates policy and administration;
- Recognizes appropriate time frames;
- Facilitates comparisons; and
- Promotes understanding without oversimplifying.

The CCAF proceeds to summarize the characteristics of effective governance as follows:

- Being comprised of people with the necessary knowledge, ability, and commitment to fulfill their responsibilities;
- Understanding their purposes and whose interests they represent;
- Understanding their objectives and strategies of the organizations they govern;
- Knowing and obtaining the information they require to exercise their responsibilities;
- Once informed, being prepared to act to ensure that the organization's objectives are met and that performance is satisfactory; and
- Fulfilling their accountability obligations to those whose interests they represent by reporting on their organization's performance.

ALTERNATIVE SERVICE DELIVERY

Alternative service delivery (ASD) can really be seen as an offshoot of the practice of business reengineering. When processes are critically examined, it frequently occurs that certain services may be delivered in a variety of alternative ways. Much of the thinking being applied to modern organizations, particularly in the context of public administration, relates to the concept of alternative service delivery.

While it takes many forms, the idea of alternative service delivery rests on the fundamental assumption that there are significantly different approaches to the delivery of all manner of services that will increase their effectiveness, efficiency, and/or economy. The range of activities associated with alternative service delivery reveals the pressing need to seek out different approaches to the "business" of government. Lindquist and Sica (1996) note the significant pressures on governments to restructure their programs and services, and have itemized some of the alternative approaches that have been taken to address these pressures:

Privatization	Commercialization	Outsourcing
Partnerships	Contracting-out	Devolution
Tax incentives	Deregulation	Re-regulating markets
Self-regulation	Special Operating Agencies	Exhortation
Grants & subsidies	Single-window service	Co-location
Loans & guarantees	Crown corporation	

Lindquist and Sica (1996) also offer a valuable summary of their survey of alternative program delivery and financing across Canada, at federal and provincial levels. They discovered a vast range of different approaches. However, in considering the substance of these different approaches, they found that:

> . . . whether or not governments emphasize one approach or another or refer to them by different names, it is increasingly useful to view all of these initiatives as part of a generic and non-ideological attempt to explore alternatives to the status quo for delivering programs; indeed, governments and citizens can no longer afford not to do so. (p. 1).

It is the notion of necessity that seems to resonate in many of the efforts made toward exploring alternative service delivery in Canada, and elsewhere. Basically, it has become an accepted piece of conventional wisdom that governments can no longer fund the full range of programs and services that they have felt compelled to offer in the past. As a result, they are looking seriously at different approaches. There is a political will that drives this search, supported by policy-makers and, in many instances, by the public.

Lindquist and Sica (1996) offer a useful grouping of known alternatives into three main categories:

1. *Finding new management*—
   - Devolution;
   - Privatization;
   - Elimination; and
   - Mergers
2. *Internal Changes*—
   - New leadership, culture, and expectations;
   - New organizational structures and processes;
   - Increased productivity with new information technologies; and
   - Shared systems or contracting out the systems.
3. *External changes*—
   - Commercialization and cost recovery;
   - Shifting administrative instruments;
   - Private or public partnerships to improve delivery; and
   - Contracting out service delivery.

The search for alternative ways of doing business within the public sector will continue to evolve over time and will involve the volunteer and private sectors in many interesting experiments.

CONTINUOUS IMPROVEMENT AND CONSTANT CHANGE

Two general themes that have issued from the whole process of organizational redesign relate to continuous improvement and the need for constant change. Both of these themes have implications for the ongoing leadership of organizations, and provide the means by which the momentum and benefits derived from corporate restructuring can be sustained (Nadler, Shaw, and Walton, 1995).

Continuous improvement, also referred to as continuous learning, corresponds to the principle that no process, policy, or practice is perfect, and that refinements may be found to extend the effectiveness or efficiency of the matter under consideration. This allows for a more beneficial and benevolent approach to organizational change, with an emphasis on seeking ways of doing things better, rather than on looking for errors. This principle implies an approach to organizational assessment that is not punitive, or fault finding, but is supportive, and problem-solving in nature. Of course, the magnitude of this change in perspective is nothing short of revolutionary in some organizations, particularly those that were formed in the tradition of the industrial or military models.

Along with the value of the continuous improvement principle is an implied acceptance of the permanence of change within the modern organization. This is something that is very difficult to come to terms with in many areas of working life, because it insists that change, sometimes extremely disruptive change, will be a constant within all organizations. This has led many management consultants and business academics to provide assistance in developing mechanisms and means to cope with this reality of organizational behaviour (Handy, 1995).

There will be a more detailed discussion of continuous learning in Chapter 7, particularly around the concept of organizational learning. However, the easiest way to understand this idea is by using the continuous work and learning model developed by Watkins and Marsick, summarized in Figure 4–3, below.

FIGURE 4–3

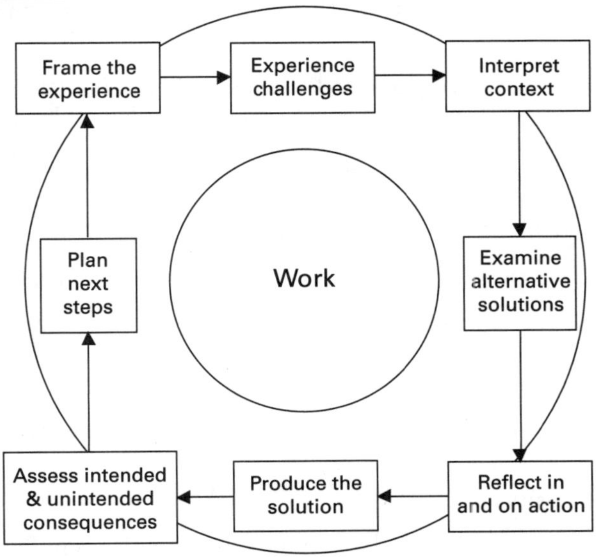

Adapted from: Watkins and Marsick (1993). *Sculpting the learning organization: lessons in the art and science of systemic change*. San Francisco: Jossey-Bass Publishers, p. 27.

POSTMODERN THEORY

While running the risk of trivializing a significant current in modern thought, it is useful, and important, to touch upon postmodern theory as it relates to modern organizations, including police organizations. It is useful because this understanding can help explain why such dramatic transformations have taken place; it is important because it illuminates the outline of possible future changes.

In the tradition of Western thought, the period known as the Age of Enlightenment marks the beginning of distinctly modern philosophy. Those who provided the foundations of the philosophy of modernity—Spinoza, Locke, Kant, and Hegel—all emphasized the pre-eminence of human rights, and thereby placed an important emphasis on the rightful focus of philosophy in the modern age. However, the spirit of postmodernism presents a different face. As Pangle notes:

> Nothing characterizes the spiritual climate of the West today so much as the pervasive disbelief in these once all-powerful philosophic pillars of modernity. Our philosophic currents are negative, skeptical, disillusioned. (p. 3)

Pangle goes on to summarize the present state of affairs:

> Here, then, is our situation in a nutshell: we in the West find ourselves in possession of fantastically powerful technological and economic resources; these resources fuel a society that is deeply unsure of its moral purpose and foundations; as an accompaniment or consequence, this society has come to be increasingly penetrated and shaped by a new, highly problematic and skeptical (not to say nihilistic) cultural dispensation known as "postmodern-ism." (p. 5) [emphasis in original]

Postmodernism is anti-rational, anti-religious, anti-philosophic, and antithetical to the view that there is some objective "truth" that can guide human action and morality. Sim (1997) notes:

> One of the best ways of describing postmodernism as a philosophical movement would be as a form of skepticism—skepticism about authority, received wisdom, cultural and political norms, etc.—and that puts it into a long-running tradition in Western thought that stretches back to Greek philosophy. Skepticism is an essentially negative form of philosophy, which sets out to undermine other philosophical theories claiming to be in possession of ultimate truth, or of criteria for determining what counts as ultimate truth. (p. 3)

Some elements of the postmodern world that have a potential impact on policing include:

- Marxism;
- The formation and advancement of the Green Party;
- Feminism;
- Ethnic and sexual minorities and activism associated with these areas;
- Environmentalism; and
- Animal rights activism.

The matter of postmodernism and its influence on politics is noted by Grant (1998):

> . . . no single story is left to hold things together; as many stories are told as there are groups to tell them. None, however, has any cultural, historical, philosophical or political priority over any other. It is in this field that, by common consent, postmodern politics finds itself: where there was unity . . . now there are only differences. Where the political will of a people, a nation or a culture used to be harnessed to long-term general goals, now fragmented groups engage in short-term struggles. The spread of identity politics over the last twenty years is testimony to this, with its emphasis on ethnicity, class, gender and sexuality replacing political credo. (p. 30)

There is also seen the strong influence of sociologists, like Daniel Bell, who wrote a number of important books, including *The end of ideology* (1960) and *The coming of post-industrial society* (1978). These works speak about the twin triumphs of Western democratic politics and capitalism. In *The coming of post-industrial society: a venture in social forecasting*, Bell saw a society with three key elements:

- Shift from manufacturing to services;
- Centrality of new science-based industries; and
- Rise of technical elites and the advent of a new principle of stratification.

Grant (1998a) also traces some of the influences of postmodernism on science and technology:

> Ever since Thomas Kuhn's The Structure of Scientific Revolutions (1962), the world of natural science has been opened for study by the social sciences. Arguing that the history of science shows a discontinuous series of breaks and radical departures (called "paradigm shifts"—a phrase that has enjoyed a certain celebrity in postmodern circles), rather than a progressive, linear accumulation of knowledge, Kuhn effectively demystified science as the 'disinterested search for facts,' and showed how scientific agendas were set by faculty squabbles, funding pressures and peer groups as much as by theoretical problems and experimental results. (p. 66)

Postmodernism is a deeply embedded trend of thought influencing all forms of culture, art, literature, criticism, education, and social experience in the Western world today. Its impact is pervasive enough to create a source of concern within law enforcement. Clearly there will be traces of the postmodern mode of thought in all parts of the criminal justice system resulting from the acceptance of, or the reaction to, this dominant trend in current thinking.

SUMMARY

This chapter has attempted to look at the recent efforts at transforming modern organizations. It has studied most closely the changes occurring within the public service across Canada. Several approaches have been considered, such as business process reengineering and organizational review and reform, including alternative service delivery. The value of exploring the "attributes of effectiveness" has been considered along with the importance of new models of governance. The significance of continuous improvement and constant change has been examined from an organizational perspective, and the prevalence of postmodern theory has been outlined.

All of these elements will serve as a useful introduction to the discussion in Chapter 5, which looks carefully at the redesign of Canadian police organizations.

QUESTIONS FOR CONSIDERATION

1. What are some of the main components of change in modern organizations?
2. How has the Public Service 2000 (PS2000) project changed the federal civil service?
3. In what ways could the eight principles of management taken from the private sector be applied to public service?
4. How might the pressures on government identified by Marson be applied to a police organization in Canada?
5. Which of the various forms of alternative service delivery (ASD) identified by Lindquist and Sica could best be applied to Canadian police organizations?
6. What are some ways in which an organization might find information relevant to the attributes of effectiveness identified by the CCAF?
7. Can the principles of continuous improvement be applied to any police organization?
8. What influence does postmodern theory have on our understanding of crime and criminality?

REFERENCES

Alberta (1994). *Measuring up*. Edmonton : Government of Alberta.

Allan, John (ed.) (1998). *Public enterprise in an era of change*. Regina, Sask.: Canadian Plains Research Center, University of Regina.

Argyris, Chris (1993). *Knowledge for action: a guide to overcoming barriers to organizational change*. New York : Jossey-Bass.

Aucoin, Peter (1990). "Administrative reform in public management: paradigms, principles, paradoxes, and pendulums." *Governance*, Vol. 3, no. 2 (April), pp. 115–137.

Aucoin, Peter (1995). *The new public management: Canada in a comparative perspective*. Montreal: Institute of Research on Public Policy.

Aucoin, Peter (1997). "The design of public organizations for the 21st century: why bureaucracy will survive in public management." *Canadian Public Administration*, Vol. 40, no. 2 (Summer), pp. 290–306.

Barzelay, Michael with the collaboration of Babak J. Armajani (1992). *Breaking through bureaucracy: a new vision for managing in government*. Berkeley, Calif.: University of California Press.

Bennis, Warren G., Kenneth D. Benne, and Bert Spector (1985). *The planning of change. 4th ed.* New York: Holt, Rinehart and Winston.

Borins, Sandford (1994). *Public sector innovation: its contributions to Canadian competitiveness*. Kingston: Government and Competitiveness, School of Policy Studies, Queen's University.

Brelin, Harvey K. et al. (1994). *Focused quality: managing for results*. Delray Beach, Fla.: St. Lucie Press.

Canada. Office of the Auditor General. (1991). *PS2000: developing a common understanding*. Ottawa: Office of the Auditor General.

Charih, Mohamed and Arthur Daniels (eds.) (1997). *New public management and public administration in Canada*. Toronto : Institute of Public Administration of Canada; Hull: École nationale d'administration publique.

Clark, Ian D. (1994). "On re-engineering the Public Service of Canada: a comment on Paul Tellier's call for bold action." *Management*, Vol. 4, no.4, pp. 20–22.

Collins, Paul and Edward Warrington (eds.) (1997). "The new public administration: global challenges—local solutions." [special issue] *Public Administration and Development*, Vol. 17, no. 1 (February).

Continuous Improvement Services Inc. and Erin Research Inc. (1992). *Best value for tax dollars: improving service quality in the Ontario government: a report to the Ontario Public Service*. [Toronto: Continuous Improvement Services Inc. and Erin Research Inc.]

Coulson-Thomas, Colin (1994). *Business process re-engineering: myth and reality*. London: Kogan Page.

Davenport, Thomas H. (1993). *Process innovation: reengineering work through information technology*. Boston: Harvard Business School Press.

Denhardt, Robert B. (1993). *The pursuit of significance: strategies for managerial success in public organizations*. Belmont, Calif.: Wadsworth.

Desveaux, James A. (1995). *Designing bureaucracies: institutional capacity and large-scale problem solving*. Stanford, Calif.: Stanford University Press.

Dinsdale, G. (1997). "The new public management and the future public service: push, pull, balance, and beyond." *Canadian Public Administration*, Vol. 40, no. 2 (Summer), pp. 370–386.

Doern, G. Bruce (1994). *The road to better public services: progress and constraints in five Canadian federal agencies*. Montreal: Institute for Research on Public Policy.

Doerr, Audrey (1996). "Delivering public services differently: some practical issues." *Management*, Vol. 7, no. 2, pp. 26–27.

Doerr, Audrey (1997). "Building new orders of government: the future of aboriginal self-government." *Canadian Public Administration*, Vol. 40, no. 2 (Summer), pp. 274–289.

Downey, T.J. and R.J. Williams (1998). "Provincial agendas, local responses: the "common sense" restructuring of Ontario's municipal governments." *Canadian Public Administration*, Vol. 41, no. 2 (Summer), pp. 210–238.

Drucker, Peter F. (1988). "The coming of the new organization." *Harvard Business Review*, Vol. 66, no. 1 (Jan.-Feb.), pp.45–53.

Drucker, Peter F. (1994). "The age of social transformation." *The Atlantic Monthly*, (Nov.), pp. 53–80.

Ford, Robin and David Zussman (eds.) (1997). *Alternative service delivery: sharing governance in Canada*. Toronto: KPMG and the Institute of Public Administration of Canada.

Fox, Charles J. and Hugh T. Miller (1995). *Postmodern public administration: toward discourse*. Thousand Oaks, Calif. : Sage Publications.

Graham, K.A. and S.D. Phillips (1998). "*Who does what* in Ontario: the process of provincial-municipal disentanglement." *Canadian Public Administration*, Vol. 41, no. 2 (Summer), pp. 175–209.

Grant, Iain Hamilton (1998). "Postmodernism and politics." In Sim, Stuart (ed.) *The Icon critical dictionary of postmodern thought*. Cambridge: Icon Books.

Grant, Iain Hamilton (1998a). "Postmodernism and science and technology." In Sim, Stuart (ed.) (1998). *The Icon critical dictionary of postmodern thought*. Cambridge: Icon Books.

Hammer, Michael and James Champy (1993). *Reengineering the corporation: a manifesto for a business revolution*. New York: Harper.

Hammer, Michael and Steven A. Stanton (1995). *The reengineering revolution: a handbook*. New York: HarperBusiness.

Handy, Charles (1995). *The empty raincoat: making sense of the future*. London: Arrow Business Books.

Hann, Robert G. and Kathryn Asbury (1992). *Change is the environment: Ontario region environmental scan, submitted to Ontario region Ministry of the Solicitor General of Canada*. Ottawa: Solicitor General Canada, Ministry Secretariat (2 vols.)

Harbour, Jerry L. (1994). *Doing it better, faster and cheaper: the power of process reengineering and information technology: NSPI conference workshop, San Francisco, Calif., April 4–5, 1994*.

Hofstede, Geert (1994). *Uncommon sense about organizations: cases, studies, and field observations*. Thousand Oaks, Calif.: Sage Publications.

Holmes, John W. (1996). "The Office of the Auditor General and public-service reform: an insider's perspective." *Canadian Public Administration*, Vol. 39, no. 4 (Winter), pp. 524–534.

Holmes, John W. and T.A. Wileman (1997). *Reform in the Australian Public Service 1983–1996*. Ottawa: Office of the Auditor General of Canada.

Ingstrup, Ole (1995). *Public service renewal: from means to ends*. Ottawa: Canadian Centre for Management Development. (Explorations; no. 4).

Jaques, Elliott and Stephen D. Clement (1991). *Executive leadership: a practical guide to managing complexity*. Arlington, Virginia: Basil Blackwell, Cason Hill & Co., Publishers.

Johnson, A.W. (1994). *What is public management?: an autobiographical view*. Ottawa : Canadian Centre for Management Development.

Kernaghan, Kenneth (1993). "Reshaping government: the post-bureaucratic paradigm." *Canadian Public Administration*, Vol. 36, no. 4 (Winter), pp. 636–644.

Kernaghan, Kenneth and M. Charih (1997). "The challenge of change: emerging issues in contemporary public administration." *Canadian Public Administration*, Vol. 40, no. 2 (Summer), pp. 218–233.

Kettl, Donald F. (1994). *Reinventing government? : appraising the National Performance Review: a report of the Brookings Institution's Center for Public Management*. Washington, D.C.: Brookings Institute.

Kotter, John P. (1996). *Leading change*. Boston: Harvard Business School Press.

Leclerc, Guy et al. (1996). *Accountability performance reporting, comparative comprehensive audit: an integrated approach*. Ottawa: CCAF-FCVI.

Lindquist, Evert and Tammy Sica (1996). "Canadian governments and the search for alternative program delivery and financing: report on a preliminary survey." *Management*, Vol. 7, no. 1, pp. 16–17.

Lyotard, Jean-François (1997). *The postmodern condition: a report of knowledge*. Translation from the French by Geoff Bennington and Brian Massumi. Minneapolis, Minn.: University of Minnesota Press.

Manion, J.L. (1994). "The challenge of the new administrative order: post-modern accountability." In *The Dewar series: perspectives on public management: values in the public service*. Ottawa : Canadian Centre for Management Development. (Explorations; no. 1).

Mintzberg, Henry (1979). *The structuring of organizations*. Englewood Cliffs, N.J.: Prentice Hall.

Mintzberg, Henry (1981). "Organization design: fashion or fit?" *Harvard Business Review*, Vol. 59, no. 1, (Jan.–Feb.), pp. 103–116.

Morgan, Gareth (1993). *Imaginization: the art of creative management*. Newbury Park, Calif.: Sage Publications.

Moynagh, W. David (1993). *Putting theory into practice: reporting and auditing effectiveness*. Ottawa: Canadian Comprehensive Auditing Foundation.

Nadler, David, Robert B. Shaw, A. Elise Walton and associates (1995). *Discontinuous change: leading organizational transformation*. San Francisco: Jossey-Bass Publishers.

Ontario. Management Board Secretariat (1993). *An introduction to business reengineering*. Toronto: Management Board Secretariat.

Ontario. Premier's Council on Economic Renewal. Task Force on the Organization of Work (1993). *The organizational change process*. [Toronto: The Task Force].

Osborne, David and Ted Gaebler (1992). *Reinventing government: how the entrepeneurial spirit is transforming the public sector from schoolhouse to state house, city hall to Pentagon*. Reading, Mass.: Addison-Wesley.

Pangle, Thomas L. (1992). *The ennobling of democracy: the challenge of the postmodern age*. Baltimore: The Johns Hopkins University Press.

Paquet, Gilles (1994). "Paradigms of governance." In *The Dewar series: perspectives on public management: rethinking government*. Ottawa: Canadian Centre for Management Development.

Pascale, Richard T. and Anthony G. Athos (1981). *The art of Japanese management: applications for American executives*. New York: Simon and Schuster.

Peters, B. Guy (1995). *The public service, the changing state and governance*. Ottawa: Canadian Centre for Management Development. (Research paper; no. 13).

Peters, Tom (1992). *Thriving on chaos: handbook for a managerial revolution*. London: Macmillan.

Price, Errol S. and Leslie P. McAdams (1996). "Bringing accountability for results to government in British Columbia." *Public Administration and Development*, Vol. 16, pp. 305–315.

Price Waterhouse Associates and The Canadian Consulting Group Inc. (1985). *A study of management and accountability in the Government of Ontario*. [Toronto]: Price Waterhouse Associates and The Canadian Consulting Group Inc.

Robbins, Stephen R. and Nancy Langton (1999). *Organizational behaviour: concepts, controversies, applications. Canadian edition*. Scarborough: Prentice Hall Canada.

Roberts, Alasdair (1996). "Worrying about misconduct: the control lobby and the PS2000 reforms." *Canadian Public Administration*, Vol. 39, no. 4 (Winter), pp. 489–523.

Rocine, Victor (1994). "Process benchmarking for superior performance: sharing, learning, improving." *Management*, Vol. 5, no. 1, pp. 27–30.

Rummler, Geary A. and Alan P. Brache (1990). *Improving performance: how to manage the white space in the organization chart*. San Francisco: Jossey-Bass.

Sancton, Andrew (1996). "Reducing costs by consolidating municipalities: New Brunswick, Nova Scotia and Ontario." *Canadian Public Administration*, Vol. 39, no. 3 (Fall), pp. 267–289.

Sim, Stuart (1998). *The Icon critical dictionary of postmodern thought*. Cambridge: Icon Books.

Tamas, Andy (1995). "The manager and the new public service." *Canadian Public Administration*, Vol. 38, no. 4 (Winter), pp. 613–621.

Taylor, James R. and Elizabeth J. Van Every (1993). *The vulnerable fortress: bureaucratic organization and management in the information age*. Toronto: University of Toronto Press.

Tellier, Paul (1994). "No time for half measures: the urgency of re-engineering of the Public Service of Canada," speech given to the Canadian Institute in Ottawa on February 21, 1994. Excerpts of his remarks appeared in the *Globe and Mail* on February 25, 1994.

Thomas, Paul (1993). "Coping with change: how public and private organizations read and respond to turbulent external environments." In Seidle, Leslie F. (ed.). *Rethinking government: reform or reinvention?* Montreal: Institute for Research on Public Policy.

Thomas, Paul (1994). "Reshaping government: now and for the future." *Management*, Vol. 5, no. 1, pp. 19–23.

Vantour, Jim (ed.) (1991). *Our story: organizational renewal in federal corrections*. Ottawa: Canadian Centre for Management Development.

Vojnovic, I. (1998). "Municipal consolidation in the 1990s: an analysis of British Columbia, New Brunswick, and Nova Scotia." *Canadian Public Administration*, Vol. 41, no. 2 (Summer), pp. 239–283.

Wagner III, John A. and John R. Hollenbeck (1992). *Management of organizational behavior*. Englewood Cliffs, N.J.: Prentice Hall.

Warrian, Peter (1995). *The end of public sector "industrial" relations in Canada? The coming challenges for labour and management*. Toronto: KPMG Centre for Government Foundation.

Waterfield, S. (1997). "The challenges facing provincial public service." *Canadian Public Administration*, Vol. 40, no. 2 (Summer), pp. 204–217.

Watkins, Karen E. and Victoria J. Marsick (1993). *Sculpting the learning organization: lessons in the art and science of systemic change*. San Francisco: Jossey-Bass Publishers.

Weisbord, Marvin R. (1992). *Discovering common ground: how future search conferences bring people together to achieve breakthrough innovation, empowerment, shared vision, and collaborative action*. San Francisco: Berrett-Koehler Publishers.

Wilson, James Q. (1989). *Bureaucracy: what government agencies do and why they do it*. New York: Basic Books.

Zuboff, Shoshana (1988). *In the age of the smart machine: the future of work and power*. New York: Basic Books.

Zussman, David (1998). "Government's new style." *Management*, Vol. 8, no. 2, pp. 21–23.

TOWARD MORE LEARNING

REENGINEERING IN THE PUBLIC SERVICE

Several sources provide excellent background on the work being done to transform the public service in Canada and other countries. The material produced by the Institute of Public Administration of Canada (IPAC) is an ongoing resource of valuable information that presents detailed analysis on the entire range of topics relevant to public administration in Canada. Through its publishing efforts, IPAC produces a reliable array of documents well suited to the student of policing who wishes to better understand the public service. Of particular value are the following periodicals:

Canadian Public Administration—this is the official journal of IPAC and is published quarterly. The journal "... is committed to the examination of the structures, processes, outputs and outcomes of public policy and public management related to executive, legislative, judicial and quasi-judicial functions at all three levels of Canadian government." It frequently features articles that address some of the major reengineering or organizational change challenges faced within the public service. For example:

Lindquist, Evert A. (1992). "Public managers and policy communities: learning to meet new challenges." *Canadian Public Administration*, Vol. 35, no. 2 (Summer), pp. 127–159.

Kernaghan, Kenneth (1992). "Empowerment and public administration: revolutionary advance or passing fancy?" *Canadian Public Administration*, Vol. 35, no. 2 (Summer), pp. 194–214.

Kernaghan, Kenneth (1993). "Reshaping government: the post-bureaucratic paradigm." *Canadian Public Administration*, Vol. 36, no. 4 (Winter), pp. 636–644.

Kernaghan, Kenneth (1994). "The emerging public service culture: values, ethics, and reforms." *Canadian Public Administration*, Vol. 37, no. 4 (Winter), pp. 614–630.

Clark, I.D. (1996). "Global economic trends and pressures on governments." *Canadian Public Administration*, Vol. 39, no. 4 (Winter), pp. 447–456.

Holmes, J.W. (1996). "The Office of the Auditor General and public service reform: an insider's perspective." *Canadian Public Administration*, Vol. 39, no. 4 (Winter), pp. 524–534.

Aucoin, P. (1997). "The design of public organizations in the 21st century: why bureaucracy will survive in public management." *Canadian Public Administration*, Vol. 40, no. 2 (Summer), pp. 290–306.

Kernaghan, K. and M. Charih (1997). "The challenges of change: emerging issues in contemporary public administration." *Canadian Public Administration*, Vol. 40, no. 2 (Summer), pp. 218-233.

Waterfield, S. (1997). "The challenges facing provincial public services." *Canadian Public Administration*, Vol. 40, no. 2 (Summer), pp. 204–217.

Public Sector Management—this is another publication of IPAC. It is also published quarterly and has shorter articles dealing with practical issues within the public service. The student of public administration looking for innovative ideas, often applied by a local police department, is well advised to seek out those issues of *Management* that catalogue the submissions made to the annual IPAC award for innovative management. These issues provide brief descriptions of projects or programs being submitted for consideration and include details about contacting those organizations. The award, which is sponsored by Coopers & Lybrand, was established to acknowledge excellence and innovation within the public sector and to recognize outstanding organizational achievements. For example:

1992—*Public Sector Management* (Fall issue);

1993—*Public Sector Management*, Vol. 4, no. 1 (Spring issue);

1994—*Public Sector Management*, Vol. 5, no. 1;

1995—*Public Sector Management*, Vol. 6, no. 2;

1996—*Public Sector Management*, Vol. 7, no. 2; and

1997—*Public Sector Management*, Vol. 8, no. 2.

Another rich source of background material may be found in the diverse publications of the Canadian Centre for Management Development (CCMD). The CCMD is an agency of the federal government in Canada that is responsible for the education and development of the senior levels of the public service. It has had a vigorous publishing program in place over the last few years that has resulted in some extremely useful documents for students, professors, and practitioners of public management. Some of the following themes have been addressed through the CCMD's publishing efforts:

- Women and work;
- Ethics and values;
- Equity and diversity;
- Negotiation and conflict management;
- Learning and executive development;
- Governance in a changing environment;
- Consultation process;
- Service and quality;
- Managing people; and
- Managing organizational change.

The CCMD may be contacted as follows:

373 Sussex Drive
P.O. Box 420, Station "A"
Ottawa, Ontario
K1N 8V4
E-mail: info@ccmd-ccg.gc.ca
Website: http://www.ccmd-ccg.gc.ca/

On an international level, the Commonwealth Association for Public Administration and Management produces a journal that frequently includes valuable articles of note for those interested in the changing world of public administration:

Public Administration and Development: The International Journal of Management Research and Practice.

ORGANIZATIONAL REVIEW AND REFORM

The topic of business process reengineering has been popular for some time. However, its most widely-known proponent is Michael Hammer. His work, in collaboration with other authors, has been responsible for guiding many organizations through the challenge of this mode of organizational review. Hammer's approach to business process reengineering is set out in two publications:

Hammer, Michael and James Champy (1993). *Reengineering the corporation: a manifesto for a business revolution*. New York: Harper.

Hammer, Michael and Steven A. Stanton (1995). *The reengineering revolution: a handbook*. New York: HarperBusiness.

The Canadian Comprehensive Auditing Foundation (CCAF) is a useful source of information on the "attributes of effectiveness." The CCAF has published several documents that develop this model.

The CCAF may be contacted as follows:
210-55 Murray Street
Ottawa, Ontario
K1N 5M3
Tel.: (613) 241-6713
Fax.: (613) 241-6900
E-mail: info@ccaf-fcvi.com
Website: http://www.ccaf-fcvi.com

NEW MODELS OF GOVERNANCE

Again, the CCAF is a good source of material on developing new models of governance. With a considerable research capacity, the CCAF has undertaken substantial study in order to assemble programs and publications that address the information and actions required for good governance. Recently, the CCAF produced a trilingual (English, French, and Spanish) lexicon of governance, accountability, and comprehensive audit terms to assist professionals in this field of study. Also, in 1994, the CCAF produced a videotape, entitled *In Search of Effective Governance*, which outlines the essential elements of this important concept through a series of discussions with leading members of governing bodies, executives, management academics, and professional auditors.

CONTINUOUS IMPROVEMENT AND CONSTANT CHANGE

The concept of continuous improvement, including continuous learning and change management, is the subject of many current organizational development textbooks. The following provides merely a sampling of works that will be useful to the student interested in this area of study:

Robbins, Stephen P. and Nancy Langton (1998). *Organizational behaviour: concepts, controversies, applications. Canadian edition.* Scarborough, Ont.: Prentice Hall Canada Inc.

This recent publication is a Canadian edition of a very popular American textbook dealing with a full range of topics relevant to organizational behaviour. It provides comprehensive coverage of the subject and offers a text-specific Internet site at http://www.prenticehall.ca/robbins that includes study guides, links to particular organizational sites, and search tools. Also included are video cases taken from the CBC's *Venture* series. An outline of the contents indicates the range of the book's coverage:

- Part I—Developing an understanding of the workplace
  - What is organizational behaviour?
  - Working in the organization of the 21st century
  - Perception and personality
  - Values, attitudes and their effects in the workplace
- Part II—Striving for performance
  - Basic motivation concepts
  - Motivation: aligning incentives to goals
  - Foundations of group behaviour
  - Understanding work teams
- Part III—Sharing the organizational vision
  - Communication
  - Decision-making, creativity, and ethics
  - Leadership
- Part IV—The uneasy sides of interaction
  - Power and politics
  - Conflict and negotiation
- Part V—Towards reorganizing the workplace
  - Foundations of organizational structure
  - Work design
  - Organizational culture
  - Organizational change and stress management

Nadler, David A., Robert B. Shaw, and A. Elise Walton (1995). *Discontinuous change: leading organizational transformation*. San Francisco: Jossey-Bass Publishers.

One of the many works available to assist in the understanding and management of change within the organizational context. This publication is helpful in that it identifies the capacity to lead change as a core competence in the twenty-first century, and discusses the fundamentals of change management.

POSTMODERN THEORY

There is a great deal written about postmodernism in virtually every subject area. Of particular value to the student wishing to learn more about the theoretical background of this approach are two publications of note:

Lyotard, Jean-François (1997). *The postmodern condition: a report on knowledge*. Translation from the French by Geoff Bennington and Brian Massumi. Minneapolis, Minn.: University of Minnesota Press. (Theory and history of literature, vol. 10).

Lyotard is considered by many to be one of the key thinkers of postmodernism. His work grows out of his disenchantment with Marxism and socialism during the popular revolt in Paris during May 1968. He develops a theme that attacks all "grand narratives" that claim to possess universal truth, including Marxism.

Sim, Stuart (ed.) (1998). *The Icon critical dictionary of postmodern thought*. Cambridge: Icon Books.

The publication offers one, accessible volume in which the key elements of postmodern thought can be found. It includes a number of critical essays that place postmodernism into various contexts. The book provides a comprehensive survey of various perspectives that fall under the general heading of postmodern thought. The contributors are experts in their subject areas and offer reflections on the impact of postmodernism in such fields as: social science, philosophy, science, politics, cinema, feminism, and the media.

CHAPTER FIVE

CREATING DYNAMIC CHANGE IN POLICE ORGANIZATIONS

> **The implementation of any complex change, such as that implied by advocates for community policing, in a resistant institutional setting, such as that identified in many police agencies, will require more attention to the organizational medium within which these changes occur. The glacial change we have witnessed in the core technology and values of policing attest to the importance of designing change strategies that are well thought out, are monitored, and that provide feedback to police administrators attempting to implement community policing.**
>
> Jack R. Greene, "Evaluating planned change strategies in modern law enforcement" (1998).

LEARNING OBJECTIVES

1. Identify the major periods of police organizational development.
2. List the predictable components of any organizational change cycle.
3. Identify significant obstacles to change in Canadian police organizations.
4. Understand the distinction between empowerment and accountability.
5. Specify the role played by governing authorities in changing police organizations.
6. Identify approaches to transforming the organizational culture of policing.
7. List ways in which police middle managers can be supported during an organizational change process.
8. Recognize the role of recruitment in supporting police organizational change.

INTRODUCTION

This chapter will consider how effective change in a police organization can contribute to dynamic growth. The word "dynamic" is used here in the sense of being aimed at effective action and providing the basis for continuous organizational change. As discussed in Chapter 4, change is a constant in modern organizational life, and police organizations should be ensuring that they are undertaking positive change projects for the right reasons.

Historically, police organizations have demonstrated a certain degree of reluctance to change (Zhao, 1996). This reality will be considered in the context of the evolution of policing, and some of the special circumstances that influence change in policing will be reviewed. This chapter will also examine the predictable components within the organizational change cycle and relate them to policing in Canada. The distinction between empowerment and accountability will be assessed as it pertains to policing, as well as the critical role played by governing authorities in the police organizational change process.

Attention will also be given to the importance of transforming the corporate culture as part of organizational transformation. This will be followed by a review of the ways in which police middle managers can be supported in ways that will help them sustain overall change initiatives. Finally, the important role of recruitment will be assessed in terms of its contribution to organizational transformation.

Each of these topics is intended to increase our understanding of the significance of organizational change in modern policing. Community policing, when it is understood as a fundamentally different way of doing police "business," necessarily requires a radical redesign of the structure and shape of the organization if it is to deliver its services in new ways (Brodeur, 1998). This chapter is designed to explain the nature and substance of such change in police organizations, and to offer certain insights into dealing with dynamic change.

"BENDING GRANITE" — A HISTORY OF RELUCTANCE?

With their origins based on a military model, police organizations have been perceived as institutions particularly resistant to change (Johnson, 1970; Guyot, 1979). The rigidity of the rank structure, the concept of unity of command, and the strict hierarchical makeup of these organizations have all contributed to that reality. Increasingly, however, police organizations have begun to explore organizational reform and, in some instances, have taken the challenge of change seriously (Ontario Provincial Police, 1995). The argument may be made that police organizations in Canada have always been involved in change initiatives to a high degree (Marquis, 1993). A counter argument has also been made that community policing is not even a new concept, derived as it is from the principles articulated by Sir Robert Peel (Braiden, 1992).

Indeed, police organizations have often been quick to adopt new technologies and tools that promise to make their operational or administrative functions more effective. As Mastrofski (1998) indicates:

> For all the hopefulness of police professionalism since August Vollmer, the state of police science at the end of the 20th century is not much advanced over what it was at the beginning, the principle difference being that today we know a bit more about what does not work. (p. 176)

The last round of substantial reform in Canadian policing follows the example set by the U.S. in the 1970s (Fogelson, 1977). While the timeframes and labels may vary from one interpreter to another, there are essentially three periods of police evolution in the modern U.S. that roughly coincide with developments in Canada:

- Political era (1830s to 1900)
- Progressive or Professional era (1900 to 1970s)
- Community-oriented or Proactive Policing era (1970s to present)

A more sophisticated treatment of the evolution of the police mandate can be found in McDougall (1988). It is important to recognize that there were significant differences between policing in the U.S. and Canada throughout this century. However, there are patterns of similarity that show a close connection in terms of mandate, organizational philosophy, and structural practice.

Zhao (1996) has examined why American police organizations change. His focus is on the recent transformation from a "bureaucratic" model to a "community-oriented policing" (COP) model. Kuykendall and Roberg (1982) identified five basic features of a bureaucratic model in relation to its organizational structure and management:

- High degree of specialization of tasks;
- Hierarchical structure;
- Top-down flow of authority;
- High degree of rule-oriented conduct;
- High incidence of position-oriented behaviour.

Throughout the political and professional eras, police departments generally displayed all of these bureaucratic features, and were singularly reluctant to modify their approach to service delivery resulting from that model. Angell (1971) observes that police organizations in the U.S. have a classic structure based on theories developed by Max Weber. The key characteristics of that structure, which may also be seen in Canadian police organizations are:

> 1 formal structures are defined by a centralized hierarchy of authority;
> 2 labor is divided into functional specialties;
> 3 activities are conducted according to standardized operating procedures;
> 4 career routes are well established and have a common entry point; promotions are based on impersonal evaluations by superiors;
> 5 management proceeds through a monocratic system of routinized superior-subordinate relationships;
> 6 status among employees is directly related to their positions (jobs) and ranks. (pp. 186–87)

It is also important to note the critical significance of technology in sustaining and supporting this organizational structure. Particularly in the case of the professional era, technological innovations allowed police departments to adhere to a static organizational structure while modifying their programs in response to crime and control issues in a way that appeared to address public safety concerns. Furthermore, the type of organizational structure outlined by Angell contributed enormously to the isolation of members of the department from the outside environment. As a paramilitary, bureaucratic organization, the police department could be seen, in its earliest forms, as being removed from the democratic approach pursued by other public institutions. Improvements would be possible if police departments applied a democratic, as opposed to a bureaucratic model of contemporary policing. Angell (1971) notes that when applying a democratic model to policing the department is compelled to provide:

> . . . a flexible, participatory, science-based structure that will accommodate change. It is designed for effectiveness in serving the needs of citizens rather than autocratic rationality of operation. It is democratic in that it requires and facilitates the involvement of citizens and employees in its processes. (p. 194) [emphasis added]

The idea that citizens, in the form of "the community" were important to policing was given expression in the 1967 U.S. President's commission report on law enforcement and administration of justice. This report also spoke to the issue of co-production between police and the community in terms of quality of life (Brudney and England, 1983; Curtis, Thurman, and Nice, 1991). Also, according to Zhao (1996):

> . . . the emergence of the COP [community-oriented policing] reform model marked the beginning of serious efforts at organizational change in bureaucratic model in terms of the key values underlying the policing function, the organizational structures required for police agencies, and the operational activities police should engage in to provide service to their community. Thus, the COP model resembles an open-system approach. That is, organizational change is influenced by both internal and external factors to such a degree that variation of these factors likely determines significantly the change process police agencies experience. (p. 8)

Zhao further distinguishes between planned change and adaptation. Planned change may also be referred to as organizational development (OD), and includes the application of group process techniques and applied psychology to achieve specific goals for the organization, including individual behaviour, group attitudes, and conflict/cooperation behaviours. Planned change has an essentially *internal* focus. Adaptation, on the other hand, tends to respond with an *external* focus, and will confront technological change, environmental variables, or political change (Sandler and Mintz, 1974; Goodman and associates, 1982; Huber et al., 1993).

In Canada, Marquis (1993) provides an excellent summary of policing during the 20th century through his treatment of the history of the Canadian Association of Chiefs of Police (CACP). The concerns of this representative organization reflect the key issues that impacted on individual Canadian police departments. Those concerns tended to focus on technology, criminal statistics, and criminal law reform. The "professionalization" of policing in Canada was a central theme for the CACP and the networking of departments and agencies across Canada (e.g., Uniform Crime Reports, inaugurated in the early 1960s and CPIC, the Canadian Police Information Centre). Training was another significant issue along with police pensions, and other such practical matters. Canadian police departments, however, tended to mirror their American cousins with regard to organizational change. Zhao (1996) notes with respect to his American study:

> The findings consistently show that contemporary organizational change in American policing is more likely forced than consciously chosen; it is more likely a response to a turbulent external environment than anything else. (p. 58)

Police departments in Canada over the last thirty years have not embraced fundamental organizational change as a top priority. This is not to suggest that these institutions have remained static. On the contrary, there have been many examples of technological innovation and interesting program development initiated with care, diligence, and ingenuity. However, the pressure for comprehensive structural change in Canadian policing has resulted largely from external pressures, not internal inspiration. Even when change has been embarked upon in Canadian policing the results are often consistent with Zhao's findings which:

> . . . suggest that little substantial organizational structural change has taken place in American policing in recent years, despite all of the discussion about COP. This means that American policing is likely engaged in a long process of evolution rather than fast-moving revolution. Although many scholars of policing might say that the direction of change is favorable, it appears to be more creeping incrementalism than decisive advancement. (p. 82)

Of course, police organizations are not the only institutions that have been somewhat slow to change. Throughout the public service, and in large pockets of the private sector, organizational change was not seen immediately as something to be pursued. Often it was a thing to be avoided if at all possible, and resisted when it appeared inevitable. The public realm held a virtual monopoly on the programs and services it supplied and did not feel an overwhelming motivation to undertake large-scale organizational change until the external factors discussed in Chapter 4 came into being.

Of particular relevance here is the influence of the professional model in creating the conditions that encouraged police departments to resist change. By removing themselves from what was seen as the undue influence of corrupt city officials in the U.S. during the political era, police leaders withdrew into a closed approach that emphasized hierarchy, rank, scientific management, and bureaucratic approaches. This "professional" alternative to the previous era of policing was pioneered by such U.S. innovators as August Vollmer (Carte and Carte, 1975), O.W. Wilson (Bopp, 1977), and William H. Parker (Bopp, 1972). The professional era enforced a principled and self-conscious detachment from the political realm that was seen as being beneficial for the police and the public, and would better preserve the rule of law. This approach was adopted in Canada in an effort to avoid the abuses and excesses of the political era in the U.S. However, by the 1970s, it was becoming painfully apparent that many police organizations were isolated from their communities, and frequently out of touch with the needs of community members (U.S. President's Commission on Campus Unrest, 1970). This realization led to the reform movement that has taken shape around various forms of community-oriented policing (COP).

In Ontario, the 1974 Task Force on Policing recognized the dilemma created by the well-intentioned pursuit of the professional model, especially the technological component of that model:

> The introduction of technology into the police role has tended to widen the gap between the citizen and his police force. Early stages of technology reform put police officers into automobiles and tied them to dispatch centres through radio. This had the effect of taking officers

> out of direct contact with people on the streets. Perhaps of equal importance, radio concentrated the officer's attention on incidents reported to the police departments by telephone. While this improved the police ability to respond to citizens in need, it also has tended to shift their attention to cases needing immediate resolution and away from a continuity with a geographic area and its people. (p. 25)

Going back to the impact of the military tradition in Canadian policing, what was true for Ontario in 1974 was true across Canada:

> During the first half of this century, policing in Ontario came to be heavily influenced by military personalities and military structures. Ranks and disciplinary systems were styled on the military model, and the concept of command was adopted as the administrative style for most forces. (p. 20)

When the leadership of Canadian policing is examined historically it is clear that the military approach to policing resulted from the direct engagement of successful soldiers as police chiefs and chief constables in Canadian police departments. When the senior executives guiding Canadian police organizations are drawn from the apex of the military establishment, it is not surprising that their management and leadership styles will mirror their military origins (Highley, 1984; Marquis, 1993). As noted above, one of the consequences of the professional model, with its heavy reliance on technology and the perceived merits of random, motorized patrol, was the increasing isolation of the individual officer. Again, the report of the Ontario Task Force on Policing (1974) is instructive:

> The shift in role for the individual constable implied by these processes of technological change have [sic] a direct impact on his beliefs and character. As an individual, his contact with society tends to be limited to other police officers. In his daily work, he sees an unbalanced sample of his community—mostly those who live on the edge of the law. Socially, because of shift work, the anxiety surrounding the real dangers of his job, and because of the community's sensitivity to the authority of the police image, he tends to concentrate his friendships among other officers and their families. Gradually, this unbalanced exposure to his community leads him to a general cynicism about human nature. His lack of exposure to people leads him to deal with them from behind the mask of his uniform and prevents him from developing the skills which would enhance easy relationships with citizens. (p. 26)

Because police organizations in Canada generally perceived it as being beneficial to stand at arm's length from political control, especially in the area of police operations, there was a sense that the police themselves held responsibility for the nature and extent of any organizational change. As a result, there was not a great deal of attention given to fundamental organizational design. The response to any perceived service delivery need was typically the development of a new program, designed almost exclusively within the police department, staffed with selected, uniformed officers. For example, a perceived need for an underwater search and recovery service would result in the creation of such a unit with a Staff Sergeant in charge, along with a sergeant supervisor and a number of constables qualified in scuba diving and other related skills. A new box would be added to the organization chart, in the field operations side of the department, and a cost centre would be established to track the budget expenditures of this new unit. Each year the budget, staffing, equipment, and other requirements of the unit would be reviewed and funded according to need.

This approach tended to work quite satisfactorily while police departments held themselves aloof from their governing authorities. The professional model may also be characterized by its view that the police themselves were in the best position to determine the service delivery needs of the community. The police were the subject matter experts in the area of law enforcement, public safety, and order maintenance. The public was the recipient of the programs devised and delivered by the police professionals and was the source of funding for these programs through their tax dollars. But pressure to change has been mounting in most police jurisdictions. This pressure has led to a serious interest in substantive organizational change in Canadian policing. It has also led to considerable strengthening of the role of civilian governing authorities in Canada. Before examining the details of this change, it is worthwhile looking at the predictable components of the change process in order to see the patterns that typically take shape during any change initiative.

THE PREDICTABLE COMPONENTS OF THE CHANGE PROCESS

As an increasing number of organizations, including police departments, have embarked on major change initiatives, a growing literature has developed setting out some of the components that can be anticipated. By understanding these elements, individuals are better able to plan and pursue the change cycle. By recognizing a certain degree of inevitability within the process, organizations can develop strategies for moving through the change cycle with a higher degree of success. It has been noted that these elements relate to both individuals and organizations as they move through the process of change. Figure 5–1 outlines the four main phases of change and sets them into a context, occurring over time, that includes the external/environmental elements, as well as the internal or personal elements:

FIGURE 5–1

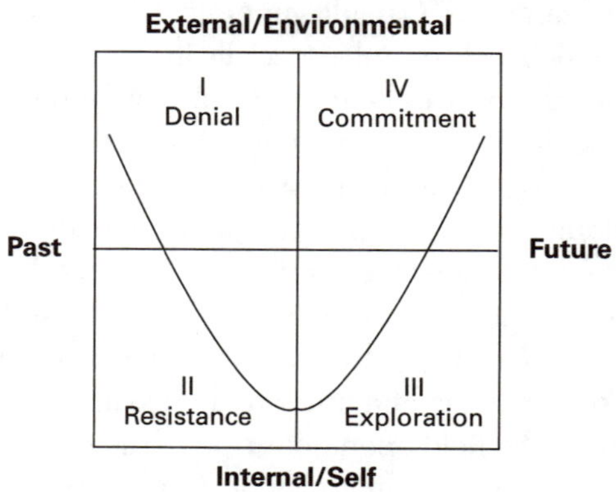

1. *Denial*—in this phase it is important to:
   - Strengthen relationships within the organization;
   - Divide the change initiative into smaller, more easily managed, steps;
   - Focus on achieving success with the first steps; and
   - Avoid confrontation.

2. *Resistance*—in this phase there will be anger directed at the change process and those perceived as being responsible for the change. Also, there will be efforts at bargaining and depression. Therefore, it will be important to:
   - Legitimize the anger;
   - Avoid taking the anger personally, if you are part of the change management team;
   - Avoid bargaining, because that will transform the nature of the change initiative and generate other responses;
   - Provide support and identify the resources available to people throughout the change process;
   - Emphasize two-way communication; and
   - Encourage people to assume responsibility for change.

3. *Exploration*—in this phase it is important to:
   - Test new options within the context of the changed organization;
   - Identify progress and the achievement of intended results;
   - Work on building confidence within the organization.

4. *Commitment*—in this phase it is important to:
   - Acknowledge and reward success within the context of the change initiative;
   - Develop an inventory of the "learnings" which have occurred throughout the change process and share those "learnings" within the organization; and
   - Develop a state of readiness for the next change intiative.

The change process itself may also be seen to have four major components. Each of these components needs to be addressed if the organization is going to successfully navigate a change initiative:

1. *Building the foundation for commitment*—
   - Recognize the need for change
   - Approach and involve other parties
   - Heighten awareness
   - Examine readiness to proceed
   - Assess what's in it for your stakeholders

2. *Getting agreement and setting direction*—
   - Establish a parallel design structure
   - Assess specific needs
   - Develop a vision statement
   - Get endorsement from key stakeholders

3. *Making changes—*
    - Analyze the organization
    - Design or redesign
    - Implement
4. *Keeping it going—*
    - Evaluate
    - Continuous improvement

SPECIAL CIRCUMSTANCES AFFECTING CHANGE IN POLICE ORGANIZATIONS

While many Canadian police organizations have gone to some lengths to engage in corporate change that will allow them to pursue a community-oriented approach, there are important circumstances that complicate the change process for police departments. Unlike a parts factory, or commercial business, that can elect to close down while it re-tools and redesigns its processes and products, a police service does not have this luxury. In fact, the organizational change process significantly impacts on the police leadership because they are compelled to deliver a critical public safety service to high standards set by federal, provincial, and municipal levels of government. This is especially difficult to accomplish when they are attempting to undertake an organizational change process that places huge demands on everyone in the department. What differentiates this from many other public service agencies that must continue to operate through change is that the police must also deal with matters of extreme, potential danger involving life and death issues, and individual freedom. The instances of bank hold-ups, domestic assaults, youth crime, and violence will not be suspended while a police department dissects its operations and management in order to design a better model.

Police organizations must also be conscious of the realities that impose certain limits on what they are able to accomplish in the context of organizational change. A number of areas are pertinent to this discussion:

- Rank structure;
- Unity of command;
- Regulations, standards, guidelines, policies, and procedures;
- Police governance and civilian oversight.

Each of these areas has an important bearing on the degree of change that is feasible within any Canadian police service, and will be discussed briefly below.

RANK STRUCTURE

The rank structure of a police organization continues to play a vital role in its administration and operation. With a quickening pace of change geared toward community-oriented policing, many police departments are beginning to assess the need for the vast array of rank levels and the sheer number of officers performing exclusively supervisory or administrative duties. Instead of having officers tasked with the job of "inspecting" the work of subordinates, it makes common sense to design systems and performance aids that are error-free, or, at least, allow the officer to self-inspect the work being done (Harbour, 1994). This is where the enabling power of technology and process reengineering have a central role to play in planned organizational change in policing in Canada (Fraser et al., 1998). There is a growing appreciation of the value of having the front-line officer work more closely with the community in a problem-solving mode. However, there are no police departments seriously contemplating the removal of all ranks. For example, the OPP dispensed with the rank of corporal several years ago and has undergone a comprehensive organizational review that reduced the number of mid-level managers within its structure.

The persistence of the rank structure has been questioned on several occasions (Guyot, 1979), yet it continues to reflect an accepted practice within police organizations.

CHAIN OF COMMAND

The chain of command is something derived directly from the military approach to organization. As a formal arrangement, it sets forth lines of control and communication that operate in a top-down manner, commonly in the form of orders (Sheehan and Cordner, 1995). Authority proceeds from the senior ranks of the organization through increasingly subordinate ranks. This format applies what is referred to as the *scalar principle* (Sheehan and Cordner, 1995). While orders originate at the top of the organization, a chain of command model also includes communication from the bottom up, typically in the form of occurrence reports, statistics, and briefings. Figure 5–2, below, presents two police organizational structures that present different chains of command.

FIGURE 5–2

VARIATIONS ON THE CHAIN OF COMMAND

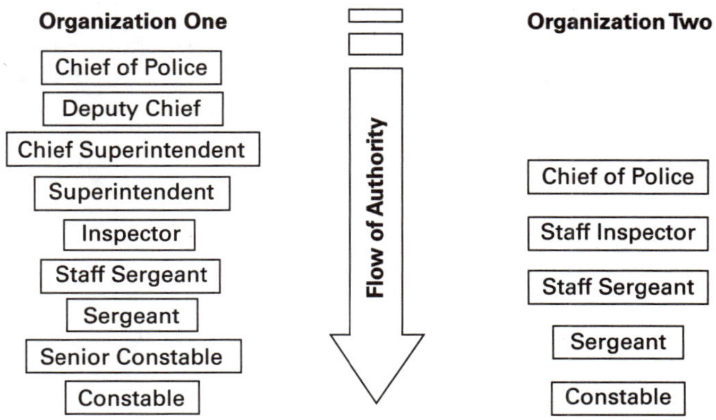

Unity of Command

The notion of unity of command in policing is an element that relates to the military tradition. In order to function effectively and safely, police officers have only one person to whom they directly report. This is a convention of organizational theory that is widely accepted. An officer receives orders and instructions from an immediate supervisor and provides feedback and status reports to that same supervisor. This one-to-one relationship is fundamental to police administrative practice because it accomplishes the following (Souryal, 1985):

- Strengthens the organization's authority and control;
- Assists in determining responsibility for action, especially in crisis situations;
- Facilitates communication and reduces confusion and contradictory orders; and
- Enhances effective supervision of staff.

This principle of unity of command has, however, been subjected to criticism because it:

- Prevents employees from gaining advice and insight from other staff members who are not immediate supervisors;
- Extends the time it takes for communications to move through the hierarchy of the police department;
- Potentially demoralizes staff when they must deal with uncooperative supervisors;
- Diminishes the value of front-line staff; and
- Promotes inflexibility within the organization.

Modern organizational design theory supports the application of a less rigid approach to command structures. In spite of this call for greater flexibility, including the innovative use of self-directed teams, it is essential in crisis situations that a strict unity of command be maintained for the purposes of operational efficiency and public accountability. Where circumstances do not warrant or require such rigorous command and control processes, the absolute application of unity of command may be accordingly relaxed. The philosophy of community policing, as an organizational approach to broader public involvement, or as an application of problem-solving principles, is particularly suited to a more flexible command structure. Police organizations must be able to distinguish circumstances and conditions in which the need for strict unity of command can be moderated for the sake of a more dynamic, consultative, and cooperative approach to service delivery.

LEADER IN COMMUNITY POLICING

David P. McKinnon, Chief of Police, Halifax Regional Police Service

What is your understanding of "community policing" and its importance to the evolution of policing in Canada?

I was hired into a very traditional and professional police organization some 29 years ago, and when I first heard the phrase "Community Policing" back in the mid 1980s I thought it was just another one of those "warm and fuzzy" concepts introduced into policing by those pressed to "do" something about the issues facing the community. My study of the subject suggested that the concepts of Community Policing were so "motherhood and apple pie" that there was no reason why it should not work . . . no reason except people. Over time I have heard Community Policing expressed in many different ways—partnerships, physician/doctor, program, project, and philosophy. Of them all, I choose the concept of philosophy—a philosophy of policing that is human and/or people-oriented, in which the police are not the sole keepers of the quality of life of the community, and where there is two-way communication of the concerns and true listening and understanding by all parties.

Community Policing will, in my opinion, keep the police from being seen as the occupiers of choice in a community, and bring them back to being those citizens entrusted by others as the custodians of society. It is without a doubt, and from a local perspective, the piece that will keep us connected to our roots, and for that reason alone it is more important than many imagine.

Outline your own contribution to the development of "community policing" in Canada.

In the mid 1980s my police organization went through a major organizational review, and I was given an opportunity to be part of that transition process. I researched, helped develop and deliver the original model of Community Policing used by the Halifax Regional Police. It was a model based on a zone team policing structure with Community Policing concepts. It was introduced along with a Constable generalist strategy and a new investigative case flow concept.

What is your view of the future of "community policing" in Canada?

Community Policing, in one form or another and with emphasis on one area or another, is here to stay. In the future I see a three-tier system of policing: 1) Community Oriented

David P. McKinnon LEADER IN COMMUNITY POLICING continued

and Problem Solving policing at the local level with a strong emphasis on crime solving; 2) Community Policing can solve crime and deal with community issues. However, the police of today are being touched within their jurisdiction by the guys from Bogotà and the Internet; hence, the scope of the policing task must have a connection between the local and the outside jurisdictional issues; and 3) the area of policing in the future is the concept of contract or private policing within the public partnering structure.

Community Policing will always suffer from the attempts of those who do not study the process to pass off projects as Community Policing, and the failures of those projects as a failure of Community Policing. The fact is that Community Policing has always been, and will always be, a part of the real policing of this country, whether we stick a label on it or just do it.

INSTITUTIONAL CHANGE AND THE TRANSFORMATION OF ORGANIZATIONAL CULTURE

When a Canadian police service has dedicated itself to major and fundamental organizational change and has dealt with the need to continue to deliver an acceptable level of policing service to the community, much more remains to be done in order to ensure success. The philosophy of community policing, however defined, cannot be simply grafted onto the organization in the form of a new and wonderful program. Once all of the decisions have been made regarding the approach to be taken for the change initiative (e.g., business process reengineering, total quality management) and a new organization has been designed to provide more effective service delivery, it is vital that the corporate culture be transformed in a way that is consistent with the goals and objectives of this new organization.

WHAT IS ORGANIZATIONAL CULTURE?

Schein (1985) offers the following excellent definition of culture:

> . . . a pattern of basic assumptions—invented, discovered, or developed by a given group as it learns to cope with its problems of external adaptation and internal integration—that has worked well enough to be considered valid and, therefore, to be taught to new members as the correct way to perceive, think, and feel in relation to those problems. (p. 9)

Schein (1985) also provides an insight into the element, or "levels," of culture. His perspective is directly relevant to the corporate culture of police organizations and is outlined in Figure 5–3, below.

In order to change the organizational culture it is necessary to consider the various levels identified by Schein and make appropriate modifications. This effort requires attention to continuous communication (both internal and external, both formal and informal), training, education, and development. Fundamental to these efforts is the

FIGURE 5-3

SCHEIN'S LEVELS OF CULTURE

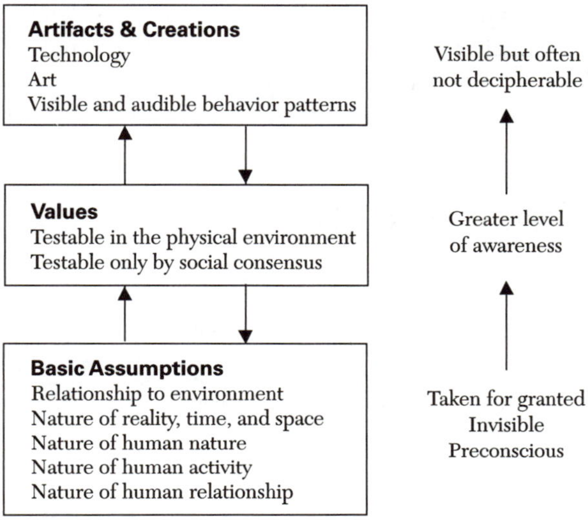

Source: Schein, Edgar H. (1985). *Organizational culture and leadership: a dynamic view*, Figure 1, page 14. Copyright 1985 Jossey-Bass Inc., Publishers. Used with permission.

re-examination and reformulation of the organizational values that guide the department. As emphasized by Sparrow, Moore, and Kennedy (1990):

> A new attention to values is important to policing for two distinct, if closely related, reasons. First, attention to values can help executives change the culture and behavior of their organizations. By articulating the values that are to guide the organization's conduct, an executive can garner outside support, establish the terms by which the organization will be held accountable, and challenge and guide employees. The specification and promulgation of values is a key managerial tool in changing an organization's culture.
>
> Second, the values that currently inhere in policing, both those explicitly articulated and those implicitly held, are in many ways pathological. They prevent the police from discovering their full potential to contribute to the communities they serve and sometimes lead to scandal and disgrace. They must be replaced with higher values that reveal the opportunities and obligations of policing and keep the police in the public's high esteem. (pp. 131–32).

It has also been identified that there are levels of change that must be taken into account:

- Environmental;
- Organizational;
- Group; and
- Individual.

Change Strategies

In order to move forward in any change initiative, it is important to make the effort at creating and sustaining the circumstances and conditions that will contribute to the successful introduction of meaningful change at the various levels identified above. In order to accomplish this effectively a great deal of planning and preparation is required within the organization. The elements of the change cycle include eight components that should be considered. These elements are presented in Figure 5–4:

FIGURE 5–4

Elements of the Change Cycle

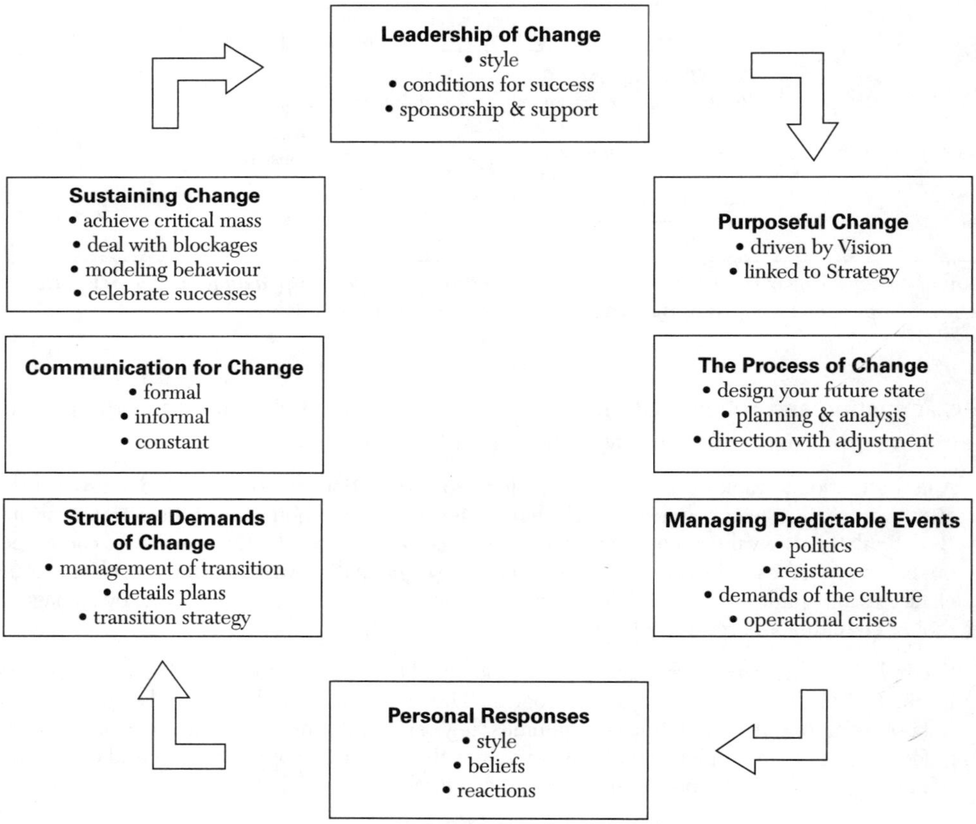

In addition to the elements of the change cycle, it is worthwhile considering the following additional aspects that have been recommended by change consultants:

Establish strong communication links—
- Embrace the need to communicate extensively during a change project
- Confront and if necessary correct the rumour mill and make it a habit to continuously send out a message on the need for change

Share perceptions—
- Allow and encourage people to relate to one another about the change process. This relies heavily on the exchange of information.

Create and sustain consistent symbols—
- This amounts to "walking the talk." Words and actions should be consistent
- Cannot lose the momentum of positive change

Plan organizational "values"—
- Make changes that are consistent with those values

Establish stewardship for the change process—
- Create teams with clear lines of responsibility and accountability

Share the burden—
- Ensure that key stakeholders are enlisted throughout the process

Create vision statements—
- Clearly articulate and frequently broadcast the organization's vision

Celebrate the moment—
- Create symbols to mark new beginnings

Communicate the vision—
- Constantly communicate on all levels, in both formal and informal ways

Recognize the pioneers—
- Acknowledge those who are early adapters of the change initiative and the progressives within the organization

Establish high-performance teams—
- Create the conditions and provide the support for successful team growth and development

Encourage experimentation and risk-taking—
- Provide the wherewithal for new initiatives to be undertaken without the need for guaranteed success

Address the burning issues—this means providing answers to the following questions;
- Why are we here? (Purpose)
- Where are we going? (Vision and Direction)
- What is expected of me? (Accountability)
- How am I doing so far? (Feedback)

- What's in it for me? (Rewards)
- Where do I go when I need help? (Support)

Have a plan—

- Attempt to anticipate possible futures and prepare accordingly.

Obstacles to Change

The following factors can present difficult obstacles to organizational change and inhibit the introduction of successful change at all levels within the police department that is attempting to transform itself:

- Poor union/management relations;
- Cynical workforce;
- Tradition;
- Poor communication;
- Failure to educate and enlist the support of employees;
- Failure to involve front-line staff and supervisors in the change process;
- Formal leadership throughout the process;
- Relying on traditional measurements;
- Lack of belief in the need for change;
- Fear of change;
- Lack of authority to make needed changes;
- Lack of commitment at all levels of management;
- Satisfaction with the *status quo*;
- Lack of improvement tools and training;
- Discouragement at the failure of previous efforts.

EMPOWERMENT, ACCOUNTABILITY, AND CIVILIAN GOVERNING AUTHORITIES

In Canadian public sector organizations there is much talk about empowerment (Kernaghan, 1995). It is often noted that meaningful organizational change will occur in police departments when police staff, particularly front-line officers, are given more genuine empowerment. However, there is a tension that exists, especially within the public service, when it comes to balancing *employee* empowerment with *organizational and individual* accountability.

Police forces in Canada are held to an exceedingly high standard of accountability, which is entirely appropriate in a liberal democracy that has enshrined constitutional principles and specifically designated rights and freedoms. Because of this reality, it is essential that the degree of empowerment enjoyed by individual police officers does not override or undermine that officer's functional accountability or jeopardize the overall accountability of the police organization the officer represents. The challenge for police leaders seeking to introduce officer empowerment is to ensure that this empowerment does not displace the accountability that resides in every police organization as a measure of its covenant with the public for the powers it wields on the public's behalf.

This tension creates a real dilemma for community policing in that it places some distinct limitations on the entrepreneurial spirit that this approach to policing tends to encourage in addressing community problems. This dilemma is not new in that it carries forward part of the debate that has centred on the question of police discretion (Grosman, 1975). Because police officers are called upon to make immediate assessments regarding circumstances that take place within their area of authority, a degree of officer discretion has always been a necessary feature of policing. Officers have a role to play in determining whether or not to invoke the legal sanctions at their disposal to handle a particular situation. For example, during a traffic stop, an officer may decide to charge the driver for an offence or violation, provide a warning, or progress to a further stage of investigation, depending on the immediate circumstances and the officer's own assessment of the situation. In a sense, police officers have always been "empowered" to make decisions based on their discretion.

In most jurisdictions, police departments are responsible to a civilian governing authority. In Canada, these bodies are referred to as police commissions or police services boards. As an extension of the community, these bodies represent the public interest in policing and have a considerable degree of authority and responsibility for setting overall goals and objectives for police departments (Stenning, 1981; Stenning, 1995).

With a level of responsibility akin to a board of directors, the police governing authorities become key partners in any organizational change process undertaken by a given police department. The issue of accountability is central to policing in Canada (Stenning, 1995). In general terms, Gordon (1986) states that:

> Accountability [is] a political principle according to which agencies and organizations, are subject to some form of external control, causing them to give a general accounting of and for their actions; an essential concept in democratic public administration. (p. 607)

Because of the web of accountability that pertains to policing in Canada, there are necessary limits to the degree of empowerment that is feasible for police officers in this context. The idea of ministerial responsibility is also something that Canadians can relate to when thinking about the scope of appropriate empowerment in policing, as a high-profile public service. This is particularly true in recent times when both federal and provincial Solicitors General have been held directly accountable (i.e., responsible) for the actions of their bureaucrats.

In looking at the overall scope of accountability, Leclerc et al. (1996: 57–58) discovered several types of relationship in their review of the relevant literature:

- Internal and external accountability;
- Political accountability—including constitutional, decentralized, and consultative aspects;
- Managerial accountability—including commercial, resource, and professional aspects;

- Legal accountability—including judicial, quasi-judicial, and regulatory aspects; and
- Procedural and consequential accountability.

Kernaghan (1992) notes the internal and external dimensions of empowerment and that both these aspects are relevant in the context of policing. The external dimension tends to relate to the clients or customers of the service and typically engages them to some extent in the decision-making process. Citizen participation is certainly a hallmark of community policing and, therefore, efforts at increased empowerment are normal in any organizational transformation of a police service.

The internal dimension relates to the scope of responsibility given to managers and employees within the organization. The individual and teams play an important part in the empowered organization. It goes beyond simple delegation. As Kernaghan (1992) indicates:

> Organizations must be prepared to overcome barriers to delegating decision-making power to their employees, to turn their managers from controllers into coaches, to promote teamwork, to keep middle managers from backsliding, to make structural changes, to allocate more resources to training, to provide increased recognition and rewards, and to resolve the labour relations issues that develop. (p. 205)

But Kernaghan (p. 208) hastens to point out the problems that confront leaders seeking to introduce a higher degree of empowerment within their organization:

> The scale and complexity of government operations, and the political, legal, and constitutional milieu in which they are conducted, complicate the task of management and organizational reform. Efforts to implement empowerment face a more formidable set of obstacles in public than in private sector organizations. The large size and the multi-purpose nature of many government departments preclude or greatly complicate successful empowerment of the whole organization.

Another obstacle to empowerment within the public sector relates to the stricter level of accountability that is imposed on government departments. In policing this is particularly true, as will be discussed in more detail below. Also the matter of ministerial responsibility is not to be diminished in this context.

Holpp (1995) offers a good conceptual framework for empowerment that can be applied within police organizations:

> Empowerment is a process for helping the right people at the right level make the right decisions for the right reasons. (p. 53)

The accountability issue is significant in police organizations and will directly influence the degree of change and the kind of change occurring in police departments. Because it occurs at various levels, accountability may be attached to:

- A superior;
- Elected officials;
- Professional norms and institutions;
- The law and regulations; and
- The public at large.

Figure 5–5 provides some of the possible specifics around accountability for a police officer in Canada:

FIGURE 5–5

LEVELS OF POLICE OFFICER ACCOUNTABILITY IN CANADA

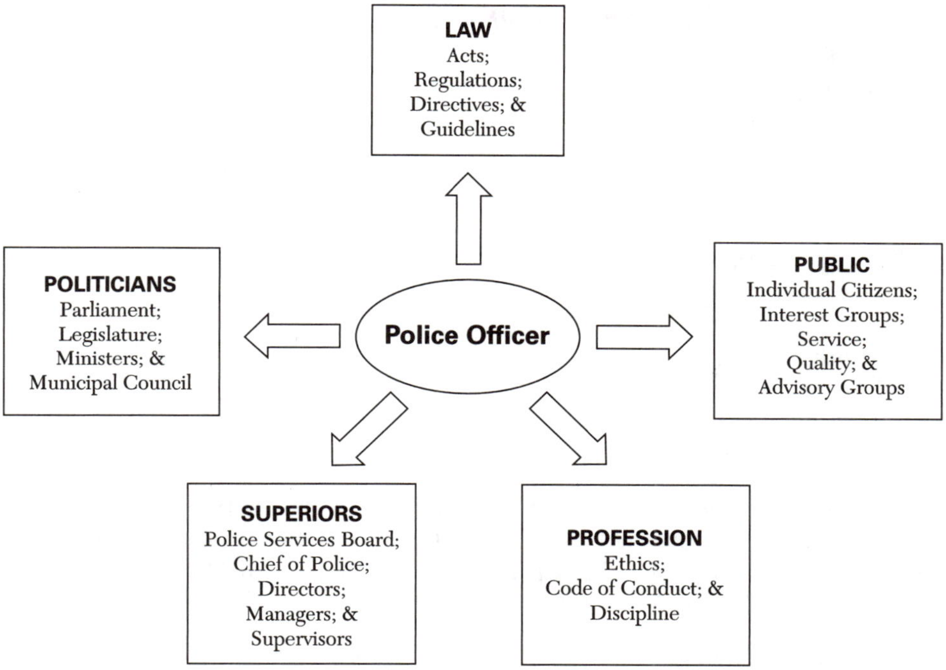

To compare public service with private business can be difficult, as Savoie (1992) indicates:

> . . . in business it does not much matter if you get it wrong 10 per cent of the time as long as you turn a profit at the end of the year. In government, it does not much matter if you get it right 90 per cent of the time because the focus will be on the 10 per cent of the time you get it wrong. (p. 115)

The challenge of empowerment in policing is complicated by the duties of an individual officer which include those that fall into two categories: public law and private law duties:

PUBLIC LAW DUTIES

Common law and statutory duties—

- Preserve the peace;
- Enforce the law;

- Protect life and property;
- Execute warrants;
- Prosecute; and
- Attend at labour disputes.

PRIVATE LAW DUTIES

Individual liability and vicarious liability—

- Use of force;
- Trespass;
- Detention and arrest;
- Use of handcuffs;
- Execution of warrants;
- Searches;
- Seizure of property;
- Malicious prosecution;
- Constitutional torts;
- Persons in custody;
- Motorists;
- Other police officers; and
- Training and supervision.

Accountability in policing includes public complaints and oversight of police activity (e.g., the Special Investigations Unit (SIU) in Ontario that investigates all incidents of death or serious injury in which the police may have used excessive force). As underlined by Normandeau and Leighton (1990):

> The Canadian public in general is more vocal and inquisitive than in the past, demanding more information from the police and desiring greater input into police practices. The movement toward community policing, too, entails greater participation by the public in policing as well as more openness on the part of police organizations. (p. 72)

MIDDLE MANAGERS

In police organizations, it is at the middle management level that change can be effectively driven and extensively sustained. The middle manager is the "glue" that binds the front-line officers to the senior management of the organization. Therefore, they are critical to the success of any change initiative. In order to understand the importance of this middle manager (i.e., staff sergeants and inspectors in Canada, lieutenants and captains in the U.S.) the following (partial) list of responsibilities has been provided by Geller and Swanger (1995), based on their review of the literature:

- Setting operational priorities for their own units;
- Supervising operations;

- Public relations with neighborhood groups;
- Coordination with other patrol and detective units;
- Determining assignments;
- Scheduling work days and shifts;
- Providing operational notices to each shift;
- Transmitting departmental/division orders to personnel within their commands;
- Communicating with senior officials and serving as a messenger between subordinates and senior officials;
- Compiling periodic crime, arrest, and other productivity statistics;
- Keeping attendance, sick leave, and other logs for personnel units;
- Preparing district or precinct budget requests;
- Preliminarily handling citizen complaints against officers;
- Running the service/administration desk in district/precinct stations;
- Procurement;
- Authorizing use of equipment and keeping track of its location and condition;
- Taking charge of complex crime scene investigations;
- Taking charge of complex operations within the middle manager's area of specialty (e.g., hostage/barricade, SWAT, containing collective violence);
- Managing special projects (dignitary protection, crowd control at public events, holiday season robbery and pickpocket prevention efforts at shopping centers, order maintenance support when a mayor temporarily moves into her city's high-rise public housing to dramatize its deplorable conditions, etc.);
- Reviewing and approving shift reports on—thus exercising some quality control over—field activity (arrests, follow-up investigations, use of force, etc.);
- Conducting roll call inspections (of weapons, the wearing of soft body armor and other uniform elements, etc.);
- Reviewing "early warning" system alerts concerning officers at risk of poor performance due to job-related or personal problems; and
- Supervising the supervisors below them. (p. 9)

Because of their importance, and because of the view that middle managers are often an impediment to transformative change, it is essential that they be taken seriously throughout any change process. When conditions of change are right, it is likelier that middle managers will become champions of meaningful change (Kanter, 1982).

RECRUITMENT AND ITS POTENTIAL FOR ORGANIZATIONAL TRANSFORMATION

Over time, the needs of a community will change. This observation is so obvious as to be banal. However, it brings into focus the question of police recruitment practices, particularly in the context of community policing and substantive organizational change.

If one considers the three main eras of policing discussed above—political, professional, and community-oriented (or proactive)—then it is possible to perceive three basic approaches to police officer recruitment. During the political era it was likely that only the most elementary qualifications were needed to qualify for a job as a police officer. In the U.S. it was widely known that political party allegiance was a critical element in being picked for a vacancy in the department. The crisis of corruption that grew out of this approach indicates the failure of this era to attract the best candidates for policing.

The professional era placed a great emphasis on specialization with the military model of strict hierarchy, chain of command, and technical expertise. This era included a degree of confidence that the police could, and should, be trained in police-directed academies removed from the influences of outside educational institutions. In spite of all the best intentions, this approach led to a style of recruitment and orientation that isolated candidates from their communities and drew them quickly and deeply into the "subculture" of policing.

The current era, which may be referred to as community-oriented, or proactive, takes a different view of recruitment. This view represents a new perspective on the mission of an effective police service. While the core functions of most Canadian police services have not changed significantly in many decades (i.e., law enforcement, crime prevention, order maintenance, assistance to victims, and emergency response), the philosophy or guiding principles for the police have gone through an evolution over the last twenty years. Consistent with the move to a community policing paradigm is a realization that police organizations need to recruit a different type of candidate if they are going to continue to function within their changing communities. This realization has taken time and has not been without its difficulties.

In Ontario, the Solicitor General began an important project in 1992 that was designed to look carefully at the strategic learning requirements of various levels within policing, including the rank of Constable (Ontario. Ministry of the Solicitor General. Strategic Planning Committee on Police Training and Education, 1992). These strategic learning requirements were then used to design a Police Learning System for Ontario, the details of which will be considered in more detail in Chapter 7 of this textbook. Using an approach known as "functional job analysis," developed by Dr. Sidney Fine, this project arrived at strategic learning requirements for constables, patrols sergeants, middle managers, senior managers, and police educators. What follows are the learning requirements identified for constables (not in priority order):

- Communication skills;
- Interpersonal and sensitivity skills;
- Knowledge of human behaviour;
- Ability to accept and work with community diversity;
- Ability to serve victims;
- Ability to initiate, promote, and facilitate community policing;
- Ability to use policing-related technology;
- Analytical skills and problem-solving ability;

- Knowledge of political systems and processes;
- Knowledge of crime prevention strategies;
- Personal and organizational development skills;
- Knowledge of other agencies;
- Team-building skills;
- Ability to use crime trend information;
- Ability to apply basic police authorities and knowledge of case preparation;
- Ability to act ethically and professionally;
- Ability to maintain a reasonable level of physical fitness and well-being;
- Ability to use force appropriately;
- Officer safety skills; and
- Conflict avoidance, resolution, and mediation skills.

The need to move into new areas of recruitment is important for enhancing the future of policing in Canada. This would also include not only more enlightened approaches to the selection and orientation of police officers consistent with the strategic learning requirements identified above, but also a broader recruitment of civilians within the structure of police leadership and management (Normandeau and Leighton, 1990; Stamper, 1992; Peak and Glensor, 1996; Thibault, Lynch and McBride, 1998).

SUMMARY

This chapter has dealt with the issues relevant to making meaningful and lasting change in Canadian police organizations. It has considered the reluctance to change that has characterized much of the history of policing, both in the U.S. and Canada. The predictable components of organizational change have been examined, along with some of the barriers to change that can be found, particularly in the context of policing.

This chapter also looked closely at the tension between empowerment and accountability and probed the question of balance between these two important concepts. The role of governing authorities in Canadian policing was briefly touched upon, particularly in terms of their impact on organizational change in policing. Tackling the corporate culture as part of a process of sustaining organizational change was considered, as was the potentially pivotal role of the middle manager in police organizations.

Finally, the topic of recruitment was considered as a means of consolidating and enhancing the achievements arrived at during a conscientious organizational transformation process.

QUESTIONS FOR CONSIDERATION

1. How have the three major eras in policing influenced the organizational structure of police departments?
2. Considering the predictable elements of the organizational change cycle, which

would require the most effort to deal with for a police organization?

3. What are the major obstacles to change in Canadian police organizations? How could each one be overcome?
4. Can empowerment be realistically adopted within Canadian police departments, particularly in light of existing accountability standards?
5. Is the role of civilian governing authorities appropriate in the context of Canadian policing?
6. Identify the major components of a police culture that need to be transformed.
7. In what ways could middle managers be instrumental in effecting organizational change in a police department?
8. How can enlightened recruitment practices advance the transformation of policing in Canada?

REFERENCES

Albrecht, K. and R. Zemke (1985). *Service America*. Homewood, Ill.: Dow-Jones Irwin.

Angell, John E. (1971). "Toward an alternative to the classic police organizational arrangement: a democratic model." *Criminology*, Vol. 9, no. 2 & 3 (August-November), pp. 185–206.

Baldridge, J. and R. Burnham (1975). "Organizational innovation: industrial, organizational and environmental impact." *Administrative Science Quarterly*, Vol. 20, pp. 165–176.

Bopp, William J. (1972). *A short history of American law enforcement*. Springfield, Ill.: Charles C. Thomas.

Bopp, William J. (1977). *"O.W.": O.W. Wilson and the search for a police profession*. Port Washington, N.Y.: Kennikat Press.

Braiden, Chris (1992). "Community policing: nothing new under the sun." In *Community oriented policing and problem solving*. Sacramento, Calif. : California Department of Justice, pp. 17–22.

Brodeur, Jean-Paul (1998). "Tailor-made policing: a conceptual investigation." In Brodeur, Jean-Paul (ed.) (1998). *How to recognize good policing: problems and issues*. Washington, D.C.: Police Executive Research Forum; Thousand Oaks, Calif.: Sage Publications.

Brodeur, Jean-Paul (ed.) (1998). *How to recognize good policing: problems and issues*. Washington, D.C.: Police Executive Research Forum; Thousand Oaks, Calif.: Sage Publications.

Brudney, J. and R. England (1983). "Toward a definition of the coproduction concept." *Public Administration Review*, Vol. 43, pp. 59–64.

Carte, Gene E. and Elaine H. Carte (1975). *Police reform in the United States: the era of August Vollmer, 1905–1932*. Berkeley, Calif.: University of California Press.

Chicago Police Department (1993). *Together we can: a strategic plan for reinventing the Chicago Police Department*. Chicago: Chicago Police Department.

Cordner, G. (1978). "Open and closed models in police organizations: traditions, dilemmas and practical considerations." *Journal of Police Science and Administration*, Vol. 6, pp. 22–34.

Crank, John P. (1990). "The influence of environmental and organizational factors on police style in urban and rural environments." *Journal of Research in Crime and Delinquency*, Vol. 27, no. 2 (May) pp. 166–189.

Crank, J., W.T. Bergman, and E.J. McLaughlin (1994). "Implementing community policing: cultural and structural change in police organizations." In Rosenbaum, Dennis P. (ed.). *The challenge of community policing: testing the promises*. Thousand Oaks, Calif.: Sage Publications.

Cummings, T.G. and C.G. Worley (1993). *Organizational development and change. 5th ed.* Minneapolis/St. Paul: West Publishing.

Curtis, C., Q.Thurman, and D. Nice (1991). "Improving legal compliance by noncoercive means: coproducing order in Washington State." *Social Science Quarterly*, Vol. 72, pp. 645–660.

Deming, W.E. (1986). *Out of the crisis*. Cambridge, Mass.: MIT Center for Advanced Engineering Study.

Durivage, A. et al. (1992). "Analysis of organizational culture: the key to strategic planning." *Canadian Police College Journal*, Vol. 16, no. 2, pp. 94–134.

Ettlie, J., W. Bridges, and R. O'Keefe (1984). "Organizational strategy and structural differences for radical vs. incremental innovation." *Management Science*, Vol. 30, pp. 682–695.

Fogelson, R. (1977). *Big-city police*. Cambridge, Mass. : Harvard University Press.

Fraser, Craig B. et al. (1998). *Challenge to change: the 21st century policing project*. Washington, D.C.: Police Executive Research Forum.

Geller, William A. and Guy Swanger (1995). *Managing innovation in policing: the untapped potential of the middle manager.* Washington, D.C.: Police Executive Research Forum and National Institute of Justice.

Goodman, P. and Associates (eds.) (1982). *Change in organizations: new perspectives on theory, research and practice*. San Francisco: Jossey-Bass Publishers.

Gordon, George J. (1986). *Public administration in America. 3rd ed.* New York: St. Martin's Press.

Greene, Jack R. (1998). "Evaluating planned change strategies in modern law enforcement: implementing community-based policing." In Brodeur, Jean-Paul (ed.) (1998). *How to recognize good policing: problems and issues*. Washington, D.C.: Police Executive Research Forum; Thousand Oaks, Calif. : Sage Publications.

Grosman, Brian A. (1975). *Police command: decisions and discretion*. Toronto: Macmillan.

Guyot, Dorothy (1979). "Bending granite: attempts to change the rank structure of American police departments." *Journal of Police Science and Administration*, Vol. 7, no. 3 (September) pp. 253–284.

Hann, Robert G. and Kathryn Asbury (1992). *Change is the environment: Ontario region environmental scan submitted to Ontario region, Ministry of the Solicitor General Canada.* Ottawa: Solicitor General Canada, Intergovernmental Affairs Regional Office. 2 vols.

Harbour, Jerry L. (1994). "Doing it better, faster, and cheaper: the power of process reengineering and information technology," presentation at the NSPI conference workshop, San Francisco, Calif., April 4–5, 1994.

Highley, Dahn D. (1984). *O.P.P.: the history of the Ontario Provincial Police Force*. Toronto: The Queen's Printer.

Holpp, Lawrence (1995). "If empowerment is so good, why does it hurt?" *Training*, Vol. 32, no. 3 (March), pp. 52–57.

Hoover, Larry T. (ed.) (1996). *Quantifying quality in policing*. Washington, D.C.: Police Executive Research Forum.

Hoover, Larry T. (1996). "Translating total quality management from the private sector to policing." In Hoover, Larry T. (ed.) *Quantifying quality in policing*. Washington, D.C.: Police Executive Research Forum.

Huber, G. et al. (1993). "Understanding and predicting organizational change." In Huber, G. and W. Glick (eds.) *Organizational change and redesign: ideas and insights for improving performance.* New York : Oxford University Press, pp. 215–254.

Institute of Public Administration of Canada. Edmonton Regional Group. Policy Committee (1996). "Accountability and public administration." Unpublished policy paper.

Jacobs, Jane (1992). *Systems of survival: a dialogue on the moral foundations of commerce and politics*. New York: Random House.

Johnson, Thomas Albert (1971). *A study of police resistance to police community relations in a municipal police department*. Ph.D. dissertation, University of California at Berkeley.

Kanter, Rosabeth Moss (1982). "The middle manager as innovator." *Harvard Business Review* (July–August), pp. 92–105.

Kelling, George L. (1996). "Defining the bottom line in policing: organizational philosophy and accountability." In Hoover, Larry T. (ed.) *Quantifying quality in policing*. Washington, D.C.: Police Executive Research Forum.

Kelling, George L. and Mary Ann Wycoff (1991). "New roles for mid-managers." In Geller, William A. (ed.). *Local government police management*. Washington, D.C.: International City/County Management Association.

Kiefer, Charles and Peter Stroh (1984). "A new paradigm for developing organizations." In Adams, John D. (ed.). *Transforming work: a collection of organizational transformation readings*. Alexandria, Virginia: Miles River Press.

Kuykendall, Jack and Roy R. Roberg (1982). "Mapping police organizational change: from a mechanistic toward an organic model." *Criminology*, Vol. 20, no. 2 (August) pp. 241–256.

Langworthy, Robert H. (1986). *The structure of police organizations*. New York: Praeger.

Leclerc, Guy et al. (1996). *Accountability, performance reporting, comprehensive auditing: an integrated perspective*. Ottawa: CCAF-FCVI Inc.

Maguire, Edward R. (1997). "Structural change in large municipal police organizations during the community policing era." *Justice Quarterly*, Vol. 14, no. 3 (September) pp. 547–576.

Mastrofski, Stephen D. (1998). "Community policing and police organization structure." In Brodeur, Jean-Paul (ed.) (1998). *How to recognize good policing: problems and issues*. Washington, D.C.: Police Executive Research Forum; Thousand Oaks, Calif.: Sage Publications.

McDougall, Allan K. (1988). *Policing: the evolution of a mandate*. Ottawa: Canadian Police College.

Miller, D. and P. Friesen (1980). "Archetypes of organizational transition." *Administrative Science Quarterly*, Vol. 25, pp. 269–299.

Moore, David and Roger Wettenhall (eds.) (1994). *Keeping the peace: police accountability and oversight*. Canberra: Centre for Research in Public Sector Management, University of Canberra and Royal Institute of Public Administration Australia.

Normandeau, André and Barry Leighton (1990). *A vision of the future of policing in Canada: police-challenge 2000: background paper*. Ottawa: Police and Security Branch, Solicitor General Canada.

Ontario. Ministry of the Solicitor General. Strategic Planning Committee on Police Training and Education (1992). *A police learning system for Ontario: final report and recommendations*. Toronto: Ministry of the Solicitor General.

Ontario. Premier's Council on Economic Renewal. Task Force on the Organization of Work (1994). *The organization change process*. Toronto: The Queen's Printer for Ontario.

Ontario. Task Force on Policing in Ontario (1974). *The Task Force on Policing in Ontario: report to the Solicitor General*. [Toronto: The Task Force.]

Ontario Provincial Police (1995). *Organizational review: a process and model for change: focus on the future*. Toronto: The Queen's Printer for Ontario.

Peak, Kenneth J. and Ronald W. Glensor (1996). *Community policing and problem solving: strategies and practices*. Upper Saddle River, N.J.: Prentice-Hall.

Reiss, A. (1992). "Police organization in the Twentieth Century." In Tonry, M. and N. Morris (eds.). *Criminal justice: a review of research*. Vol. 15, pp. 51–97. Chicago: University of Chicago Press.

Reuss-Ianni, Elizabeth (1983). *Two cultures of policing: street cops and management cops*. New Brunswick, N.J.: Transaction Books.

Roberg, Roy R. (1994). "Can today's police organizations effectively implement community policing?" In Rosenbaum, Dennis P. (ed.). *The challenge of community policing: testing the promises*. Thousand Oaks, Calif.: Sage Publications.

Rosenbaum, Dennis P. and Arthur J. Lurigio (1994). "An inside look at community policing reform: definitions, organizational changes, and evaluation findings." *Crime & Delinquency*, Vol. 40, no. 3 (July) pp. 299–314.

Sandler, Georgette and Ellen Bennett Mintz (1974). "Police organizations: their changing internal and external relationship." *Journal of Police Science and Administration*, Vol. 2, no. 4 (December) pp. 458–463.

Savoie, Donald J. (1995). "What is wrong with the new public management?" *Canadian Public Administration*, Vol. 38, no. 1 (Spring), pp. 112–121.

Schneider, Frank W. (1991). "Police organization effectiveness: the manager's perspective." *Canadian Police College Journal*, Vol. 15, no. 3, pp. 153–165.

Schein, Edgar H. (1985). *Organizational culture and leadership: a dynamic view*. San Francisco: Jossey-Bass Publishers.

Seagrave, Jayne (1992). "Community policing and the need for police research skills training." *Canadian Police College Journal*, Vol. 16, no. 3, pp. 204–211.

Sheehan, Robert and Gary W. Cordner (1995). *Police administration. 3rd ed.* Cincinnati, Ohio: Anderson Publishing.

Souryal, Sam S. (1985). *Police organization & administration*. Cincinnati, Ohio: Anderson Publishing.

Sparrow, Malcolm K., Mark H. Moore, and David M. Kennedy (1990). *Beyond 911: a new era for policing*. New York: Basic Books.

Stamper, Norman H. (1992). *Removing managerial barriers to effective police leadership: a study of executive leadership and executive management in big-city police departments*. Washington, D.C.: Police Executive Research Forum.

Stenning, Philip C. (1981). *Police commissions and boards in Canada*. Toronto: University of Toronto, Centre of Criminology.

Stenning, Philip C. (ed.) (1995). *Accountability for criminal justice: selected essays*. Toronto: University of Toronto Press.

Thibault, Edward A., Lawrence M. Lynch, and R. Bruce McBride (1998). *Proactive police management. 4th ed.* Upper Saddle River, N.J.: Prentice Hall.

Tomovich, V.A. and D.J. Loree (1989). "In search of new directions: policing in Niagara Region." *Canadian Police College Journal*, Vol. 13, no. 1, pp. 29–54.

U.S. President's Commission on Campus Unrest (1970). *The report of the President's Commission on Campus Unrest*. Washington, D.C. : U.S. Government Printing Office.

Wycoff, Mary Ann and Wesley G. Skogan (1994). "The effect of a community policing management style on officers' attitudes." *Crime & Delinquency*, Vol. 40, no. 3 (July) pp. 371–383.

Zhao, Jihong (1996). *Why police organizations change: a study of community-oriented policing*. Washington, D.C.: Police Executive Research Forum.

TOWARD MORE LEARNING

CHANGE MANAGEMENT

A series of clearly written booklets has been produced by Pritchett & Associates which contains material relevant to the general topics dealt with during a change project. This is a growing area of research and there is no shortage of popular items that offer guidance for individuals, teams, and organizations facing the challenge of change. Listed below are several recommended publications that may be useful.

Pritchett, Price (1996). *Resistance: moving beyond the barriers to change.* Dallas, Texas: Pritchett & Associates, Inc.

Pritchett, Price and Ron Pound (1990). *The employee handbook for organizational change*. Dallas, Texas: Pritchett & Associates, Inc.

Prichett, Price and Ron Pound (1995). *A survival guide to the stress of organizational change*. Dallas, Texas: Pritchett & Associates, Inc.

Ontario Provincial Police (1995). *Organizational review: a process & model for change*. Toronto : Queen's Printer for Ontario.

A thorough guide to the entire process of large-scale organizational change within a Canadian police department. This particular project used the techniques of business process reengineering to accomplish its organizational change objectives, and examines the lessons learned as a result of the change initiative.

Kotter, John P. (1996) *Leading change*. Boston, Mass.: Harvard University Press.

Kotter has condensed his exposure to many successful, and unsuccessful, change projects into an easy-to-read, and succinct summary for a wide audience of readers. He concentrates on successful change and has developed an eight-stage process that highlights vision, strategy, communication, empowerment, and lifelong learning.

EMPOWERMENT AND ACCOUNTABILITY

Kernaghan, Kenneth (1992). "Empowerment and public administration: revolutionary advance or passing fancy?" *Canadian Public Administration*, Vol. 35, no. 2

Leclerc, Guy et al. (1996). *Accountability, performance reporting, comprehensive audit: an integrated perspective*. Ottawa : CCAF-FCVI Inc.

This is a very detailed publication that traces the historical roots of several interrelated topics that are pertinent to accountability in public and private organizations, namely: accountability, governance, performance reporting, and comprehensive auditing. The text builds on the substantial research undertaken by the CCAF-FCVI in developing a framework based on the twelve "attributes of effectiveness" of a well-performing organization.

ORGANIZATIONAL CULTURE

Robbins, Stephen P. and Nancy Langton (1998). *Organizational behaviour: concepts, controversies, applications*. Scarborough, Ont.: Prentice Hall Canada Inc.

This is a recent publication that includes a complete chapter (Chapter 16) on organizational culture. It offers a Canadian perspective on this topic and includes a self-assessment questionnaire that allow readers to diagnose the type of organizational culture they prefer.

Schein, Edgar H. (1985). *Organizational culture and leadership: a dynamic view*. San Francisco: Jossey-Bass Publishers.

MIDDLE MANAGERS

Geller, William A. and Guy Swanger (1995). *Managing innovation in policing: the untapped potential of the middle manager*. Washington, D.C.: Police Executive Research Forum, sponsored by the National Institute of Justice.

A practical approach to bringing middle managers into the process of successful organizational change within police departments. It draws on the experience of police practitioners who have dealt directly with the difficulties of making change happen in traditional police agencies. It acknowledges that change is difficult but also that community policing requires new ways of doing business.

POLICE RECRUITMENT

Ontario. Ministry of the Solicitor General. Strategic Planning Committee on Police Training and Education (1992). *A police learning system for Ontario: final report and recommendations*. Toronto: The Ministry.

Hay Management Consultants and Dennis Strong & Associates (1993). *Police constable selection: community recruiting, selecting constables for the future: phase II, final report recommended design of the new selection system*. [Toronto: Ontario Ministry of the Solicitor General].

CHAPTER SIX

> As in all their actions it will be seen here how wise they were to depart from generally accepted methods in order to pave the way to greatness of the highest order.
>
> Niccolo Machiavelli, Discourses on the First Ten Books of Titus Livy, Bk. II, Discourse 6 (Leslie Walker, S.J., translation).

PROBLEM-SOLVING—NEW MODES AND ORDERS

LEARNING OBJECTIVES

1. Identify a working definition for problem-solving.
2. Identify the limitations of incident-driven policing.
3. Identify key principles relating to the application of problem-oriented policing.
4. State the components of the conceptual framework for problem-oriented policing.
5. Identify and explain the components of the S.A.R.A. model of problem-solving.
6. Distinguish "leadership" from "sponsorship" in policing as it relates to problem-oriented policing.
7. List several applications of the problem-solving approach to policing being applied in Canada and other jurisdictions.
8. Apply problem-solving to contemporary issues in Canadian policing.

INTRODUCTION

Beginning with the pivotal scholarly work of Herman Goldstein, in 1979, the problem-oriented approach to policing has been gaining understanding, acceptance, and prominence throughout North America. It has been noted that there may be some value in distinguishing problem-oriented policing (POP) from the broader category of community policing in order to avoid the taint of confusion that has attached itself to the whole concept of community policing (Brodeur, 1998). However, for the purposes of this study, it is appropriate to include the problem-oriented, or problem-solving, approach under the general rubric of community policing as part of a fundamental rethinking of the entire policing enterprise. Certainly, even the less rigorous applications of community policing reveal, if only on a superficial level, an understanding that contemporary policing requires new tools and concepts. While problem-oriented policing may, as argued by Brodeur, be seen as a more profound commitment to "... the need to reconceptualize the *general* police function," it fits into a larger context that can usefully be categorized as community policing.

This chapter will provide a working definition of problem-solving, the key ingredient in any recipe for problem-oriented policing (POP). It will rely heavily on the significant theorizing undertaken by Herman Goldstein and the initial, practical application of Professor Goldstein's theories in Newport News, Virginia. The limitations of an incident-driven approach to policing are outlined in order to introduce the problem-oriented approach that works to transform the entire process of policing.

Next there will be a consideration of the problem analysis guide inspired by the work of Barry Poyner (1980; 1983). This guide offers assistance to the problem-oriented police officer in gathering sufficient and relevant data to undertake a thorough analysis of the problem at hand. The model of problem-solving featured in this chapter involves four simple components: scanning, analysis, response, and assessment (S.A.R.A.). The S.A.R.A. model will be outlined in some detail for further application, and the principles relating to the application of POP will be discussed. The instances in which POP has been applied will be considered briefly, with particular reference to Canadian examples.

This chapter will draw a distinction between leadership and sponsorship in relation to problem-oriented policing, and will encourage the reader to see the possibilities that exist for applying the theory of problem-solving to any range of contemporary issues in Canadian policing.

DEFINITIONS

In order to understand the concepts that constitute problem-oriented policing, it is useful to begin with a definition of the word "problem" and the term "problem-solving." Eck and Spelman (1987) define a problem as:

> A group of incidents occurring in a community, that are similar in one or more ways, and that are of concern to the police and the public. (p. 42)

With this understanding as the basis for a perspective on problem-solving, the Crime Prevention Center within the California Department of Justice (1992) proceeds to state that:

> Problem-solving refers to a process of *identifying problems/priorities* through coordinated community/police needs assessment; *collecting and analyzing* information concerning the problem in a thorough, though not necessarily complicated manner; *developing or facilitating responses* that are innovative and tailor-made with the best potential for eliminating or reducing the problem; and finally *evaluating* the response to determine its effectiveness and modifying it as necessary. (p. 3) [emphasis in original]

To quote Herman Goldstein (1990) himself on the subject of problem-oriented policing:

> While problem-oriented policing builds on the best of the past, it is obviously much more than just a new tactic or program to be added on to prevalent forms of policing. It entails more than identifying and analyzing community problems and developing more effective responses to them. In its broadest context, it is a whole new way of thinking about policing that has implications

> for every aspect of the police organization, its personnel, and its operations . . . Fully implemented, it has the potential to reshape the way in which police services are delivered. (p. 3)

And, finally, Murphy (1991) makes this clear observation about problem-oriented policing (POP):

> One of the most promising alternative strategies to incident driven policing is called **Problem Oriented Policing (P.O.P.)**. Problem Oriented Policing is a style of policing that emphasizes the strategic use of problem solving techniques as a basis for more effective police response to repetitive or related crime and other problems. While problem solving has always been an informal part of policing, problem oriented policing programs systematically and formally introduced problem solving as basic police philosophy, training and practice. (p. 2) [emphasis in original]

To assist police officers in selecting problems to focus on in the context of a problem-oriented approach, Goldstein (1990) proposes that the following factors be considered:

- *Impact*—what is the effect of the problem on the community?
- *Seriousness*—what is the degree of severity of the problem?
- *Complexity*—how many factors are at play in the context of this problem?
- *Solvability*—what is the probability that this problem can be resolved?
- *Interest*—what is the extent of concern regarding this problem within the community?

FOCUS OF ACTION FOR A PROBLEM-ORIENTED POLICE SERVICE

With this relatively straightforward framework for understanding the function and format of problem-oriented policing, it is possible to provide more detail on the focus of this approach. In outlining a specific approach for any police service dedicated to the application of a problem-oriented approach to policing, Goldstein (1979) offers the following suggestions:

1. Focus on problems of concern to the public.
2. Zero in on effectiveness as the primary concern.
3. Be proactive.
4. Be committed to systematic inquiry as first step in solving substantive problems.
5. Encourage use of rigorous methods in making inquiries.
6. Make full use of the data in police files and the experience of police personnel.
7. Group like incidents together so that they can be addressed as a common problem.
8. Avoid using overly broad labels in grouping incidents so separate problems can be identified.
9. Encourage a broad and uninhibited search for solutions.
10. Acknowledge the limits of the criminal justice system as a response to problems.
11. Identify multiple interests in any one problem and weight them when analyzing the value of different responses.
12. Be committed to taking some risks in responding to problems.

Cited in: Eck, John E. and William Spelman (1987). *Problem solving: problem-oriented policing in Newport News.* Washington, D.C.: Police Executive Research Forum, p. 3.

LIMITATIONS OF INCIDENT-DRIVEN POLICING

At the root of the movement toward Goldstein's problem-oriented policing model was a keen sense of dissatisfaction and frustration with what is referred to as the incident-driven approach (Eck and Spelman, 1987: 11–16). While it is true that an "incident," however defined, will always be an important factor in the equation of policing, the traditional practice of treating that incident in isolation has come under attack on both theoretical and practical grounds (Goldstein, 1979; Cahn and Tien, 1981; Toch and Grant, 1991). Police scholars and practitioners have cited several limitations to an incident-driven approach to policing. Basically, such an approach is:

- *Reactive*—officers are dispatched after a call for service has been received; the response is triggered by an occurrence;
- *Limited in the information available*—officers rely on information gathered, again, after the event, and are focused on a narrow range of sources, typically victims, witnesses, and potential suspects;
- *Reliant on the criminal justice system*—this approach does not consider possible alternative measures for problem resolution; and
- *Reliant on aggregate statistics*—police performance, both individual and organizational, is measured on the basis of jurisdictional statistics that gather together total crime rates, arrest rates, and clearance rates. This is an emphasis on quantity over quality.

Figure 6–1 shows the typical structure of incident-driven policing.

FIGURE 6–1

INCIDENT-DRIVEN POLICING

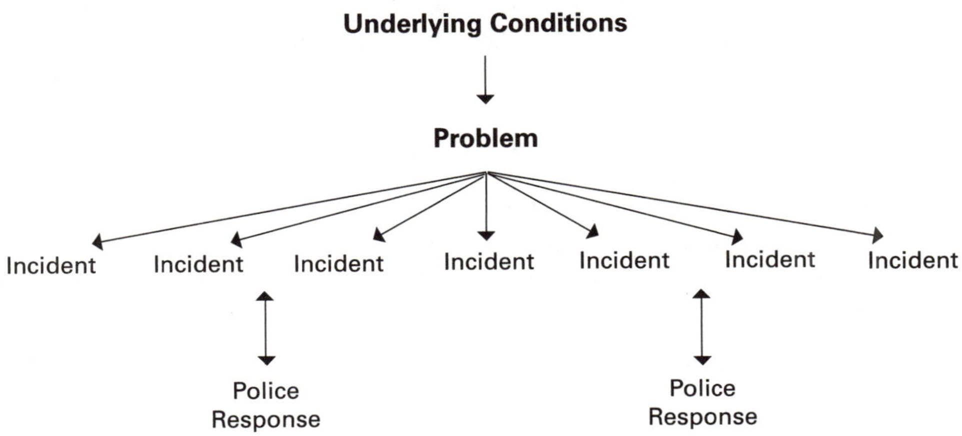

Source: Eck, John E. and William Spelman (1987). *Problem-solving: problem-oriented policing in Newport News*. Washington, D.C.: Police Executive Research Forum, p. xvi. Used with permission.

It would be unfair to suggest that the incident-driven approach was devoid of value as a response to crime and disorder problems (Toch and Grant, 1991: 7). Certainly, a traditional police response has often involved problem-solving, and officers have frequently been engaged in the perceptive gathering of relevant information. What makes the POP model different is that the problem-solving process is systematically applied, it draws upon other public and private responses, and it consciously looks to the underlying conditions that have contributed to the incident(s) being addressed.

GUIDING PRINCIPLES FOR PROBLEM-SOLVING

When the problem-solving approach was being developed for introduction as an experimental model, the U.S. National Institute of Justice insisted that any system would have to adhere to the following five principles (Eck and Spelman, 1987):

1. Officers of all ranks and from all units should be able to use the system as part of their daily routine.
2. The system must encourage the use of a broad range of information, including but not limited to conventional police data.
3. The system should encourage a broad range of solutions, including but not limited to the criminal justice process.
4. The system should require no additional resources and no special units.
5. Finally, any large police agency must be able to apply it.

These guiding principles are significant because they work toward an approach that can be consistently and broadly applied in virtually any police organization. It serves to stretch the creativity and ingenuity of police officers, and it encourages them to look "outside the box" of conventional sources of information. It creates a common language and practice that can advance shared learning within, and between, police officers and their respective departments. Figure 6-2 shows that the simple addition of the problem-solving approach substantially transforms incident-driven policing into something more effective as a tool for innovative policing.

Figure 6–2 represents the elements of a problem-oriented policing process.

British researcher, Barry Poyner (1983), has developed a process model for street attacks that was extremely helpful to the work of the Problem-Oriented Policing Project Task Force in Newport News. Poyner's model provided a straightforward means of describing the problem, and inspired the Task Force to formulate a Problem Analysis Guide which amounts to a checklist of issues an officer should consider when analyzing a problem.

Figure 6–3 summarizes Poyner's guide for problem analysis.

FIGURE 6-2

PROBLEM-ORIENTED POLICING

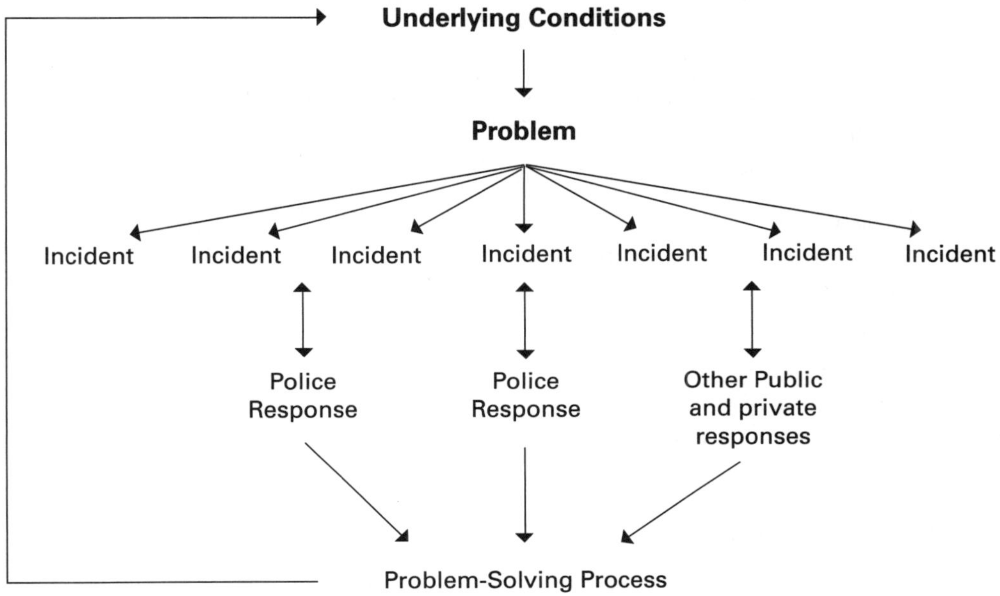

Source: Eck, John E. and William Spelman (1987). *Problem-solving: problem-oriented policing in Newport News.* Washington, D.C.: Police Executive Research Forum, p. xvii. Used with permission.

FIGURE 6-3

PROBLEM-ANALYSIS GUIDE

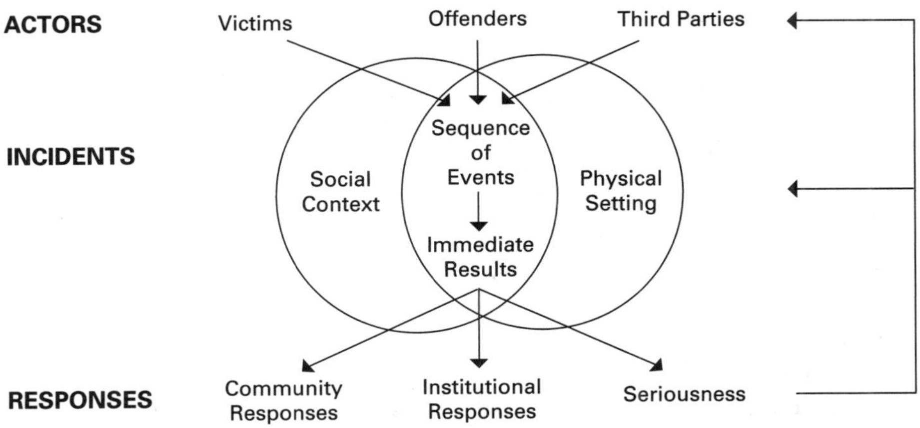

Source: Poyner, Barry (1983). *Design against crime: beyond defensible space.* London: Butterworths. Used with permission.

The Task Force's guide, based on Poyner's model, is presented below:

PROBLEM ANALYSIS GUIDE—LIST OF TOPIC HEADINGS

ACTORS

Victims—
- Life style
- Security measures taken
- Victimization history

Offenders—
- Identity and physical description
- Life style, education, employment history
- Criminal history

Third parties—
- Personal data
- Connection to victimization

INCIDENTS

Sequence of events—
- Events preceding act
- Event itself
- Events following criminal act

Physical context—
- Time
- Location
- Access control and surveillance

Social context—
- Likelihood and probable actions of witnesses
- Apparent attitude of residents toward neighborhood

Immediate results of incidents—
- Harm done to victim
- Gain to offender
- Legal issues

RESPONSES

Community—
- Neighborhood affected by problem
- City as a whole
- People outside the city

Institutional—
- Criminal justice system
- Other public agencies
- Mass media
- Business sector

Seriousness—
- Public perceptions
- Perceptions of others

Clearly, the problem analysis guide, reproduced above, reveals the enormous effort that is required in assembling all of the information that may be pertinent to the analysis phase. But there is more to problem-analysis than mere data (Eck and Spelman, 1987: 63). Poyner's presentation is a more elaborate rendering that builds on another model shown in Figure 6–4, below, referred to as the problem analysis triangle:

FIGURE 6–4

PROBLEM ANALYSIS TRIANGLE

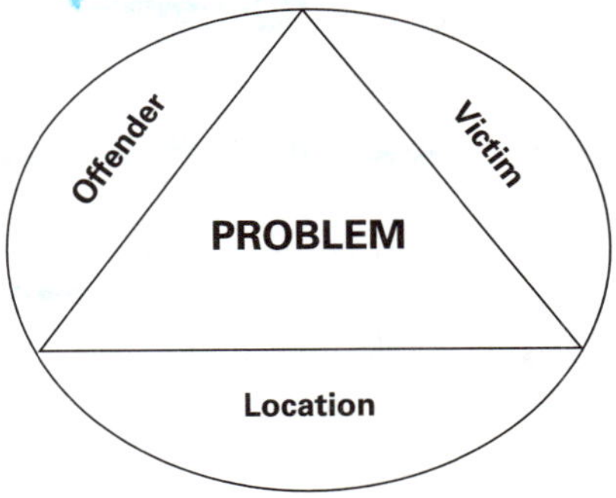

Source: Bureau of Justice Assistance, U.S. Department of Justice. *Comprehensive gang initiative: operations manual for implementing local gang prevention and control programs.* (Draft, October, 1993) pp. 3–10. Used with permission.

THE S.A.R.A. MODEL OF PROBLEM-SOLVING

The S.A.R.A. model that resulted from the extensive efforts of the POP Project Task Force in Newport News has been broadly applied in dozens of police agencies in North America and has proven to be a galvanizing force in modern policing. Its components are simple yet directly relevant to the challenging task of addressing the fullest range of crime and social disorder problems.

Figure 6–5 presents the basic S.A.R.A. model for a problem-solving process.

FIGURE 6–5

A PROBLEM-SOLVING MODEL

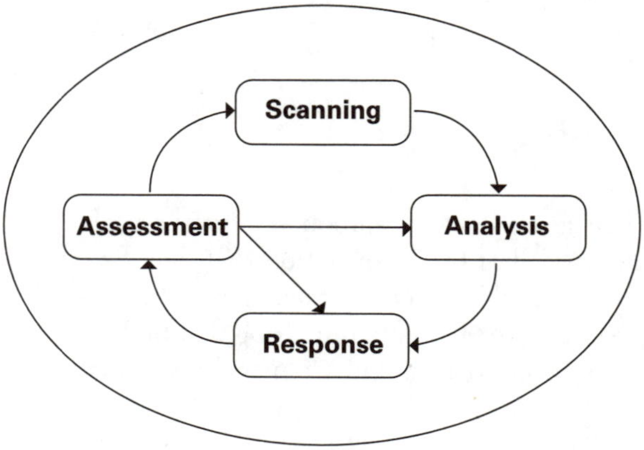

Scanning—this is also referred to as the problem identification stage, and involves officers in pinpointing specific community concerns and understanding what the priorities are for resolution;

Analysis—in this phase, officers collect detailed background information relevant to the problems which have been identified in the earlier stage. There is strong support in the literature for ensuring that information is gathered from the broadest range of possible sources. It is critical that officers acquire a deep understanding of the nature and extent of the problem. Watson, Kenney, and Lusardi (1995) suggest that this stage includes four steps:

1. Analyzing the forces impinging on the problem;
2. Determining the events that precede, accompany, and follow the problem;
3. Analyzing the problem's strength and the harms caused; and
4. Determining the assets, or resources, available to address the problem. (p. 5)

Response—in this stage the officers are intended to develop and implement a response, or set of responses, that will address the problem and are consistent with the information received. Here there may be a need to call upon additional resources outside of the police service to facilitate the implementation of the response, e.g., businesses, other government services, or local citizens. This broadens the scope of the action of the police organization and works to assist in the challenge of making the problem more manageable.

Assessment—also called the evaluation stage, when officers are tasked with collecting further data and information in order to gauge the effectiveness of the response(s) applied to the problem. The results of this stage will make it possible for the police to know if their planned response to the problem is working, or if it is not working, to possibly diagnose areas in which improvements may be made in the future.

THE EVALUATION CUBE

The evaluation cube presented below provides an excellent overview of the range of criteria for assessing the implementation of a problem-solving model within any given police department. Figure 6–6 represents the evaluation cube presented by Oettmeier and Wycoff (1997) to assess performance in police organizations across several key dimensions.

FIGURE 6–6

THE EVALUATION CUBE

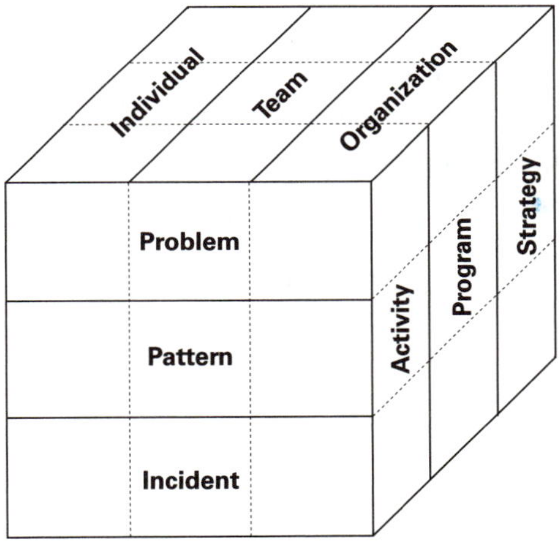

Source: Oettmeier, Timothy N. and Mary Ann Wycoff (1997). *Personnel performance evaluations in the context of community policing*. Washington, D.C.: Police Executive Research Forum, p. 5. Used with permission.

The evaluation cube, above, can be used as an effective tool in gauging the effectiveness of a police department's approach to problem-solving. Chapter 10 will provide more detail on measuring overall success in community policing, and will further explain the application of the evaluation cube. However, it can quite easily be seen how this evaluation cube logically relates to the whole range of organizational actions relating to problem-solving (Oettmeier and Wycoff, 1997).

TRAINING FOR PROBLEM-SOLVING

In order to prepare officers for the challenge of working with the community on the basis of a problem-solving method, it is essential that proper training, education, and development be offered which are tailored to this purpose. Recently, there has been a substantial growth in the materials available for this type of instruction. In Chapter Seven, the educational challenge of community policing will be considered in more detail. However, specific to the matter of problem-solving, the following learning objectives are offered by Watson, Kenney, and Lusardi (1995):

Objectives For The Problem-Solving Training

1. To understand and be able to complete each of the four broad stages of problem-solving.
2. To know and be able to practice the individual steps in each stage of problem-solving.
3. To use the problem-solving model as an effective tool for interacting with members of the community.
4. To be able to explain and mentor the problem-solving process to other officers and members of the community. This involves working collaboratively with members of the community instead of working alone as the expert.
5. To develop the communication skills required during the problem-solving process.
6. To develop leadership skills necessary to elicit support from the community.
7. To encourage initiative and creativity from other problem-solving officers. (p. 11)

Drawing back from the close inspection of the problem-oriented approach at the level of the front-line officer, it is possible to gain a better understanding by examining matters from a broader viewpoint. Figure 6-7 shows the nature of the changes in the role and structure of policing in the context of individual officer decision-making and the scope of police organizational activity.

FIGURE 6-7

ROLE AND STRUCTURE OF POLICING

		RANGE OF ORGANIZATIONAL ACTIVITY	
		Limited	**Broad**
RANGE OF INDIVIDUAL DECISION-MAKING	**Limited**	I Incident-driven policing	II Focus on disorders, services
	Broad	III Participative management	IV Problem-oriented policing

Adapted from: Eck, John E. and William Spelman (1987). *Problem-solving: problem-oriented policing in Newport News.* Washington, D.C.: Police Executive Research Forum, p. 17. Used with permission.

When a police department evolves beyond the realm of conventional incident-driven policing (represented in quadrant I) to tackle larger issues, it moves into the area shown in quadrant II. However, this change will not cause any broadening in the range of decision-making for individual officers. If the police service retains its limited, incident-driven approach to policing, but introduces a degree of latitude for officer decision-making, the organization moves into quadrant III, which has been characterized as participative management. Only when both areas, i.e., organizational activity and individual officer decision-making, have been systematically expanded does the problem-oriented approach come into place, as shown in quadrant IV.

A PROBLEM-ORIENTED POLICING DATABASE

With the realization that the Internet offered a powerful and cost-effective tool for broadly disseminating information on law enforcement, the Police Executive Research Forum has developed POPNet for the Community Policing Consortium, in Washington, D.C. POPNet is a free database containing information on problem-solving that allows officers to gather ideas and insight about this strategy. It also serves as a clearinghouse for the sharing of experiences that may have value for practitioners, educators, and researchers around the world.

POPNet offers a selection of cases in which individual departments have applied problem-oriented policing to particular operational situations. The database has coded the following types of problems:

Abandoned buildings	Fondling/handling	Other
Abandoned vehicles	Fraud	Panhandling
Animal issues	Gambling	Parking
Arson	Gangs	People with mental illness
Assault/battery	Graffiti	Phone calls
Auto theft	Hate crimes	Prostitution
Bribery	Health hazards	Robbery
Burglary	Homicide	Sex offenses
Calls for service	Illegal dumping	Smuggling
Child abuse	Indecent exposure/lewd	Stalking
Code/zoning violations	behavior	Stolen property
Community decline	Juveniles	Suicide
Counterfeit/forgery	Litter	Theft
Disturbance	Loitering	Transients/street people
Drunks	Missing person/runaway	Traffic
Elder abuse	Neighborhood eyesore	Trespassing
Extortion	Noise	Vandalism
False alarms	Off road activity	Weapons violations

The PERF database also allows users to search their files on the basis of specific locations where the problem might occur, including any of the following:

Commercial business	Other	Public transportation
Construction site	Parking lot (business)	Public space
Graveyard	Private property	Religious institution
Marina	Private vehicle (car, boat, truck)	School
Money-matic or automatic teller machine (ATM)	Public building	Street/alley/sidewalk
Neighborhood block	Public park/recreational area	Vacant building/house
		Warehouse
		Wooded area or field

LEADER IN COMMUNITY POLICING

W.I. James Harding, Executive Officer, Special Investigations Unit & ex-Chief of Police, Halton Regional Police Service

What is your understanding of "community policing" and its importance to the evolution of policing in Canada?

The absence of a nationally accepted definition of Community Based Policing is a sanctuary for all those who, in their heart of hearts, still believe that if we continue to pour additional staffing and resources into a reactive policing process then we will, eventually, win the war against crime and disorder.

While I recognize that an effective and competent reactive response to committed crime and incidents of social disorder will always be necessary, it must be equally recognized that as a total, singular policing strategy, it has failed.

Empirical studies show that reactive policing does not enjoy the public confidence that it provides a safe environment in which to live. However, such studies show also that Community Based Policing has done much to secure the public confidence which stems, in part, from the ability to participate with the police in securing their own protection and security. It is my personal belief that this is the cement which binds and strengthens Canadian democracy.

If it is true that ". . . the more people who participate in a democracy the stronger it will become . . . ," then Community Based Policing must stand as one of the great foundation stones upon which a democracy is built.

W. I. James Harding LEADER IN COMMUNITY POLICING continued

As a believer in democracy and in effective policing I made the decision to change from the traditional method of delivering a police service, and embarked upon the policing philosophy of Community Based Policing which for me means:

Community Based Policing is a voluntary Cooperative partnership between the police and the community in which problems and concerns are identified and for which a mutually satisfactory solution is found and collectively acted upon.

If Canadian democracy is to be strengthened, then the public confidence and trust in the Justice System and all its components will greatly assist in the process. The expressed public concern for safety and the need to feel free from the threat of crime is addressed in a very significant way by the practice of Community Based Policing.

Outline your own contribution to the development of "community policing" in Canada.

I came to the position of Chief of Police with the conviction that, based upon the mountain of evidence available, new strategies in the administration and operation of police services were required if the public confidence in our ability to protect them was to be maintained or restored.

When I took command of the Halton Regional Police on May 19th, 1989, the existing circumstance was such that it afforded me an opportunity to embark upon two significant initiatives critical to preparing the service to more efficiently and effectively meet the demands being made of it.

The first was to break from the traditional autocratic and regulatory style of management to move to a more participatory style of management. This was done for a variety of reasons, not the least of which was the fact that levels of staffing were critically low and there was a need to create an environment in which people could rise to be the very best that they could be, in order to achieve any of the tasks that lay before us. It was also based upon my personal belief that it was the responsibility of each and every member of the service to contribute toward its efficiency. Participatory management and a policy of inclusion made such a contribution possible. The second decision was to adopt Community Based Policing as the policing philosophy of the Halton Regional Police Service.

Decentralizing decision making authority to the line Constable was recognized as critical to the success of Community Based Policing, and that this could only be made possible, in a meaningful way, by a participatory management environment. Many initiatives were undertaken to achieve a competent management environment in order that the PRACTICE of the philosophy of Community Based Policing could be properly supported.

The move to Community Based Policing was not an easy sell to the rank and file, many of whom were still seized with the traditional police "Hunter" mentality they had been urged to adopt and develop since the day they joined the service. However, having

W. I. James Harding LEADER IN COMMUNITY POLICING continued

committed the service to "a new adventure in policing," I was determined that we would not merely pay it lip service and, as a consequence, committed 59% of all available patrol resources to the concept.

It became my personal task to explain to the Police Services Board, Regional Council members, Municipal Council members, associated services and agencies, and the public at large that by doing so, even though the time to respond to calls for service would likely increase, the public would be better protected and the environment would be a safer one in which to live. Crime statistics for the next 12 years would confirm this.

Community Based Policing will not succeed just because you wish it to . . . To ensure its initial and continuing success within the Regional Municipality of Halton, two initiatives were embarked upon:

- Within the Halton Regional Police, "Management Teams" were established at every level, section, and platoon, with the intent that they be the means by which any idea, concern, or issue affecting the delivery of our service or the welfare of any staff member would be brought to the attention of Supervisory, Managerial, or Administrative staff. Once an issue is raised, it must be responded to and dealt with. If such an issue was not resolved at a lower level then it became the responsibility of the Chief of Police to do so.
- Within the Regional Municipality of Halton, "Community Consultation Committees" were established to work in concert with local police officers in identifying community issues and concerns and developing a mutually acceptable response to those issues. This was the stuff and substance which made the concept of "needs driven policing" a reality in the Regional Municipality of Halton.

A word of caution is warranted here. Incident driven policing was the accepted policing philosophy for a very long time, so, make no mistake about it, if we reduce, in any way, our commitment to and faith in Community Based Policing, our delinquent policing past will return, and all those who still have the "Hunter" mentality will warmly welcome it.

What is your view of the future of "community policing" in Canada?

One of the realities with which Canada is now dealing (and will continue to do so for a considerable time) is the explosion of racial and cultural diversity within the nation. In addition to the advantage which flows from the richness such racial and cultural diversity brings, comes a series of problems. Many of those problems have been identified, often through the torture of bad experience, and have been well responded to by governments at the federal, provincial, and municipal levels.

It is clear that if Canadians, in all the richness of their racial and cultural diversity, are to be strengthened in their faith and confidence in the government elected to serve them, especially when it relates to Justice, then their participation in the democratic

W. I. James Harding LEADER IN COMMUNITY POLICING continued

process is critical. That democratic participation is accommodated in a constructive, beneficial way by the *practice* of Community Based Policing.

Within the Province of Ontario, Community Based Policing is now mandatory as a consequence of the requirements of the *Police Services Act*. Other provinces should be mandated to follow the practice. Community Based Policing should be adopted as the philosophy of policing of the federal government.

What other comments would you like to make on this topic?

One of the greatest benefits to flow from the practice of Community Based Policing and, perhaps the area receiving the least recognition, is the fact that volunteerism flourishes in a community environment that advocates and practises the concept of inclusion. The notion that a safe community is best achieved when individual and corporate citizens play an active role in its pursuit, is readily accepted and willingly responded to.

With regard to Community Based Policing, it can be said that: Lip service is worse than no service. If you are going to do it, do it right . . . if you are not prepared to make a real commitment to this way of policing, it's better to leave it to someone else.

LEADERSHIP VS. SPONSORSHIP

In order to make an explicit distinction between police practice of the past and the future promise of problem-oriented policing, an understanding of the need for new approaches to "leadership" in police organizations is required. In fact, it may be simpler to use an entirely different term to express the nature of the distinction intended in this area. When we think of "leaders," the literature presents the strong, often charismatic, person who sets the direction, articulates the vision, and inspires all those below to new heights of achievement and excellence (Peters and Waterman, 1982; Donaldson and Lorsch, 1983; Janis, 1989). Of course there is nothing wrong with this presentation. Leadership constitutes a positive organizational value. It has worked to the advantage of many private and public corporations to move toward significant goals. However, in the context of problem-oriented policing, that places an enormous amount of responsibility and reliance on the individual officer within the police service. It may be more fruitful to consider a different model of leadership. While it requires an enlightened leader to serve as the prime mover in any significant transition into the stream of innovation, such as that represented by problem-oriented policing, the real work of this approach happens in the day-to-day engagement of the various communities that make up the police department's jurisdiction. The term "sponsorship" is being suggested as a way of symbolizing the fact that police executives and senior managers who are truly dedicated to the goals of problem-oriented policing will understand that their role is to function as sponsors of the essential work of the people at the "leading" edge of their organizations, the front-line officers who must act on the principles and practices of problem-oriented policing. A sponsor is defined as: "one who binds himself to answer for another; one who is responsible for another's default." The "real work" of a police organization has always occurred at the front line. The nature of policing, regardless of the core functions it pursues, consists of the interaction of officers with citizens in circumstances in which the officers operate to maintain, or secure, order in their capacity as functionaries of the state vested with extensive powers and considerable discretion. This responsibility, with its myriad of possibilities, could never be held to a definitive set of rules and regulations. The officer requires leadership that sets the strategic direction for the organization. However, once that strategic direction has been set, the leadership of the organization should operate to ensure that everything within the organization serves to support and enhance the front-line officers' capacity to do their job. The enlightened organization moves to generate leadership at every level (McKenna, 1998).

Given the past, this type of transformation is incredibly difficult for many police organizations. Sparrow, Moore, and Kennedy (1990) note that there are several police chiefs who have welcomed change.

> They are no longer what their predecessors largely were: modern major generals presiding over departments so established that they did not so much evolve as simply age. They are, instead, something new to the field, police executives, who believe their departments' missions, operational methods, and administrative arrangements may all be adjusted to adapt to new demands and to exploit new opportunities. Their privilege, and their burden, is to point the way, not only for themselves and their departments, but for all the other police executives who so sorely need leave to search for better ways of policing. (p. 59–60).

By viewing the front line of the police organization as the leading edge, it becomes readily apparent that the function of police executives and senior managers is to "sponsor" the efforts of the front-line service deliverers (commonly known as constables). This is truly revolutionary thinking and poses one of the key organizational challenges for problem-oriented policing. As the sponsors of change, police chiefs and senior officers will need to work differently. Sparrow, Moore, and Kennedy (1990) offer further assistance when they recommend the need for:

> . . . understanding and promoting innovation and freeing up the creativity of ordinary officers, one of policing's great untapped resources. If policing can learn to promote innovation, it will be a long way toward winning the adaptability it so badly needs. If it can learn to learn from its officers, it will unleash a vast reservoir of energy and insight. (p. 198).

With appropriate consideration for the layers of police accountability discussed in Chapter Five, there is something valuable to be learned from Wheatley's (1992) observation:

> We are beginning to see organizations that tap into [the] property of self-organizing or self-renewing systems. Some theorists have termed these "adaptive organizations", where the task determines the organizational form...In a separate but related example are those corporations structured around core competencies, as described by C.K. Prahalad and Gary Hamel (1990). Both types of organizations avoid rigid or permanent structures and instead develop a capacity to

> respond with great flexibility to external and internal change. Expertise, tasks, teams, and projects emerge in response to need. When the need changes, so does the organizational structure. (p. 91).

Many experts on organizational development have written about this need for greater flexibility and more effective use of the intellectual capital possessed by the people within those organizations (Senge, 1990; Drucker,1994; Dalla Costa,1995). However, for this to take place in Canadian police organizations there needs to be greater emphasis on developing a new mindset at the upper levels of police departments. As Braiden (1992) notes:

> If a conventional police agency is to adopt the Community Policing way of doing things, then there first has to be a re-tooling of the heads of the brass before you can re-tool the feet of the grunts. (p. 24).

Part of that retooling relates to the concept of leadership and what it means for an innovative, modern police service. This chapter suggests substituting the term sponsorship for leadership to highlight the significant amount of responsibility and accountability the leader bears for those who function at the front line of the organization. While still using the term "leadership," Braiden (1995) is nonetheless correct when he makes the following assertion:

> Leadership is the one thing we cannot have unless we are prepared to give it away by sharing power and ownership. True leaders have many talents but the most important one is the ability to recognize talent in others, and then liberate it. True leaders are forever creating new heroes around them. I have read that the biggest difference between Winston Churchill and F.D. Roosevelt was that Churchill was always championing young politicians while FDR was busy silencing those around him. (p. 15).

PRACTICAL APPLICATION OF PROBLEM-SOLVING IN POLICE ORGANIZATIONS

The principles and practice of problem-solving have been applied in many police organizations, resulting in an approach that may be termed problem-oriented policing (POP). While the approach may carry a different label, the essential ingredients are consistent with the details outlined above. Typically, there is a conscious effort to have an impact on crime and social disorder through an application of the kinds of strategic techniques pioneered by Herman Goldstein and those first police organizations that have embraced this methodology. What follows is a brief survey of police departments that have experimented with the problem-solving approach.

CANADA

MONTREAL URBAN COMMUNITY POLICE

In November 1995, the Montreal Urban Community Police received approval from the Montreal Urban Community Council to proceed with a Neighbourhood Policing model made up of five components:

- A problem-solving approach;
- Geographical responsibility;
- A customer approach;
- Partnership; and
- Staff recognition.

The Montreal Urban Community Police adopted a model similar to that outlined in Eck and Spelman (1987) and made a commitment to this approach, fully aware of the organizational implications of this decision as a departure from previous practice.

> The problem-solving approach is essentially a new organizational strategy. It modifies the Police Service's main operational and administrative methods to enable it to react and adapt to its environment. It touches all aspects of the organization, including human resources and operations. For the police, it is a new way of thinking and acting.
>
> The SPCUM's [Service de police de la communauté urbaine de Montréal] organization will thus be subjected to a complete metamorphosis. The intention is obviously not to stop responding to urgent calls under the pretext that this strategy does not allow the resolution of problems. The problem-solving approach will, however, enable a better balance between the proactive and reactive aspects of the service. (p. 23).

Ontario Provincial Police

The Ontario Provincial Police (OPP) has endorsed a philosophy of community policing since the early 1980s as an integral part of its strategic plan (Ontario Provincial Police, 1995). Through its major reorganization, commenced in 1993, the OPP renewed its commitment to community policing, including the creation of a new organizational entity entitled the Community Policing Development Centre (CPDC). This area was designed to take responsibility for driving the research agenda necessary to support the organization's commitment to community policing.

The CPDC prepared, in consultation with various stakeholders, a manual intended to assist local detachments and community policing committees in understanding and delivering community policing services. A portion of this manual is devoted to the problem-solving approach applied within the OPP. Known as the P.A.R.E. model, this approach is consistent with the elements of the S.A.R.A. model, outlined above, and has the following components:

- *Problem identification*—includes a listing of the perceptions and symptoms of the problem and a clear identification of the perceived problem(s);
- *Analysis*—includes an examination of the problem to determine answers to the who, what, where, when, why questions, a consideration of the victim, the offender, and the situation, a determination of the impact, seriousness, complexity, and solvability of the problem along with some prioritization of the issue, and an identification of the goal in this instance (e.g., eliminate, reduce, or redefine responsibility for the problem);
- *Response*—includes the identification of an appropriate response strategy and the development of a plan, including tasks and an evaluation mechanism; and

- *Evaluation*—includes the actual conduct of a summary evaluation and the analysis, documentation, and sharing of the results.

Edmonton Police Service

Under the enlightened "sponsorship" of Superintendent Chris Braiden (see, "*Leaders in Community Policing in Canada*" box, this chapter) an experiment was undertaken within the Edmonton Police Service that applied much of the thinking developed under the auspices of problem-oriented policing. The approach will be discussed again in Chapter Twelve, and is recounted in several other sources (Koller, 1990; Edmonton Police Service, 1995; Hawkins, Jr., 1996). However, it amounted to a dramatic departure from previous practice within this Canadian police service.

Mount Pleasant, Vancouver

The project launched in this part of the City of Vancouver by the Vancouver Police Department was a brief attempt to apply the POP approach in an area that had been plagued by serious social problems (i.e., drug dealing and prostitution). It is seen as a "highly successful project" (Goldstein, 1990). However, it was not implemented throughout the organization.

The project involved the formation of the Mount Pleasant Liaison Team (MPLT) which included officers drawn from across the local community. The members of the MPLT were trained in the S.A.R.A. approach to problem-solving and applied those skills to the analysis of identified problems. They targeted problem buildings and worked with property owners and local health agencies to deal with the pressing concerns about prostitution, disturbances, and drug activity. This three-year experiment in problem-oriented policing is described by Rossmo and Fisher (1993).

Peel Regional Police

Peel Regional Police has the honour of being one of two Canadian police services that have been awarded the Herman Goldstein Award for Excellence in Problem-Oriented Policing (the other being the Calgary Police Service in 1993). The individual project that was recognized for this award in 1996 dealt with a problem of trespassing in one of the local high schools and involved an application of Crime Prevention Through Environmental Design (CPTED), discussed in Chapter Two. The principal of the school identified the problem and sought advice from the police on dealing with this matter. By using the CPTED approach, the constable assigned to this problem was able to pinpoint several design and use deficiencies at the site, including difficulties with the layout and location of the school parking lot, lack of appropriate signage, lack of suitable walkways, among others.

The constable was able to develop a detailed CPTED study that outlined a number of external changes that would assist in alleviating the identified deficiencies. In order to prepare this report, the constable received assistance from the police department's crime analyst, as well as technical assistance in the areas of desktop publishing and computer graphics. When the plan was presented to the local Board of Superintendents it was approved, with some amendments, and construction work began in a matter of months. In the three years that following these CPTED-based modifications, police occurrences at this school (with some 2000 students) have dropped by an average of 87%. This may be considered a successful application of both problem-solving and CPTED.

UNITED STATES

Newport News (Virginia) Police Department

This is one of the key sites for the early application of problem-oriented policing. The project began in 1983 and is thoroughly detailed in several presentations (Eck and Spelman, 1987; Goldstein, 1990; Peak and Glensor, 1996). The careful planning that went into the design, development, and deployment of the S.A.R.A. model in this police department ensured that it would provide a solid foundation for future work in the area of problem-solving.

New York City Police Department

In 1984, the New York City Police Department was led by Commissioner Benjamin Ward. Commissioner Ward, with technical support from the Vera Institute of Justice, began a project known as the Community Patrol Officer Program (Ward, 1985; Farrell, 1986, New York City Police Department, 1987). As part of this project, 9 or 10 officers in the participating precincts would be assigned to a specific beat. These officers were expected to provide a full range of police services within their beat area. However, they were not required to respond to "routine" calls. The community patrol officer was challenged to work with community residents in the identification and the resolution of specific problems within the beat area. The assessment of this program, conducted by staff from the Vera Institute, revealed that the officers' role as problem-solver was the most significant feature of their function within the beat. It was in their capacity as concentrated problem-solving agents that the officers distinguished themselves from regular foot patrol officers (Goldstein, 1990).

GREAT BRITAIN

Metropolitan London Police

In 1983, Sir Kenneth Newman, commissioner of the Scotland Yard, began a project to determine the feasibility of adopting the techniques of problem-oriented policing within the Metropolitan London Police. Newman's interest in this approach was derived from his exposure to the positive results of the experiments conducted in Madison, Wisconsin (see Goldstein and Susmilch, 1981, 1982, 1982a). The London project consisted of four pilot initiatives that were to be examined for their possible overall organizational benefit in operations, training, resources, research and management (Hoare, Stewart, and Purcell, 1984). The problems identified for this experimental application of problem-oriented policing were: Asian gangs, shopper victims in Oxford Street, Bedford Hill prostitutes, and motor vehicle crime.

The assessment of these initiatives found that the problem-oriented approach held potential to improve the performance of the police. However, the project was structured in a manner that made it difficult to administer in a coherent fashion (Hoare, Stewart, and Purcell, 1984, summary). There was a growing awareness of the need to create an environment in which a problem-oriented approach could function more effectively. Commissioner Newman understood this need and expressed a desire to make fundamental changes in the Metropolitan Police, unprecedented in over 150 years. In reworking the principles of the Metropolitan Police, Newman (1985) articulated a commitment to problem-oriented policing in the following words:

> Many of the isolated incidents to which police are called are symptoms of more general substantive problems with roots in a wide range of social and environmental conditions. The aim of the Metropolitan Police will, therefore, be to work with other agencies to develop what is known as a "situational" or "problem-solving" approach to crime prevention, where, rather than merely dealing with individual acts of law-breaking, careful analysis is made of the total circumstances surrounding the commission of types of crime, taking account of wide-ranging social and environmental factors, in order better to understand—and counter—the causes of those acts. (p. 12).

OTHER JURISDICTIONS

Several authors have begun the important work of documenting the instances of problem-oriented policing in other jurisdictions (Goldstein, 1990; Peak and Glensor, 1996; Skolnick and Bayley, 1988). Something similar to problem-oriented policing is beginning to take shape in countries such as Germany, Japan, Singapore, South Australia, and the Scandinavian countries (Norway, Sweden, Denmark, and Finland).

SUMMARY

This chapter has examined the application of problem-solving to the issues of crime and disorder confronting police departments. When this technique is implemented in a systematic manner throughout a police service it is normally referred to as problem-oriented policing and has become an important component within the range of options commonly referred to as community policing. The limitations of the traditional incident-driven approach to policing have been considered, and definitions and key principles relating to problem-oriented policing have been presented.

The S.A.R.A. model of problem-solving has been outlined in some detail, as well as the training requirements that are recommended in order to make the transition to this consistent approach. This chapter presents a perspective on leadership within the problem-oriented policing model and suggests that true leaders should be seen, and see themselves, as sponsors of the front-line officers who will actually lead the organization to successful implementation of the problem-solving technique in their communities.

Finally, this chapter cites some of the various police departments that have attempted to apply the problem-oriented approach to policing, and offers some examples of where it has worked to resolve specific operational issues.

QUESTIONS FOR CONSIDERATION

1. What are possible sources for information on an identified police/community problem?
2. Are there positive aspects associated with an incident-driven approach to policing?
3. Can the guiding principles identified for problem-solving policing be applied to Canadian police services?
4. What are some community problems that would **not** usually be addressed by problem-oriented policing?
5. What are some approaches that a police department could use in the assessment/evaluation stage to determine if a particular problem response has been effective?
6. What other police services in Canada have introduced problem-oriented policing?

REFERENCES

Alpert Geoffrey P. and Roger G. Dunham (1988). *Policing mulit-ethnic neighborhoods: the Miami study and findings for law enforcement in the United States.* New York: Greenwood Press.

Braiden, Chris (1986). "Bank robberies and stolen bikes: thoughts of a street cop." *Canadian Police College Journal*, Vol. 10, pp. 1–30.

Braiden, Chris (1992). "Community policing: nothing new under the sun." In *Community oriented policing & problem solving*. Sacramento, Calif.: California Department of Justice, Crime Prevention Center.

Braiden, Chris (1993). "Ownership: who paints a rented home?" paper prepared by Chris Braiden Consulting Inc.

Braiden, Chris (1994). "Policing—from the belly of the whale." Paper prepared by Chris Braiden Consulting Inc.

Cahn, Michael F. and James M. Tien (1981). *An alternative approach in police response: Wilmington management of demand program.* Washington, D.C.: U.S. Government Printing Office.

Capara, David and Bill Alexander (1989). *Empowering residents of public housing: a resource guide for resident management*. Washington, D.C.: National Center for Neighborhood Enterprise.

Clarke, Ronald V. (1983). "Situational crime prevention: its theoretical basis and practical scope." In Tonry, Michael and Norval Morris (eds.) *Crime and justice: an annual review of research, Vol. 6*. Chicago: University of Chicago Press.

Clarke, Ronald V. (1992). *Situational crime prevention: successful case studies*. Albany, N.Y.: Harrow and Heston, Publishers.

COPPS: community oriented policing & problem solving. Sacramento, Calif.: California Department of Justice, Attorney General's Office, Crime Prevention Center, 1992.

Cordner, Gary W. (1985). *The Baltimore County citizen oriented police enforcement (COPE) project: final evaluation*. Baltimore: Criminal Justice Department, University of Baltimore.

Cordner, Gary W. (1988). "A problem-oriented approach to community-oriented policing." In Greene, Jack R. and Stephen D. Mastrofski (eds.) *Community policing: rhetoric or reality*. New York: Praeger.

Couper, David C. and Sabine H. Lobitz (1991). "The customer is always right: applying vision, leadership and problem-solving method to community policing." *Police Chief*, Vol. 58, pp. 16–23.

Couper, David C. and Sabine H. Lobitz (1991). *Quality policing: the Madison experience*. Washington, D.C.: Police Executive Research Forum.

Dalla Costa, John (1995). *Working wisdom: the ultimate value in the new economy*. Toronto: Stoddart.

Donaldson, G. and J.W. Lorsch (1983). *Decision making at the top: the shaping of strategic direction*. New York: Basic Books.

Drucker, Peter F. (1994). "The age of social transformation." *Atlantic Monthly*, Vol. 274, no. 5 (November), pp. 53–80.

Eck, John E. and William Spelman (1987). *Problem-solving: problem-oriented policing in Newport News*. Washington, D.C.: Police Executive Research Forum.

Eck, John E. William Spelman (1987). "Who ya gonna call?: the police as problem-busters." *Crime & Delinquency*, Vol. 33, pp. 31–52.

Eck, John E. and William Spelman (1989). "A problem-oriented approach to police service delivery." In Kenney, Dennis Jay (ed.) *Police and policing: contemporary issues*. New York: Praeger.

Farmer, Michael T. (ed.) *Differential police response strategies*. Washington, D.C.: Police Executive Research Forum.

Farrell, Michael T. (1986). "C.P.O.P. (community patrol officer program)," *Interim progress report no. 2*. New York City Police Department and Vera Institute of Justice. May.

Goldstein, Herman (1977). *Policing a free society*. Cambridge, Mass.: Ballinger.

Goldstein, Herman (1979). "Improving policing: a problem-oriented approach." *Crime & Delinquency*, Vol. 25, pp. 236–258.

Goldstein, Herman (1990). *Problem-oriented policing*. New York: McGraw-Hill Inc.

Goldstein, Herman and Charles E. Susmilch (1981). *The problem-oriented approach to improving police service: a description of the project and an elaboration of the concept*. Madison, Wis.: Law School, University of Wisconsin.

Goldstein, Herman and Charles E. Susmilch (1982). *Experimenting with the problem-oriented approach to improving police service: a report and some reflections on two case studies*. Madison, Wis.: Law School, University of Wisconsin.

Goldstein, Herman and Charles E. Susmilch (1982a). *The repeat sexual offender in Madison: a memorandum on the problem and the community's response*. Madison, Wis.: Law School, University of Wisconsin.

Goldstein, Herman, et al. (1981). *The drinking driver in Madison: a study of the problem and the community's response*. Madison, Wis.: Law School, University of Wisconsin.

Guyot, Dorothy (1991). "Problem oriented policing shines in the stats." *PM: Public Management*, Vol. 73, no. 9, pp. 12–16.

Hatry, Harry P. and John M. Greiner (1986). *Improving the use of quality circles in police departments*. Washington, D.C.: Urban Institute.

Hawkins, Jr., Carl W. (1996). "Ready, fire, aim: a look at community policing in Edmonton, Alberta, Canada." In *Themes and variations in community policing: case studies in community policing*. Washington, D.C.: Police Executive Research Forum.

Hayward (City) Police Department (1991). *Community oriented policing and problem solving, "C.O.P.P.S.": implementation plan*. Hayward, Calif.: City of Hayward Police Department.

Hoare, M.A., G. Stewart, and C.M. Purcell (1984). *The problem oriented approach: four pilot studies*. London, U.K.: Metropolitan Police, Management Services Department.

Janis, Irving L. (1989). *Crucial decisions: leadership in policymaking and crisis management*. New York: The Free Press.

Jeffrey, C.R. (1971). *Crime prevention through environmental design*. Beverly Hills, Calif.: Sage Publications.

Koller, Katherine (1990). *Working the beat: the Edmonton neighbourhood foot patrol*. Edmonton: Edmonton Police Service.

McKenna, Paul F. (1998). *Foundations of policing in Canada*. Scarborough, Ont.: Prentice Hall Canada.

Mitchell, William P. (1990). "Problem-oriented policing and drug enforcement in Newport News." *Public Management*, Vol. 72, pp. 13–16.

Murphy, Chris (1991). "Problem oriented policing: a manual for the development and implementation of problem oriented policing." In Ontario Ministry of the Solicitor General (ed.) *Community policing: shaping the future*. Toronto: Queen's Printer for Ontario.

New York City Police Department (1987). "The community patrol officer program: implementation guide," New York: N.Y.P.D.

Newman, Kenneth (1985). *The principles of policing and guidance for professional behaviour*. London: The Metropolitan Police.

Oettmeier, Timothy N. and Mary Ann Wycoff (1997). *Personnel performance evaluation in the community policing context*. Washington, D.C.: Police Executive Research Forum.

Ontario Provincial Police. Community Policing Development Centre (1997). *"How do we do it?": manual of community policing*. Orillia, Ont.: Ontario Provincial Police, Community Policing Development Centre.

Parker, Patricia A. (1990). "POP vs. drugs." *Police*, Vol. 43, pp. 34–37.

Parker, Patricia A. (1991). "Herman Goldstein: the man who made problem-oriented policing a reality." *Police*, Vol. 15, nos. 10–12, p. 75.

Peters, Thomas J. and Robert H. Waterman (1982). *In search of excellence*. New York: Harper & Row.

Police Executive Research Forum (1989). *Taking a problem-oriented approach to drugs: an interim report*. Washington, D.C.: Police Executive Research Forum.

Poyner, Barry (1980). *Street attacks and their environmental settings*. London: Tavistock Institute of Human Relations.

Poyner, Barry (1983). *Design against crime: beyond defensible space*. London: Butterworths.

Rossmo, Kim and Doug Fisher (1993). "Problem-oriented policing: a co-operative approach in Mount Pleasant, Vancouver." *Royal Canadian Mounted Police Gazette*, Vol. 55, pp. 1–9.

Seagrave, Jayne (1997). *Introduction to policing in Canada*. Scarborough, Ont.: Prentice Hall Canada Inc.

Senge, Peter M. (1990). *The fifth discipline: the art & practice of the learning organization*. New York: Doubleday Currency.

Sherman, Lawrence W., Catherine Milton, and Thomas Kelly (1973). *Team policing: seven case studies*. Washington, D.C. : Police Foundation.

Skolnick, J.H. and D.H. Bayley (1988). *Community policing: issues and practices around the world.* Washington, D.C.: National Institute of Justice.

Spelman, William and John E. Eck (1987). "The problem-oriented policing bureaucracy," paper presented to the annual meeting of the American Political Science Association.

Spelman, William and John E. Eck (1989). "The police and delivery of local government services: problem-oriented approach." In Fyfe, James J. (ed.) *Police practice in the 90's: key management issues*. Washington, D.C.: International City/County Management Association.

Spergel, Irving (1969). *Community problem-solving*. Chicago, Ill.: University of Chicago Press.

Talarico, Susette M. and Charles R. Swanson, Jr. (1980). "The limits of team policing?" *Police Studies*, Vol. 3, pp. 21–29.

Toch, Hans and J. Douglas Grant (1991). *Police as problem solvers*. New York: Plenum.

Ward, Benjamin (1985). *The community patrol officer program*. New York: New York City Police Department.

Watson, T. Steuart, Dennis Jay Kenney, and John Lusardi (1995). *Problem-solving: intermediate curriculum guide*. Washington, D.C.: Police Executive Research Forum.

Weisbord, Marvin R., Howard Lamb, and Allan Drexler (1974). *Improving police management through problem-solving task forces: a case study in organization development.* Reading, Mass.: Addison-Wesley.

Weisel, Deborah Lamm (1990). "Playing the home field: a problem-oriented approach to drug control." *American Journal of Police*, Vol. 9, pp. 75–95.

Weisel, Deborah Lamm (1990). *Tackling the drug problem in public housing: a guide for police*. Washington, D.C.: Police Executive Research Forum.

Wheatley, Margaret J. (1992). *Leadership and the new science: learning about organization from an orderly universe*. San Francisco: Berrett-Koehler Publishers, Inc.

TOWARD MORE LEARNING

The most valuable way to develop a deeper understanding of the problem-solving, or problem-oriented, approach to policing is to practice applying the S.A.R.A. model to various law enforcement situations. A great deal of assistance can be found in the website of the Police Executive Research Forum (PERF) in the United States. This organization is dedicated, as its name implies, to encouraging and sponsoring relevant research in several areas of police studies, including problem-solving and problem-oriented policing. To that end, PERF has established a feature, known as POPNet, which provides a database dedicated to collecting case studies from police departments in Canada, England, Australia, and the U.S. The Police Executive Research Foundation's website can be found at:

http://www.policeforum.org

Because of its broad mandate, PERF has taken on responsibility for providing research and training materials in several areas of policing, including problem-oriented policing (POP). To that end, PERF has installed some free training curricula for use by police departments wishing to better understand and apply this approach. Currently, PERF is making available the following training programs that can be downloaded from the following website:

- The Mechanics of problem-solving;
- Supervising the problem-solving process; and
- Collaborative problem-solving.

http://www.policeforum.org/home/public/curricula.html

PERF also publishes a newsletter, *Problem Solving Quarterly*, that includes examples of successful problem-solving cases for consideration by police organizations dealing with similar issues. Also of value from a training perspective is material prepared by the Community Policing Consortium (PERF is a member of this organization). The Consortium has produced a number of useful publications dealing with community policing, as well as a series of training modules geared to the transition to a community policing model. One of these modules is entitled "Community policing problem-solving: taking a problem-solving approach to tackling crime, fear and disorder." This module has the following goal and can be downloaded from the Consortium website, also listed:

Module Goal: Problem-solving and community partnerships are the key components to community policing. An in-depth focus on crime, fear, and disorder problems is essential to the success of community policing. This module will introduce problem-solving to street-level officers, their supervisors and/or community volunteers who are interested in addressing crime, fear, and disorder.

http://www.communitypolicing.org/mod3.html

Eck, John E. and William Spelman (1987). *Problem-solving: problem-oriented policing in Newport News*. Washington, D.C.: Police Executive Research Forum.

This should be considered essential reading for anyone interested in the fundamentals of problem-solving. It traces the development of the first major problem-oriented policing project in Newport News, Virginia. What makes this early treatment so valuable is the detailed presentation of the research background that guided this early project, and the considerable support provided by the Police Executive Research Forum. This is an important project, and the summary of the work undertaken provides an extremely helpful insight into the promise of problem-oriented policing.

Goldstein, Herman (1979). "Improving policing: a problem-oriented approach." *Crime & Delinquency*, Vol. 25, pp. 236–258.

This is the seminal article that has been recognized as the source of original thinking around the problem-oriented policing movement. It should be considered a "must read" for those seeking to understand the theory behind the practice of problem-oriented policing.

Goldstein, Herman (1990). *Problem-oriented policing*. New York: McGraw-Hill Inc.

Goldstein's most recent refinement of his earlier work on problem-oriented policing, this publication offers a detailed assessment of the application of this approach. Goldstein provides a critique of the present state of problem-oriented policing, reviewing the basic elements of the model, and revisiting some of the early practical experiences with this model in several jurisdictions (including: Madison, Wisconsin; Baltimore, Maryland; Metropolitan London Police, Newport News, Virginia, and Edmonton, Alberta). He goes on to offer insight into the important need for significant change in police management that will support and sustain ongoing success in problem-oriented policing.

Murphy, Chris (1991). "Problem oriented policing: a manual for the development and implementation of problem oriented policing." In Ontario Ministry of the Solicitor General (ed.) *Community policing: shaping the future*. Toronto: Queen's Printer for Ontario.

An excellent Canadian source on the application of problem-oriented policing. Professor Murphy has contributed a great deal to the development of community policing in Canada, and this contribution to the early phases of community policing in Ontario is helpful for students, researchers, and practitioners alike.

Watson, T. Steuart, Dennis Jay Kenney, and John Lusardi (1995). *Problem-solving: intermediate curriculum guide*. Washington, D.C.: Police Executive Research Forum.

CHAPTER SEVEN

THE CHALLENGES RAISED BY COMMUNITY POLICING

Every art and every inquiry, and similarly every action and pursuit, is thought to aim at some good; and for this reason the good has rightly been declared to be that at which all things aim.

Aristotle, Nicomachean Ethics, Bk. 1, 1094a (Translated by W. D. Ross).

LEARNING OBJECTIVES

1. Recognize the importance of a "system" of police learning in the context of community policing.
2. Identify the key elements of a new paradigm for police learning in Canada.
3. Identify the four stages of team development and relate them to the implementation of community policing.
4. List major stakeholders that would benefit from exposure to a police learning system.
5. Identify ethical challenges resulting from the introduction of community policing.
6. State the main race relations issues connected with community policing in Canada.

INTRODUCTION

This chapter is designed to examine, in some detail, the major challenges raised by the introduction of community policing into Canadian police departments. While Chapter Two attempted to grapple with the first significant challenge, i.e., defining what is actually meant by the term "community policing," this chapter assumes that community policing involves some new way of doing business that directly involves the community in a substantive manner. The concept of the "co-production of order" is relevant here because it captures the idea that the public and the police form some kind of partnership aimed at transforming communities into places that are safe, secure, and stable.

The first challenge to be addressed in this chapter relates to education. What police personnel are taught, from their initial recruitment phase through to senior executive development, will have a profound bearing on the success or failure of community policing. It is worth noting that the organization's responsibility for education, however, does not end with police officers. Civilian staff, and the fullest range of external stakeholders, must also be provided with sound educational content in order to help them understand their vital role within the new paradigm of policing, referred to as community policing.

Next, this chapter looks at the ethical challenge of community policing. With a more direct engagement of the police with the community, questions about integrity become ones of paramount importance. Of course, ethical policing has always been a concern. However, the nature of community policing has raised many pertinent questions for consideration. Finally, there will be a review of the special challenge of race relations in Canadian policing. The composition of our country makes it imperative that community policing "fit" all the various communities that comprise our cultural and ethnic mosaic. Is community policing achieving this "fit"?

THE EDUCATIONAL CHALLENGE OF COMMUNITY POLICING

In years gone by, it was not uncommon for police managers to complain about the time "lost" to such things as sick leave, workers' compensation, holidays, court, and *training*. While this type of reaction may be understood on some level with regard to sick leave (especially that strange malady known as the "blue flu"), it betrays a terrible misunderstanding about the proper role of training within a police organization. It was not long ago that responsible police organizations began placing more of an emphasis on the design, development, delivery, and evaluation of their training programs. Much of this attention was the result of significant liability on the part of police organizations when it was demonstrated that officer training was deficient. Training departments have become a point of focus in several police organizations and appropriate attention is being paid to making an "investment" in training, as opposed to seeing it as time "lost" to the other more important activities, like crime-fighting and law enforcement (Oettmeier, 1992).

The movement from the concept of "training" to that of "education" is a subtle, but significant, one and it may be helpful in this regard to consider some definitions. Training may be defined as:

> The process of developing specific skills to perform immediate job functions to a level of acceptable competence using the equipment, processes, and information available for those skills.

Examples of police training could include the use of approved firearms, techniques of interview and interrogation, accessing computerized police databanks, CPR/first aid, handcuffing techniques, and a whole range of specific skills that must adhere to established standards of performance set forth by law, or directive, and using authorized and specialized equipment. The emphasis here is typically on the acquisition of skills. Training does not usually involve much independent critical thinking on the part of the learner. Normally something called a "course training standard" is developed that fixes the specific objectives for the training program, and course candidates (i.e., the trainees) work through the exercises until they have achieved an acceptable (i.e., identified) level of skill to qualify them in this particular area of job performance. The challenge here relates to the fact that such an approach is not suited to the higher order problem-solving skills anticipated in a problem-oriented approach to policing, as formulated by Herman Goldstein and his successors. However, it remains the case that officers must be able to demonstrate the basic skills expected of them in doing their jobs. Officers who cannot use the issued firearms, who do not possess the skills for interviewing witnesses and suspects, cannot properly handcuff a suspect, or operate a police cruiser are of limited overall value to the department, regardless of their possession of higher order problem-solving aptitudes. The educational challenge for Canadian police organizations, in the era of community policing, is that they must balance highly effective *training* programs, geared to skills acquisition, with suitable *learning* strategies that will foster and focus problem-solving and critical thinking capacities within their members.

A SYSTEMS APPROACH TO TRAINING (S.A.T.)

One of the most effective approaches to developing sound training programs, in any organization, is referred to as the "systems approach to training." Essentially, this approach places a great deal of emphasis on "front-end analysis" (Harless, 1988) and relies on the preparation of clear and specific behavioural objectives, along with directly relevant evaluation tools that will ensure that a skill has been acquired by the student. An overview of the systems approach to training (S.A.T.) is presented below:

FIGURE 7-1

THE SYSTEMS APPROACH TO TRAINING MODEL

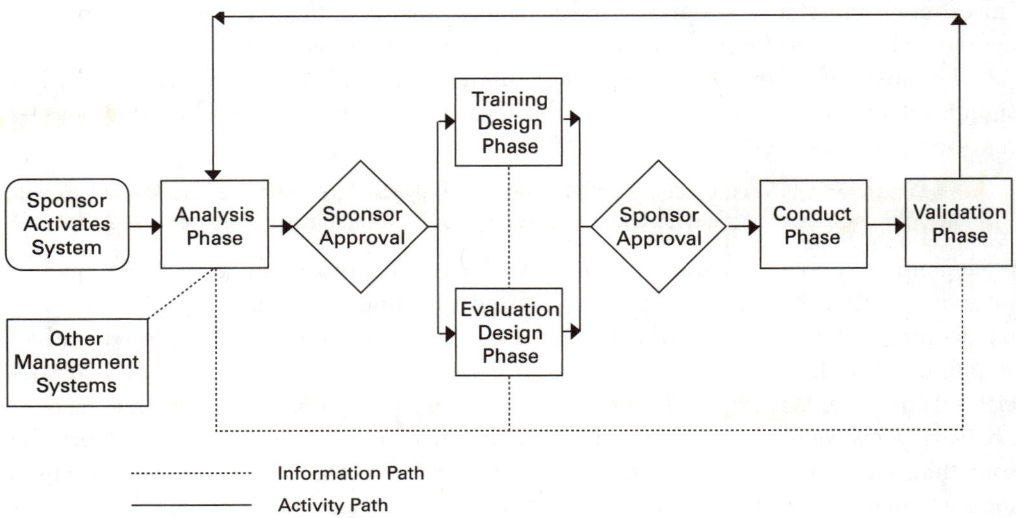

This approach has proven successful in many police departments. However, it is not ideally suited to the theory and practice of organizational learning, encouraged by many authorities on innovative management and leadership (Senge, 1990; Watkins and Marsick, 1993; Dixon, 1993; Kotter, 1996).

EDUCATION AND ORGANIZATIONAL LEARNING

Moving beyond the realm of straightforward training, which continues to be vitally important and relevant to the many skills-based activities of policing in Canada, it is valuable to turn attention to the concepts of education and organizational learning. Education may be defined as:

> . . . a complex collection of actions, procedures, and predictable results through which deliberate instruction is provided to a designated group of learners, usually pupils or students . . . learning is the act of an individual whereas education is a relatively coherent group of social activities usually associated with a particular institution . . . (Alan Thomas, 1991)

As a concept, organizational learning has many definitions, but at its most basic, it rests on the significant insight that organizations are composed of individuals, and it is only through the efforts of those individuals that organizations function, or fail to function. Therefore, organizational learning is a continuous process of generating and regenerating knowledge at the following levels:

- Individual;
- Team; and
- Organization or system-wide.

A model of organizational learning is one that most closely parallels the problem-oriented policing paradigm articulated by scholars, such as Herman Goldstein, and practitioners in numerous police jurisdictions. Organizational learning is also a tool that can be used to facilitate and drive strategic change within the police department (Canadian Centre for Management Development, 1994). Organizational learning, as a process, must be seen as something new and different, not simply another "course" being offered through the Training Branch.

Some of the key characteristics of a "learning organization" are as follows:

- Openness to new ideas;
- Tolerance of error and risk-taking;
- Culture that encourages and provides opportunities for learning and development;
- Clear and shared vision;
- Capacity for systems thinking;
- Understanding of mental models;
- Application of "action learning" and "action research";
- Encouragement of personal growth and mastery;

- Capacity for creating, containing, and retrieving organizational memory;
- Capacity to question fundamental assumptions and/or first principles; and
- Practice of "dialogue."

Although many police departments have been very effective at designing and delivering excellent training programs, it has become increasingly clear that the "reactive" approach typical of police training is not really satisfactory given the constantly changing demands of modern policing. Just as police departments have been encouraged to take a different approach to the delivery of services—indeed the whole "philosophy" of policing has undergone a change to a more "proactive" model—so too a more proactive approach to learning is required in order to prepare police organizations for this new world.

If community policing, in it most reasonable form, implies a different mode of operation within the community, and calls forth significant organizational change, it should come as no surprise that the whole approach to police learning must also be transformed. This area, like that of organizational change, discussed in Chapter Five, is not easy. Commentators on community policing note both the importance, and the difficulty, of different approaches to learning in making the transition to this mode of operation (Goldstein, 1990; Normandeau and Leighton, 1990; Seagrave, 1992; Peak and Glensor, 1996).

An excellent example of the attention given to this issue is to be found in Ontario. Beginning in 1989, the Ministry of the Solicitor General undertook a project to develop short-term solutions for pressing problems in the area of police training. In 1990, a Strategic Planning Committee on Police Training and Education was formed to systematically address the long-range issues in this area (Ontario. Ministry of the Solicitor General. Strategic Planning Committee on Police Training and Education, 1992). This broadly representative group embarked on an ambitious project that amounted to a "paradigm shift" in terms of police learning and applied organizational learning principles and practices developed through the project's considerable research efforts.

One of the primary results of this research was the introduction of a new Police Learning System in Ontario. The system recognized that police personnel, at all levels, required an array of knowledge, skills, and abilities that had not been fully addressed in earlier approaches to police training and education. While the Ontario Police Learning System, inaugurated on March 20, 1995, anticipated a complete redesign of training, education, and development for police personnel, progress has been slow in bringing this undertaking into reality. Part of the structural change at the Ontario Police College, which coordinates training and education for police in the province, has included staffing of two key units:

- *Research and Evaluation Unit*—to conduct ongoing educational research and evaluation of training programs and to provide research in support of Ontario's policing community; and
- *Race Relations and Adult Education Unit*—to provide training and support within the college, ensuring that course content and policies apply appropriate theory and practice in these critical areas.

With a recognition that learning is central to the success of community policing, especially the varieties that feature an adherence to problem-solving, it is not surprising that organizational learning and other related organizational development initiatives have come into place (McKenna, 1998, pp. 221-227). Certainly the framework for community policing introduced by the Ontario Ministry of the Solicitor General and Correctional Services and the Ontario Association of Chiefs of Police reveals the central importance of police learning. That framework is outlined below:

FIGURE 7-2

OACP & MSGCS FRAMEWORK FOR COMMUNITY POLICING

Adapted from: *Community Policing . . . Our Future Today.* © 1996 Queen's Printer for Ontario. Reproduced with permission.

The acceptance of a new paradigm for police learning does not relieve police executives of the responsibility for providing the more traditional, skills-based training programs. Policing continues to be an occupation fraught with danger, and police officers in Canada continue to have a legal authority to use force, include lethal force, in the execution of their duties. The levels of accountability and the layers of oversight in place to ensure that police operate within the law, require that police departments have detailed and definitive training curricula, as well as intensive requalification programs to verify that officers are maintaining a high level of proficiency in some of the following areas:

- Use of force options;
- Search and seizure;
- Arrest techniques; and
- Techniques of interview and interrogation.

Also, because the law is changing with great regularity, and only some predictability, officers require training in the substance of legislative change and on the impact of court rulings and decisions in areas of police operation. With this often bewildering array of new information, police officers rely on their organizations to provide them with the training and tools necessary to understand, practise, and absorb the knowledge, skills, and abilities they must have to be effective in their law enforcement role. The challenge embedded in this observation is that police leaders will need to be especially creative in developing approaches to learning that bridge the gulf between traditional training and more progressive organizational learning. Unlike other enterprises, such as IBM, Hewlett-Packard, General Electric, and others that can recruit people already equipped with the essential competencies to perform various functions, police organizations in Canada typically recruit candidates with no specific police training. They are required to send these new candidates through a basic orientation program and then place them on a period of probation during which they normally work in tandem with a coach officer who guides and monitors their early performance.

A change in mindset is called for in police learning. It requires that higher level skills be established for entry into policing and yet ensures that the basic competencies are met by new candidates. The available research does not prove with any certainty that police officers with university degrees perform better than those without post-secondary education (Worden, 1990; Buckley, 1991; Krimmel, 1996). However, there has also been no clear demonstration that police organizations have created the conditions to ensure that highly educated police personnel are able to thrive and apply their knowledge, skills, and abilities without interference or opposition. What is capable of demonstration is that police personnel require continuous learning if the organizations they represent are going to survive in the face of significant complex change.

The approach that needs to be taken is not simple; it implies a whole new way of "looking" at the development of individuals within the organization (McKenna, 1995). There is certainly an opportunity for human resource developers in this new paradigm, as expressed by Watkins and Marsick (1993):

> Human resource developers will be asked to take on new roles in the learning organization. Human resource developers have often functioned as trainers who were working against time to ensure that the workforce had the requisite skills. They are not being asked to drop this challenge. But they are being asked to in addition, help the workforce face an unknown future that is changing at unbelievable rates in an unpredictable fashion . . . Human resource developers can extend their impact in [an] almost exponential fashion by working with all employees to make self-managed learning more effective and by creating ways for people to share what they learn. (p. 21)

Another view asserts that for organizational learning to take shape in a real and profound way, it must be driven by individuals *outside* of the training or human resource departments. It is something too important to entrust to the human resource realm, with its preference for forms, procedures, conventions, and rigidity (Driver, 1985).

As a point of reference, the following Canadian police organizations have made serious efforts aimed at embedding continuous learning into their strategic plans:

- Calgary Police Service;
- Royal Canadian Mounted Police; and
- Ontario Provincial Police.

Calgary Police Service

In Calgary, the arrival of a new chief of police in 1995 brought about innovative and fundamental changes in the organizational structure, as well as the department's approach to learning. There was an initial interest in the Massachusetts Criminal Justice Training Council and its "values-based approach to police education." This was seen as a useful model to assist the Calgary Police Service in moving away from the traditional, military model with its heavy reliance on command and control.

Mr. Howard Liebowitz was engaged by the Massachusetts Criminal Justice Training Council to develop a model that emphasized community-based values and citizen input, i.e., a model directly linked to the principles of community policing. The program resulting from Mr. Liebowitz's research was a twenty-two week training curriculum. Central to this curriculum was the identification of underlying values upon which learning would be based. The model has four key organizing principles:

- *Integrity*—includes the need to operationalize integrity in every situation and a concentration on ethics and character;
- *Law*—includes an understanding of the rule of law, and the philosophical foundations of law;
- *Fitness*—with an emphasis on wellness as opposed to endurance, including the consideration of a "fitness for life" contract; and
- *Community & Neighbourhood Policing*—with continuous reference to the police role as being guided by the expectations of the community.

The basic curriculum that emerged from the research done for the Massachusetts Criminal Justice Training Council is presented below:

FIGURE 7-3

A Values-Driven Educational Model of Police Training

The Calgary Police Service built upon this research and proceeded to develop their own competency-based human resource management system, based upon extensive consultations. These competencies are to be used in recruitment, developmental assessment, lateral transfer promotions, and planning (Ko, 1997; Juschka, 1998).

The Calgary Police Service organized all of their formal learning under five key disciplines:

- Police science;
- Law;
- Social science;
- Management/administrative studies; and
- Health education.

Learning in the Calgary Police Service is linked to every stage of an employee's career and, therefore, promotes lifelong interest in continuous growth. Additional learning initiatives include the development of university programs, the enhancement of a "learning lab" for testing, distance learning, the implementation of computer-based training, and computerized arms training for the assessment of judgment in realistic use of force scenarios (Jushka, 1998). The process of transforming the learning culture is a long-term investment. However, the Calgary Police Service appears to be making considerable headway as a Canadian leader in the area of police learning.

Royal Canadian Mounted Police

As pioneers in the application of the systems approach to training (S.A.T.) outlined earlier in this chapter, the RCMP has long been concerned about the quality and effectiveness of its educational programs (Public Service Commission of Canada, 1984). The leadership shown by the Canadian Police College has been significant throughout Canada, and many police departments in other jurisdictions have sent their police personnel to the College in order to gain the benefits of its extensive course offerings. However, consistent with a clearly identified need to transform their approach to learning in order to support innovative policing, the RCMP has established its own model for a "learning organization."

To a large extent, the RCMP has adopted much of the thinking expressed by Peter Senge and his colleagues (Senge et al., 1993). There was a clear recognition within the RCMP that there needed to be renewal in that organization's approach to training based on their vision of community policing, problem-solving policing, and an analysis of core competencies for organizational effectiveness (Himelfarb, 1995). In 1990, the RCMP began an integrated review of the following areas of training:

- Cadet training program;
- Field coach program;
- Management/executive training program; and
- In-service training program.

An initial step involved the development of a framework that would reflect the RCMP's mission and lead to a clearer understanding of core competencies. This involved broad consultation within the RCMP as well as an assessment of work done in other jurisdictions, especially the work undertaken by the Strategic Planning Committee on Police Training and Education in Ontario. The resulting CAPRA model developed through this process now guides all training within the RCMP and links with a new paradigm for organizational change, that sets out leadership responsibilities throughout the RCMP. The CAPRA model is presented below:

FIGURE 7–4

THE RCMP'S CAPRA MODEL

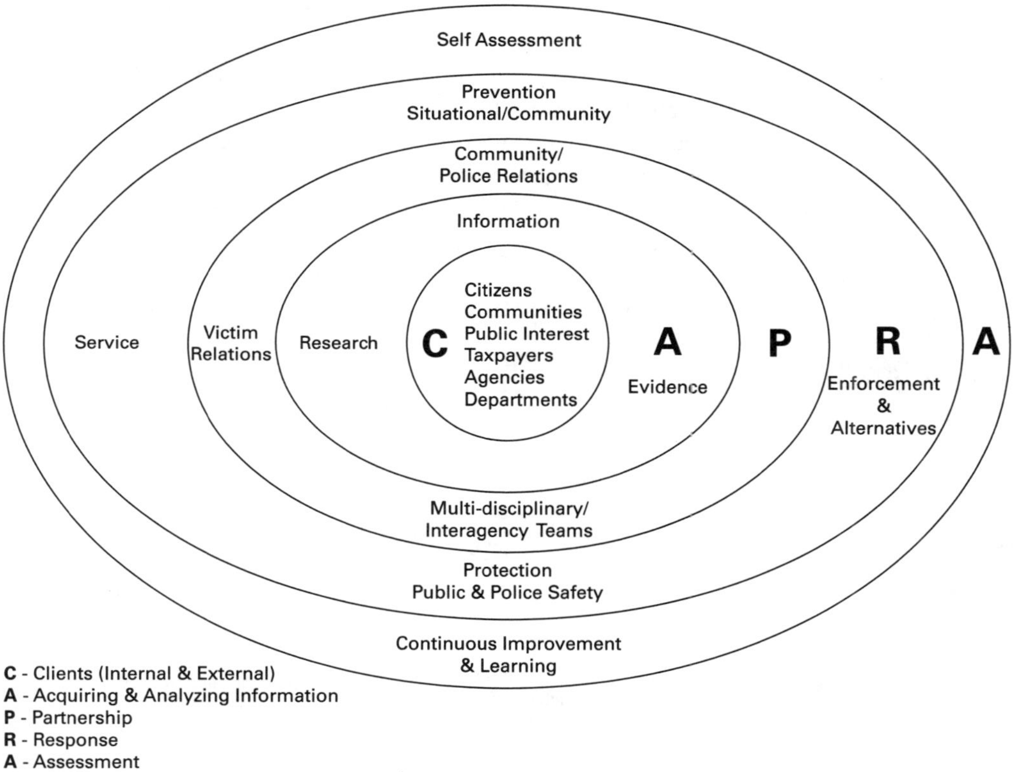

Adapted from: Himelfarb, Frum (1995). "Training and executive development in the RCMP." In *Management Challenges in 21st century policing conference proceedings, Ottawa, Canada, September 22/23/24, 1995*, [p. 217]. Reproduced with permission.

As noted by Himelfarb (1995):

> The reform of training in the RCMP is part of the larger transformation to a new philosophy of policing and a learning culture in which all members of the RCMP share responsibility for their own development and for the development of the organization . . . The core competencies which shape training throughout the RCMP are derived not from traditional task analysis or discipline-based approaches, but from an analysis of what constitutes quality service to the community—client-centred problem solving partnerships. (p. 216)

Ontario Provincial Police

The Ontario Provincial Police (OPP) has always been conscientious in its approach to the design and development of police training (Highley, 1984). In the 1980s, it incorporated the systems approach to training (S.A.T.) model developed by the RCMP and applied this methodology to the preparation of course training standards for all aspects of police education, from advanced law enforcement, through counterfeit investigations and explosives disposal, to supervisory management and administration. Good training has always been a hallmark of the success of the OPP.

However, following the groundbreaking work of the Strategic Planning Committee on Police Training and Education in 1992, and especially in the context of the OPP's major Organizational Review Project in 1994-95, the OPP began to seriously re-examine and reformulate its strategy with regard to police training and development (OPP, 1995). As one of the ten "businesses" identified as central to the OPP, and therefore, earmarked for the application of process reengineering, the OPP focused on "employee selection and competency development" as an area for major organizational improvement. In the overall context of community policing, the OPP realized that police training and education had to be enhanced. What resulted from the OPP's efforts was a full realization of the value of organizational learning as a model for the development of a police curriculum. It placed this understanding squarely within its overall business strategy for the future, as outlined below:

FIGURE 7–5

OPP's Business Strategy for the Future

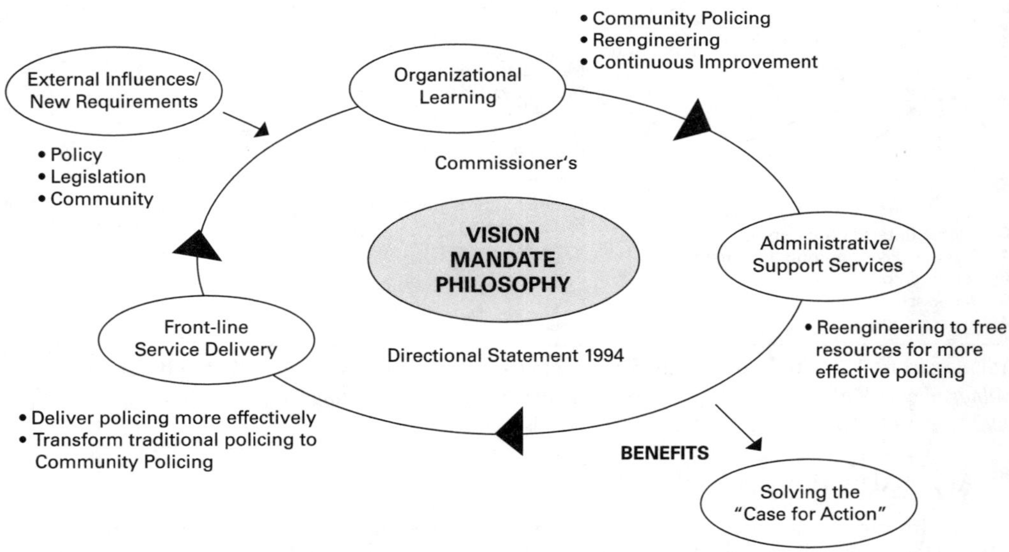

Source: Ontario Provincial Police (1995). *Organizational review: a process & model for change: focus on the future*. Toronto: The Queen's Printer for Ontario, p. 27. © 1995 Queen's Printer for Ontario. Reproduced with permission.

As a result of the OPP's investment in the positive value of organizational learning, it proceeded to apply relevant principles to the implementation phase that followed the successful conclusion of the Organizational Review Project. As part of the subsequent Organizational Renewal Project, the OPP approved and supported the development of a "Transition Through Learning" strategy. The overall goals of this strategy were:

- To develop and enhance senior management team's capabilities to skillfully lead the way toward the new vision and goals of the OPP; and
- To provide a set of programs which will begin to support and enhance the work of all OPP members in realizing the OPP's vision and goals and which will encourage continuous learning and positive adaptation to change.

The "Transition Through Learning" strategy recognized the need for deep insight and substantial assistance in developing skills and knowledge across the entire new organization. Building on the considerable wisdom already available within the Ontario Ministry of the Solicitor General and Correctional (1994), the senior management of the OPP approved the following series of learning modules:

FIGURE 7-6

OVERVIEW OF THE OPP'S "TRANSITION THROUGH LEARNING" STRATEGY

Learning Module # & Title	Level of Delivery			
	Executive Group	*Senior Management Group*	*Managers & Supervisors*	*Constables & Support Staff Group*
Module #1—Orientation to the "new" OPP & the Dynamics of Change	*	▲	▲	▲
Module #2 – Leadership Styles & Managing Change	*	▲	▲	▲
Module #3 – Addressing Organizational Priorities Through "Action Learning" Teams	*	*	* ▲	* ▲
Module #4 – Remote Management Technology & Methods	*	*	* ▲	* ▲
Module #5 – Best Practices (Benchmarking) & the Fundamentals of Research (Problem-solving)	*	* ▲	* ▲	*

* = Immediate, short-term implementation (i.e., 2 months to one year)

▲ = Long-term implementation (i.e., one + years)

The introduction of the "Transition Through Learning" strategy was an experiment in organizational learning within the reorganized OPP, and several important lessons were learned as a result of this undertaking (Richter and McKenna, 1995). From this point of departure, the OPP is attempting to reformulate its organizational learning strategy and continues to make efforts at embedding this paradigm within the overall approach to training, education, and development. It remains a challenge in this organization to fully integrate the principles of organizational learning into the everyday practice of uniform and civilian members. There is certainly a general understanding of the importance of Peter Senge's "disciplines" (i.e., systems thinking, personal mastery, mental models, shared vision, and team learning). However, the process of incorporating these "disciplines" and the collective consciousness that it represents is very difficult to achieve without continuous effort and high-level executive support.

The higher goal of moving to a knowledge organization is expressed extremely well in the work of several authors (Drucker, 1994; Davenport and Prusak, 1998). However, it is important to make a start at the level of understanding and acceptance appropriate to the organization and its current needs.

TEAM BUILDING AND TEAM LEARNING

A major component of the new paradigm organizational learning is the greater emphasis on teams (Dixon, 1993; McKenna, 1998, pp. 226-227). In the past, policing certainly did not ignore teamwork, indeed there was a great deal of emphasis on the effective "squad," the emergency (or tactical) response team, and other closely-knit groups of officers with highly complex patterns of performance in which allegiance and mutual trust were well-established. However, the typical police team has been a grouping of officers possessing a similar set of skills and abilities, tasked with a discrete function (e.g., underwater search and rescue, hold-up investigations, forensic identification). Teams were rarely composed of cross-functional members (both uniform and civilian) brought together to address organization-wide problems or challenges. Community policing has done a great deal to encourage a more broadly-defined approach to teams within police departments.

However, rewards within the typical Canadian police service continue to be granted on an individual basis. Promotions (probably *the* most significant events in a police officer's career) are normally based on highly competitive processes that pit one officer against another for a limited number of vacancies within that rank level. But in overall organizational terms, teams have become extremely important when we are looking for better ways of working (Katzenbach and Smith, 1993; Senge et al., 1994; Isgar, 1995; Bennis and Biederman, 1997). In order for teams to work effectively they must be developed properly. There is a considerable body of literature that now offers assistance and advice in this regard, much of it based on a shared understanding of the phases, or stages, that each team goes through during its life cycle (Phillips and Elledge, 1989; Elledge and Phillips, 1994). Holpp (1995, p. 56) provides an excellent overview of those phases of team development, represented in Figure 7–7, below:

FIGURE 7-7

TEAM DEVELOPMENT: WHO AM I AND WHO ARE YOU?

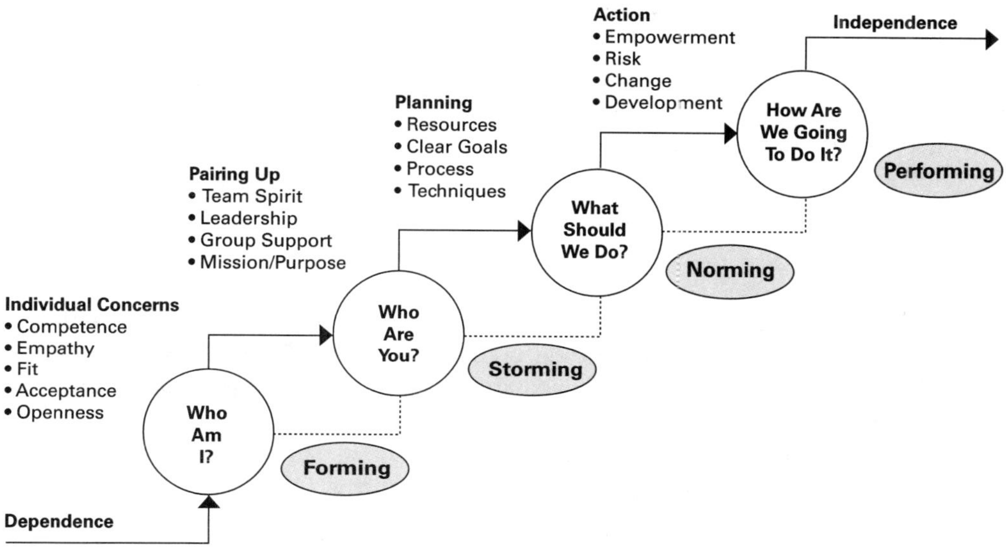

Adapted with permission from the March 1995 issue of *Training Magazine*. Copyright 1995. Lakewood Publications, Minneapolis, MN. All rights reserved. Not for resale.

The emphasis on team building and team learning may create new approaches unheard of before through unique combinations of talent, skill, perspective, and experience. The promise of team policing is considerable (De Paris, 1997). Again, it is necessary to remember that effective teams rarely come into being unassisted; they must be carefully constructed and deliberately supported at all levels of the organization. It must make sense to have teams, and the teams themselves must be able to make sense of what they are expected to do.

STAKEHOLDER EDUCATION

Because an important premise of community policing, as an authentic approach to service delivery, is the engagement of the several communities being served, it is worthwhile considering how various stakeholders may be educated with regard to this fundamental change in the philosophy and practice of policing in Canada.

It is essential that police departments transform their internal training functions, as seen in the promising efforts undertaken by the Calgary Police Service, the RCMP, and the OPP. However, this is not sufficient given the *external* nature of community policing and its underlying intention to create a new dynamic for public safety and security. One facet of the educational challenge of community policing is the need to facilitate understanding across the full range of the department's stakeholders. This includes, but is not limited to, the following categories:

- Politicians;
- Civilian governing authorities;
- Citizens;
- Police executives;
- Middle managers; and
- Front-line officers.

POLITICIANS

It is contrary to common sense to suggest that policing is, or even should be, completely apolitical. The Greek philosopher, Aristotle, noted that man is, by nature, a political animal. Police departments everywhere in Canada are created by law to uphold the law, and in order to do this they are guided by several streams of legislation and regulation that flow from the federal Parliament, or from provincial legislatures, that are led by elected politicians. Policing, therefore, is inescapably a political undertaking. Of course, it is essential that policing be held at arm's-length from what may be referred to as partisan politics. We do not expect, and would not condone, the RCMP to do the bidding of the Prime Minister for purely party purposes. The Canadian public would not tolerate the perception, not uncommon in earlier days, that provincial police forces were being used as strike-breakers on behalf of the party in power. In order to ensure that politicians comprehend the thrust, and limitations, of community policing it is essential that they be schooled in the philosophy and purpose of this approach. This may not be an easy task given the reluctance of politicians to deal with experimentation and the particularly intense scrutiny that police departments are placed under (Mastrofski, 1998).

CIVILIAN GOVERNING AUTHORITIES

In all Canadian jurisdictions, police departments report to some form of civilian governing authority (Stenning, 1981; Hann et al., 1985). In many instances this body is referred to as a police commission, or, a police services board. In other jurisdictions, the police are directly accountable to an elected official (e.g., a Minister of Justice, or a Solicitor General). Regardless of the specific local arrangements, it is critical that the governing authority be well-grounded in the principles of community policing adhered to within the police department. The nature of their understanding may be focused on the strategic (i.e., high-level) application of the preferred community policing model. However, it is important that these individuals, essentially the "board of directors" of the police service, have a sound grasp of the department's approach and be aware of the operational and administrative results expected from the implementation of community policing principles.

CITIZENS

In order for people to know that something different is taking place in the delivery of policing services there must be some mechanism in place for informing the public about the intent and substance of community policing as a new approach (McGahan, 1992). This can happen in a number of ways:

- Public "town hall" meetings;
- Publications outlining the police department's philosophy of community policing;
- Public surveys that include an information and educational component;
- Local cable programs featuring community policing topics and speakers;
- Citizen Police Academies in cooperation with local community colleges or through the police department;
- Service club presentations by members of the police service, the police governing authority dealing with community policing; and
- Feature on community policing initiatives and topics in the local newspapers.

POLICE EXECUTIVES

The people who are in executive positions in Canadian policing today have, in many instances, been operating out of what may be termed the "professional" model of policing. Frequently they will be individuals who have "come up through the ranks" and have tailored their careers within the context of a strict, hierarchical, paramilitary structure. They have had their organizational habits formed in departments that placed the highest value on obedience to the chain of command, discipline, and technical competence. Many of those police executives have willingly and eagerly made the transition to the community policing paradigm, with its emphasis on proactive problem-solving and community-based partnerships. Indeed, if not for those innovative police executives who seized on the need for a new approach to the whole policing enterprise, it is doubtful that community policing, in any form, would have secured a foothold in Canadian policing.

Regardless of the approach being taken to community policing, it is essential that the layer of police executives be supportive of the philosophy and its implementation. This implies that the organization will make an effort to ensure that police executives comprehend the nature of the change and have an opportunity to internalize its impact on their working life. If the "top" layer of a police organization does not have a deep understanding of community policing it will not succeed in transforming the organization. If this level of the department does not in fact champion the cause of community policing it will be extremely difficult for this approach to be sustained in ways that will embed the philosophy in the organization and practices over time. Accordingly, efforts must be made to promote executive learning about community policing. This can be accomplished through workshops, newsletters, guest speakers, and site visits to other departments applying the principles of community policing, all geared to the learning needs of these police executives.

MIDDLE MANAGERS

The importance of the police middle manager has already been addressed in Chapter Five. Because of the positive role they can play in making meaningful change occur in police departments, it is essential that educational efforts be made to inform these members about community policing as it is applied within their organization. Of course, the ideal circumstances would be to enlist their efforts in arriving at the "right" understanding of community policing within that jurisdiction. The effectiveness of having broad involvement throughout the police service in creating the approach to community policing cannot be overstated.

FRONT-LINE OFFICERS

Education in this arena needs to emphasize the practical value of community policing as an approach to the demands of police work. It is not enough to present the grand vision of community policing and expect that front-line service delivery personnel will accept responsibility for its implementation. There needs to be a planned process for engaging the entire organization in the delivery of this philosophy as a functional way of doing business. It must be meaningful to front-line officers, as well as to other police personnel, who support the front-line services. Everyone in the organization must understand why, and how, this new approach is being undertaken.

LEADER IN COMMUNITY POLICING

Ross Hastings, Professor, Faculty of Social Sciences, University of Ottawa

What is your understanding of "community policing" and its importance to the evolution of policing in Canada?

There is a great divergence of views on what constitutes the philosophy and the practice of community policing. At their heart, though, almost all versions of community policing share three common features:

- A commitment to a proactive approach, and a recognition of the importance of the prevention of crime and victimization as a counterbalance to reactive law enforcement. Prevention holds the promise of demand reduction in an era of competition for limited resources;
- A recognition of the need to include a wide array of partners in the mission of public safety and security. The fact of the matter is that the inclusion of prevention in the

Ross Hastings LEADER IN COMMUNITY POLICING continued

police mission broadens the work beyond the legal mandate and the resources of police organizations, and requires that others become involved. The key to success in this area is the willingness of the police to share or relinquish control over problem-solving partnerships;
- A recognition of the responsibility of the community to become involved in assuring public safety and security, and an awareness of the right of the community to exercise some control over decisions relating to problem definition, the selection of priorities and the design, implementation, and evaluation of solutions.

In recent years, community policing has become the rallying cry for police organizations around the country. Unfortunately, there are mixed motives for this enthusiasm. The good news is that progressive police leaders have been open to exploring and implementing ideas which have proven successful in helping communities cope with their crime and victimization problems. The bad news is that, too often, the enthusiasm for the philosophy has been little more than a mask for an attempt to recruit the public as support workers in the delivery of traditional police services. This is little more than a new way of delivering the types of police services whose limits have been recognized.

Outline your own contribution to the development of "community policing" in Canada.

The major focus of my work has been on the development of a typology of the different and competing approaches to community policing. I argue that there are significant variations in the goals pursued, the tactics used, the organizational changes required, and the types of individuals modern police organizations must recruit and train. The challenge is to combine these different approaches in a manner that maximizes the benefits of each, while recognizing that they are somewhat contradictory. The reality is that community policing cannot be an add-on—police leaders will have to face the responsibility of directing their organizations and their communities in difficult choices about what they hope to accomplish, and about the prices they are willing to pay to do so.

I have had the privilege of being involved in four major "change projects" in recent years. The first was the Strategic Planning Committee on Police Training and Education (Ontario, 1991–1992), which looked at the question of how to prepare police officers for the challenge of the new orientation to policing. The other three were major reorganization projects in large police organizations: the Ontario Provincial Police, the Service de Police de la Communauté Urbaine de Montréal, and the Ottawa-Carleton Regional Police Service. In each of these, my role was to try to contribute to the design of alternative models to guide decisions about organizational change.

Ross Hastings LEADER IN COMMUNITY POLICING continued

What is your view of the future of "community policing" in Canada?

In my view, the major challenge at this point in time is to come to grips with the problem of translating the philosophy of community policing into clear job descriptions and sets of expectations for all individuals in a police organization. It is one thing to commit to community policing. It is another to be able to explain to police officers in clear terms exactly how their workday will change, and exactly how they will be evaluated in the future. Until we can do this, and until we can align recruitment, evaluation, and promotion procedures with these new expectations, we have no right to expect the front-line to replace cynicism with enthusiasm.

THE ETHICAL CHALLENGE OF COMMUNITY POLICING

The question of ethics has always been an important one for policing in Canada. There has been an effort within the Canadian police leadership to avoid the scandals and corruption frequently associated with American police departments, particularly during the "political" era. While Canadian police have certainly not been free of scandal, or charges of unethical behaviour, the record of performance is reasonably sound.

Along with the introduction of community policing, there has been renewed interest in matters of integrity, ethics, and proper police conduct. Such an interest in the ethical dimension of policing seems entirely consistent with the general spirit of community policing initiatives. Because community policing brings individual officers into closer contact with members of various communities, it becomes terribly important that those officers have the character and integrity necessary to function in such close quarters with the problems and complexities of any given community. Because community policing endeavours to give greater levels of empowerment to police officers in order for them to be more flexible in their response to problems, it is vital that this empowerment is not abused or misdirected. Finally, because community policing often requires officers to operate directly with various groups in any given community, over an extended period of time (perhaps for several years), it is critical that these officers do not become so closely identified and attached to a particular group within the community that it impairs their ability to deliver services in an impartial and equitable manner.

Since close ties to the community characterized the "political" era of policing, and are often seen as the source of significant corruption within police departments, community policing as an approach had to overcome the stigma that attached as a result of this kind of perception. There are many influences that impact on an individual officer's ethical world, and they may be categorized into three key groups:

- Personal forces;
- Organizational or departmental forces; and
- External or environmental forces.

Figure 7-8, below, provides an excellent perspective on these various forces:

FIGURE 7-8

DYNAMICS OF POLICE INTEGRITY

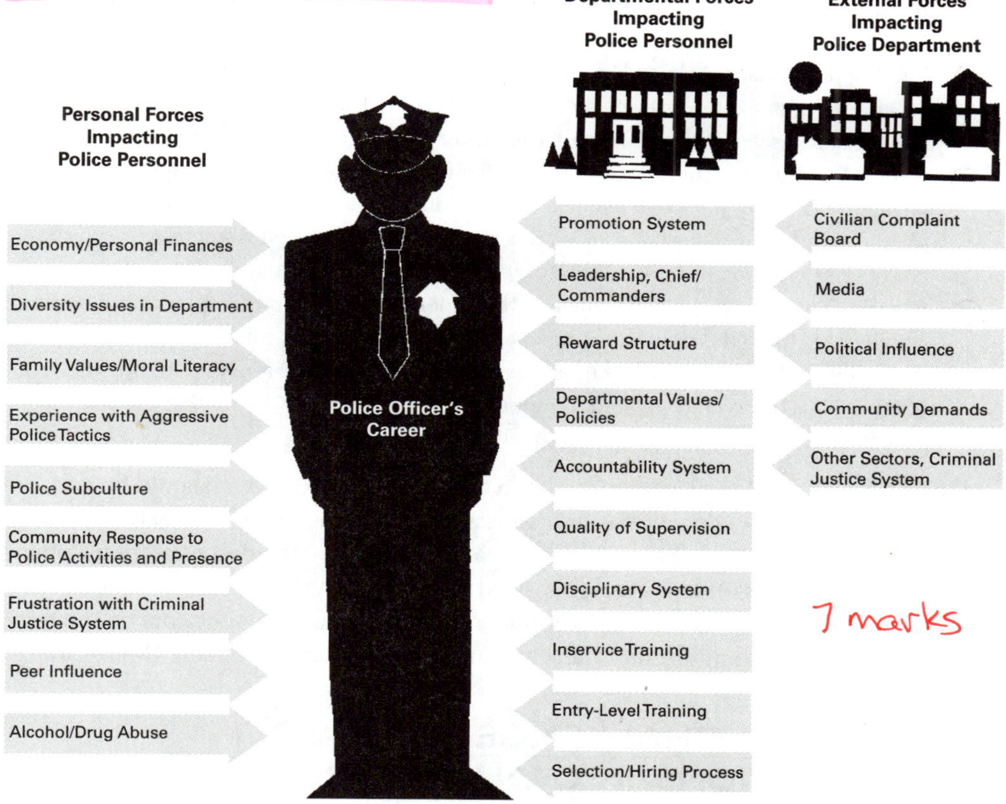

Graphic: Jennifer S. Novecio
Adapted from: Gaffigan, Stephen J. and Phyllis P. McDonald (1997). *Police Integrity: public service with honor; a partnership between the National Institute of Justice and the Office of Community Oriented Policing Services*. Washington, D.C.: U.S. Dept. of Justice. p. 92. Used with permission.

In attempting to identify the "core virtues" required in a police officer, Vicchio (1997) lists the following:

- *Prudence.* Practical wisdom, the virtue of deliberation and discernment. The ability to unscramble apparent conflicts between virtues while deciding what action (or refraining from action) is best in a given situation.

- *Trust.* This virtue is entailed by the three primary relationships of the police officer: the citizen-officer relationship, the officer-officer relationship, and the officer-supervisor relationship. Trust ought to engender loyalty and truthfulness in these three contexts.
- *Effacement of self-interests.* Given the "exploitability" of citizens, self-effacement is important. Without it citizens can become a means for advancing goals of the department other than those to protect and serve.
- *Courage.* As Aristotle suggests, this virtue is a golden mean between two extremes: cowardice and foolhardiness. There are many professions—surgery and police work, to name two—where the difference between courage and foolhardiness is extremely important.
- *Intellectual honesty.* Acknowledging when one does not know something and being humble enough to admit ignorance is an important virtue in any professional context. The lack of this virtue in police work can be very dangerous.
- *Justice.* We normally think of justice as giving the individual what he or she is due. But taking the virtue of justice in a police context sometimes requires the removal of justice's blindfold and adjusting what is owed to a particular citizen, even when those needs do not fit the definition of what is strictly owed.
- *Responsibility.* Again Aristotle suggests that a person who exhibits responsibility is one who intends to do the right thing, has a clear understanding of what the right thing is, and is fully cognizant of other alternatives that might be taken. More importantly, a person of integrity is one who does not attempt to evade responsibility by finding excuses for poor performance or bad judgment. (p. 15)

Given the importance placed on this area of policing in Canada, it should be no wonder that there is a major educational, or organizational learning, component attached to its successful implementation. Many police departments have included courses on police ethics in their course calendars. Workshops and seminars with the theme of "police ethics" sponsored by police departments are not uncommon. Several commentators tie the topic of ethics very closely to any attempts at achieving a "profession" of policing (Marin, 1991).

THE SPECIAL CHALLENGE OF RACE RELATIONS IN CANADIAN POLICING

Canada has never experienced the intense racial or ethnic unrest found in the worst moments of American history. There have never been flashpoints like the aftermath of the trial of police officers involved in the beating of Rodney King. However, police organizations in Canada have not been complacent simply because they have been spared this kind of confrontation. They have been challenged to improve their sensitivity and awareness of racial and ethnic issues. They have also been prompted to improve their efforts at recruiting, educating, and retaining police officers that represent the broadest spectrum of the Canadian population. By and large, they have made some good beginnings in these directions.

Community policing as a philosophy of inclusion, has driven much of this improvement in race relations. In the Canadian context, there are problems that earlier styles of policing did not adequately address, particularly with regard to visible ethnic minorities. Community policing rests on an assumption that communities deserve, indeed are owed, the best policing available. As a theoretical model it does not discriminate among communities. However, the history of Canada, let alone Canadian policing, is not devoid of examples of racism, both official and systemic (Walker, 1997).

There have been instances in Canadian policing in which questions have been raised about the way in which police have handled an investigation, a victim, or a suspect. Often these questions focus on issues of race and unequal treatment. In Canada, it is frequently a matter concerning Aboriginal people and the concern over systemic racism in our criminal justice system (Sunahara, 1992).

A great deal has been accomplished as a result of the focus on race relations as a special feature of community policing. The downside of the "professional" era of policing, that placed an emphasis on rapid response and highly specialized operations, was an estrangement from the various communities served. When the demographics of these communities began to transform in fairly dramatic ways with respect to ethnic and racial composition, the police were often ill-equipped to understand the citizens they were attempting to serve and protect. This included an inability to even communicate with certain segments of the community because of language and other cultural barriers.

The issue of race relations has accelerated much of the progress made on the community policing front. The self-interest argument has been made that police officers require the skills necessary for dealing with a variety of racial and ethnic backgrounds in order to properly function "on the street." From an organizational perspective, its stands to reason that the police department will serve all of its communities, without fear or favour.

The most intense period of activity around race relations and policing can be found between 1975 and 1995 in Ontario, during which time there were not less than seven Royal Commissions, task forces, or special studies conducted on such matters. In 1989, as a result of serious concerns raised on behalf of the visible minority communities in that province, a task force was established to examine police training, policies, practices, and attitudes. One of the strongest precipitating factors for the creation of the task force was the fatal shooting of Michael Wade Lawson. The Race Relations and Policing Task Force spent a considerable amount of time and energy examining the key issues related to its "terms of reference" and conducted exhaustive public hearings in order to arrive at its recommendations. The main issues addressed by the Task Force were as follows:

- *Monitoring*—the need for establishing a specialized system to independently monitor the issues involved in race relations and policing ;
- *Hiring and promotion*—addressing the need for specific programs to recruit more broadly from within the visible minority community and to promote visible minority police personnel in a more equitable manner ;
- *Race relations training*—improve the state of existing race relations training in police departments and establish a sound training foundation for the design and delivery of all levels of police training to ensure that they are appropriate throughout the system;

- *Use of force*—addressing the need to collect statistics on the police use of force in order to identify deficiencies in police training and to improve the relationship between the police and minority groups;
- *Community relations*—developing approaches that emphasize community-based policing principles. Work at breaking down the barriers that separate the police from the visible minority communities and move toward more partnerships. Addressing the need for a "new breed" of police officer who is willing to work on a more cooperative basis;
- *Public complaints*—emphasize the need for a publicly credible, accountable and independent civilian mechanism for public complaints in order to respond to allegations of racial intolerance and other police misconduct;
- *First Nations peoples*—while not a visible minority in Canada, the Task Force acknowledged the importance of the broader issue of a separate justice system for Aboriginal people. Encouraged the creation of a tri-partite (federal government, Ontario government, and First Nations representatives) task force to study the feasibility of native justice system;
- *Police commissions*—emphasis on the accountability of civilian governing authorities to the communities they represent and the need to be situated in non-police facilities with non-police staff to fulfill their statutory obligations; and
- *Police associations*—encourage these representative groups to become agents for positive change in the area of race relations and policing. They cannot defend the *status quo* without diminishing the quality of policing and destroying claims to professionalism claimed by association leaders.

The Task Force was convinced, on the strength of its hearings, that a problem did exist in the area of race relations and policing. Accordingly, they offered the Ontario Solicitor General fifty-seven recommendations aimed at improving the quality of policing in the province. The Task Force (1989) encouraged immediate change and observed that:

> Police must understand that their reputation will not founder nor will police forces be brought into disrepute if they accept criticism as helpful in eradicating problems. Public approval will be the result of visible and credible efforts by police to assess the nature and degree of racism in police forces. If the police are seen to eliminate organizational factors which perpetuate prejudice and to punish discriminatory behaviour they will be assured of public confidence. (p. 25)

SUMMARY

This chapter has examined several of the challenges raised by the philosophy of community policing. The educational challenge has been addressed by several police organizations in Canada, and a considerable amount of research has been undertaken to determine the framework for a police learning system that will address the future requirements of police personnel at various levels.

The ethical challenge of policing in the context of community policing has also been considered in some detail. The impact of this dimension is far-reaching, touching as it does recruitment, training, education, promotion, and other areas of police management and operation.

Finally, this chapter has looked at the issue of race relations and policing. The need to address the broadest range of neighbourhood problems requires a degree of cultural sensitivity that challenges police departments to develop detailed policies and programs that will address the concerns raised through several race relations studies.

QUESTIONS FOR CONSIDERATION

1. Why is a "system" of police learning important for the successful introduction of community policing in Canada?
2. What are some important differences between police "training" and police "learning"?
3. What are the components of the systems approach to training?
4. What are the four guiding principles of a learning system for police leaders identified by Hastings?
5. Who are the stakeholders that require varying degrees of exposure to the police learning system?
6. What are the key virtues identified for police officers in this chapter?
7. What are the main forces that impact on a police officer's ethical situation?
8. What is the connection between education and ethics in Canadian policing?
9. How do race relations impact on policing in Canada?
10. What are some of the major race relations issues that have confronted police organizations in Canada?
11. How do race relations in policing impact on the development of police training and education programs?

REFERENCES

Abbey-Livingston, Diane and David Kelleher (1988). *Managing for learning in organizations : the fundamentals*. Toronto: Ontario Ministry of Tourism and Recreation.

Ackoff, Russell L. (1981). *Creating the corporate future : plan or be planned for*. New York: John Wiley & Sons.

Alleyne, Noreen and Ron Hoffman (1992). *High impact learning methodologies*. Toronto: Ontario Ministry of the Solicitor General, Strategic Planning Committee on Police Training and Education.

Argyris, Chris (1964). *Integrating the individual and the organization*. New York: John Wiley & Sons.

Argyris, Chris (1965). *Organization and innovation*. Homewood, Ill.: Richard D. Irwin & the Dorsey Press.

Argyris, Chris (1977). "Double loop learning in organizations." *Harvard Business Review*, Vol. 55, no. 9.

Argyris, Chris (1990). *Overcoming organizational defenses: facilitating organizational learning*. Boston: Allyn and Bacon.

Argyris, Chris (1991). "Teaching smart people how to learn." *Harvard Business Review*, Vol. 69, no. 5.

Argyris, Chris (1992). *On organizational learning*. Cambridge, Mass.: Blackwell Publishing.

Argyris, Chris (1993). *Knowledge for action: a guide to overcoming barriers to organizational change*. San Francisco, Calif.: Jossey-Bass Publishers.

Argyris, Chris and Schon, Donald A. (1978). *Organizational learning: a theory of action perspective*. Reading, Mass.: Addison-Wesley Publishing Co.

Bennis, Warren and B. Nanus (1985). "Organizational learning and management of the collective self." *New Management*, Vol. 3.

Bennis, Warren and Patricia Ward (1997). *Organizing genius: the secrets of creative collaboration*. New York: Addison-Wesley Publishing.

Booth, Ralph and Wes Gerrish (1994). "Performance improvement through team alignment: group decision making to focus project objectives." Paper presented at the 1994 Annual conference of the National Society of Performance and Instruction, San Francisco, April 4–8, 1994.

Brown, John Seeley (1991). "Research that reinvents the corporation." *Harvard Business Review*, Vol. 69, no. 1, pp. 102–111.

Brown, John Seeley (1991a). "Organizational learning and communities of practice: toward a unified view of working, learning and innovation." *Organization Science*, Vol. 2, no. 1, pp. 40–56.

Buckley, Leslie Brian (1991). "Attitudes toward higher education among mid-career police officers." *Canadian Police College Journal*, Vol. 15, no. 4, pp. 257–274.

Canadian Association of Chiefs of Police (1992). *Police race relations: raising your effectiveness in today's diverse neighbourhoods through community policing*. Ottawa: CACP.

Canadian Centre for Management Development (1994). *Continuous learning: a CCMD report*. Ottawa: Canadian Centre for Management Development. (CCMD Report No. 1).

Capra, F. (1988). *Uncommon wisdom: conversations with remarkable people*. Toronto: Bantam Books.

Cohen, Howard S. and Michael Feldberg (1991). *Power and restraint: the moral dimension of police work*. New York: Praeger.

Cook, Janice A., Derek Staniforth, and Jack Stewart (eds.) (1997). *The learning organization in the public services*. Aldershot, Hampshire: Gower.

Dalla Costa, John (1995). *Working wisdom: the ultimate value of the new economy*. Toronto: Stoddart.

Davenport, Thomas H. and Laurence Prusak (1998). *Working knowledge: how organizations manage what they know*. Boston: Harvard University Press.

De Geus, A.P. (1988). "Planning as learning." *Harvard Business Review*, Vol. 66, no. 3.

De Paris, Richard J. (1997). *Neighborhood team policing: organizational opportunities and obstacles*. Washington, D.C.: Police Executive Research Forum.

Delattre, Edwin J. (1994). *Character and cops: ethics in policing. 2nd ed*. Washington, D.C.: AEI Press.

Delattre, Edwin J. (1998). "Leadership: a position? An activity? And . . . ?: ethics in policing." Reprint of a paper delivered at the *Leadership* '97 conference sponsored by the Police Leadership Forum, November 6, 1997, in Sydney, Nova Scotia. Ottawa: The Police Leadership Forum. (Occasional papers collection).

Dixon, George (1988). *What works at work: lessons from the masters*. Minneapolis, Minn.: Lakewood Books.

Dixon, Nancy M. (1993). *Organizational learning*. Ottawa : Conference Board of Canada. (Report 111-93).

Dixon, Nancy M. (1999). *The organizational learning cycle: how we can learn collectively. 2nd ed.* Brookfield, Vt.: Gower.

Dodge, R. Bruce (1991). *Learning in an organizational setting: the public service context*. Ottawa: Canadian Centre for Management Development.

Driver, Michael J. (1985). "Response and commentary." In Lawler III, Edward E. et al. (eds.) *Doing research that is useful for theory and practice*. San Francisco: Jossey-Bass.

Drucker, Peter (1994). "The age of social transformation." *The Atlantic Monthly*, Vol. 274, no. 5 (November) pp., 53–80.

Elledge, Robin L. and Steven L. Phillips (1994). *Team building for the future : beyond the basics*. Toronto: Pfeiffer & Co.

Elliston, Frederick A. and Michael Feldberg (eds.). *Moral issues in police work*. Totawa, N.J.: Rowman & Allanheld.

Ellul, Jacques (1964). *The technological society*. Translated by J. Wilkinson. New York: Knopf.

Ellul, Jacques (1990). *The technological bluff*. Grand Rapids, Mich.: William B. Eerdmans Publishing Co.

Gaffigan, Stephen J. and Phyllis F. McDonald (1997). *Police integrity: public service with honor: a partnership between the National Institute of Justice and the Office of Community Oriented Policing*. Washington, D.C.: National Institute of Justice.

Garvin, David A. (1993). "Building the learning organization." *Harvard Business Review*, Vol. 71, no. 4, pp. 78–91.

Geller, William A. and Hans Toch (eds.) (1995). *And justice for all : understanding and controlling police abuse of force*. Washington, D.C.: Police Executive Research Forum.

Gow, James Iain (1994). *Learning from others: administrative innovations among Canadian governments.* Toronto: Institute of Public Administration of Canada. (Monographs on Canadian Public Administration; No.16).

Greater Toronto Regional Working Group on Policing in Multicultural, Multiracial Urban Communities (1988). *Guidelines for police/minority community relations and liaison*. [Toronto]: Greater Toronto Regional Working Group on Policing in Multicultural, Multiracial Urban Communities.

Hann, Robert G. et al. (1985). "Municipal police governance and accountability in Canada: an empirical study." *Canadian Police College Journal*, Vol. 9, no. 1, pp. 1–85.

Hastings, Ross (1993). "Leadership for community policing in Canada." Unpublished report prepared for the Canadian Police College (RCMP).

Heck, William P.(1990). *Creating a cadre of police leaders: a study of police executive development programs in the United States: a dissertation presented to the Faculty of the College of Criminal Justice, Sam Houston State University in partial fulfillment of the requirements for the degree of Doctor of Philosophy*. Huntsville, Texas: Sam Houston State University.

Hefferman, William C. and Timothy Stroup (eds.). *Police ethics: hard choices in law enforcement*. New York: The John Jay Press.

Hegelsen, Sally (1995). *The web of inclusion: a new architecture for building great organizations*. New York: Currency Doubleday.

Highley, Dahn D. (1984). *O.P.P.: the history of the Ontario Provincial Police Force*. Toronto: The Queen's Printer.

Himelfarb, Frum (1995). "Training and executive development in the RCMP." In *Management challenges in 21st century policing: conference proceedings*. Published proceedings of a conference held in Ottawa, Canada, sponsored by the Solicitor General of Canada, organized through the Centre for Police Management and Research, Staffordshire University (U.K.) and the Canadian Research Institute for Law and the Family, University of Calgary.

Holpp, Lawrence (1995). "If empowerment is so good, why does it hurt?" *Training*, Vol. 32, no. 3 (March) pp. 54–56.

Ingstrup, Ole (1995). *The strategic revolution in executive development : what does it mean for you and your organization?* Ottawa: Canadian Centre for Management Development.

Janis, Irving (1989). *Crucial decisions : leadership in policymaking and crisis management*. New York The Free Press.

Jushka, Barbara (1998). "Living a learning model." *Subject to Debate*, Vol. 12, nos.3 & 4 (March/April), pp. 6–8.

Katzenbach, Jon R. and Douglas K. Smith (1993). *The wisdom of teams: creating high-performance organization.* Boston: Harvard Business School Press.

Kets de Vries, Manfred F.R.(1980). *Organizational paradoxes : clinical approaches to management.* London: Tavistock Publications.

Klein, David A. (ed.) (1998). *The strategic management of intellectual capital*. Boston: Butterworth-Heinemann.

Kleinig, John (1990). "Teaching and learning police ethics: competing and complementary approaches." *Journal of Criminal Justice*, Vol. 18, no. 1 (January), pp. 1–18.

Kleinig, John (1996). *The ethics of policing*. New York: Cambridge University Press.

Kline, Peter and Bernard Saunders (1993). *Ten steps to a learning organization*. Arlington, Va.: Great Ocean Publishers.

Ko, Yvonne (1997). *An alignment of internal and external systems: a competency-based approach*. Calgary, Alberta: Calgary Police Service, unpublished.

Kofman, Fred and Peter M. Senge (1993). "Communities of commitment: the heart of learning organizations." *Organizational Dynamics*, Vol. 22, no. 2, pp. 5–22.

Kotter, John P. (1996). *Leading change*. Boston: Harvard Business School Press.

Krimmel, John T. (1996). "The performance of college-eductated police: a study of self-rated police performance measures." *American Journal of Police*, Vol. 15, no. 1, pp. 85–96.

Laiken, Marilyn (1992). *The evaluation of adult learning in the workplace: a review of current research and field applicatons*. Toronto: Ontario Ministry of the Solicitor General, Strategic Planning Committee on Police Training and Education.

Larson, Peter and Robert Mingie (1993). *Leadership for a changing world : developing executive capability : detailed findings*. Ottawa: Conference Board of Canada, 1993.

Lawler, Edward E. et al. (eds.) (1985). *Doing research that is useful for theory and practice*. San Francisco: Jossey-Bass.

Leeuw, Frans L., Ray C. Rist, and Richard C. Sonnichsen (1994). *Can governments learn?: comparative perspectives on evaluation and organizational learning.* New Brunswick, N.J.: Transaction Publishers.

Mancuso, Maureen et al. (1998). *A question of ethics: Canadians speak out*. Toronto: Oxford University Press.

Marin, René J. (1991). "Professionalism and ethics in policing." *Canadian Police College Journal*, Vol. 15, no. 4, pp. 291–309.

Mastrofski, Stephen D. (1998). "Community policing and police organization structure." In Brodeur, Jean-Paul (ed.). *How to recognize good policing: problems and issues*. Thousand Oaks, Calif.: Sage Publications and Police Executive Research Forum.

McGahan, Peter (1992). "Public awareness of policing initiatives." *Canadian Police College Journal*, Vol. 16, no. 1, pp. 24–46.

McKenna, Paul F. (1995). "Policing in 2020 & the application of organizational learning: grinding new lenses to view the development of human resources: a prolegomenon." Paper prepared for the Society of Police Futurists International, undelivered.

McKenna, Paul F. and Donald G. Evans (1995). "Leadership as learning: learning as leadership." *Police Governor*, Vol. 5, no. 2 (Fall/Winter), pp. 13–18.

Mager, Robert F. (1988). *Making instruction work : or, skillbloomers*. Belmont, Calif.: Lake Books.

Morgan, Gareth (1993). *Imaginization: the art of creative management*. Newbury Park, Calif.: Sage Publications.

Nadler, David A., Robert B. Shaw, A. Elise Walton and associates (1995). *Discontinuous change: leading organizational transformation*. San Francisco, Calif.: Jossey-Bass Publishers.

Normandeau, André and Barry Leighton (1990). *A vision of the future of policing in Canada: police-challenge 2000*. Ottawa: Police and Security Branch, Solicitor General Canada.

Oettmeier, Timothy N. (1992). "Matching structure to objectives." In Hoover, Larry T. (ed.). *Police management: issues & perspectives*. Washington, D.C.: Police Executive Research Forum.

Ontario. Ministry of Municipal Affairs. Subsidies Management Branch (1993). *On becoming a learning organization*. Toronto: Queen's Printer.

Ontario. Ministry of the Solicitor General. Strategic Planning Committee on Police Training and Education (1992). *A police learning system for Ontario: final report and recommendations*. Toronto: Ministry of the Solicitor General.

Ontario. Ministry of the Solicitor General and Correctional Services (1994). *Learning for change: report of the staff training and recruitment review*. Toronto : MSGCS.

Ontario. Ministry of the Solicitor General and Correctional Services. Police-Race Relations Monitoring Board (1996). *Good beginnings: a catalogue of race relations initiatives in Ontario*. Toronto: Police-Race Relations Monitoring Board.

Ontario. Race Relations and Policing Task Force (1989). The report of the Race Relations and Policing Task Force. [Toronto]: The Task Force.

Ontario. Race Relations and Policing Task Force (1992). The report of the Race Relations and Policing Task Force. [Toronto]: The Task Force.

Ontario Provincial Police (1995). *Organizational review: a process & model for change*. Toronto: Queen's Printer.

Pearn, Michael et al. (1994). "Becoming a learning organization: how to as well as why." In Burgoyne, John, Mike Pedlar and Tom Boydell. *Towards the learning company*. London: McGraw-Hill.

Peak, Kenneth J. and Ronald W. Glensor (1996). *Community policing & problem solving: strategies & practices*. Upper Saddle River, N.J.: Prentice Hall.

Pedler, M., J. Burgoyne, and T. Boydell (1997). *The learning company: a strategy for sustainable development. 2nd ed.* London: McGraw-Hill.

Phillips, Steven L and Robin L. Elledge (1989). *The team-building source book*. San Diego, Calif.: University Associates, Inc.

Public Service Commission of Canada. Staff Development Branch (1984). *Systems approach to training. Rev. ed.* [Ottawa]: Public Service Commission of Canada.

Richter, Ingrid and Paul F. McKenna (1995). "Organizational learning and policing: paradox or paradigm?" Unpublished paper presented at the joint American Society for Public Administration and the Institute of Public Administration of Canada meeting in Toronto, October 15–18, 1995.

Schein, E.H. (1985). *Organizational culture and leadership*. San Francisco: Jossey-Bass.

Seagrave, Jayne (1992). "Community policing and the need for police research skills training." *Canadian Police College Journal*, Vol. 16, no. 3, pp. 204–211.

Senge, Peter M. (1990). *The fifth discipline: the art & practice of the learning organization*. Toronto: Doubleday/Currency.

Senge, Peter M. et al. (1994). *The fifth discipline fieldbook: strategies and tools for building a learning organization*. Toronto: Doubleday/Currency.

Sims, Jr., Henry P., Dennis A. Gioia and associates (1986). *The thinking organization: dynamics of organizational social cognition*. San Francisco, Calif.: Jossey-Bass.

Snetsinger, Douglas (1993). *Learning leaders: perspectives from Canadian CEOs*. Toronto : Institute of Market Driven Quality, Faculty of Management, University of Toronto and Skill Dynamics Canada.

Sunahara, David F. (1992). "Public inquiries into policing." *Canadian Police College Journal*, Vol. 16, no. 2, pp. 135–156.

Toffler, Alvin & Heidi (1995). *Creating a new civilization: the politics of the third wave*. Atlanta: Turner Publishing Inc.

Vicchio, Stephen J. (1997). "Ethics and police integrity: some definitions and questions for study." In Gaffigan, Stephen J. and Phyllis F. McDonald (1997). *Police integrity: public service with honor: a partnership between the National Institute of Justice and the Office of Community Oriented Policing*. Washington, D.C.: National Institute of Justice, pp. 11–21.

Walker, James W. St. G. (1997). *"Race," rights and the law in the Supreme Court of Canada.* Toronto: The Osgoode Society for Canadian Legal History and Wilfrid Laurier University Press.

Watkins, Karen E. and Victoria J. Marsick (1993). *Sculpting the learning organization: lessons in the art and science of systemic change*. San Francisco, Calif.: Jossey-Bass Publishers.

Weick, Karl E. and Michael G. Bougon (1986). "Organizations as cognitive maps: charting ways to success and failure." In Sims, Jr., Henry P., Dennis A. Gioia and Associates. *The thinking organization: dynamics of organizational social cognition*. San Francisco, Calif.: Jossey-Bass.

Wheatley, Margaret J. (1994). *Leadership and the new science: learning about organization from an orderly universe*. San Francisco, Calif.: Berrett-Koehler Publishers, Ltd.

Worden, Robert E. (1990). "A badge and a baccalaureate: policies, hypotheses, and further evidence." *Justice Quarterly*, Vol. 7, no. 3 (September), pp. 565–592.

TOWARD MORE LEARNING

EDUCATIONAL CHALLENGE

It's difficult to know where to begin in suggesting additional resources for the study of the educational challenge of community policing. Certainly, one of the best starting points would be the final report of the Strategic Planning Committee on Police Training and Education, briefly discussed in this chapter. In addition to the report and recommendations of the committee, there are available a series of background research papers and studies prepared for this project that contain a wealth of useful and instructive insight. These items have been listed below and will provide a solid foundation upon which to build a better understanding of police learning in Canada.

Ontario. Ministry of the Solicitor General. Strategic Planning Committee on Police Training and Education (1992). *A police learning system for Ontario: final report and recommendations*. Toronto : Ministry of the Solicitor General.

Ontario. Ministry of the Solicitor General. Strategic Planning Committee on Police Training and Education (1992b). *Report on high impact learning methodologies*. Toronto: The Ministry.

Ontario. Ministry of the Solicitor General. Strategic Planning Committee on Police Training and Education (1992c). *Report on police training and education in other jurisdictions*. Toronto: The Ministry.

Ontario. Ministry of the Solicitor General. Strategic Planning Committee on Police Training and Education (1992d). *Report on consultations with the community on future policing and police training issues*. Toronto: The Ministry.

Ontario. Ministry of the Solicitor General. Strategic Planning Committee on Police Training and Education (1992e). *Report on Ontario police community initial consultation*. Toronto: The Ministry.

Ontario. Ministry of the Solicitor General. Strategic Planning Committee on Police Training and Education (1992f). *Report on private sector learning initiatives*. Toronto: The Ministry.

Ontario. Ministry of the Solicitor General. Strategic Planning Committee on Police Training and Education (1992g). *Report on strategic learning requirements for police personnel*. Toronto: The Ministry.

Ontario. Ministry of the Solicitor General. Strategic Planning Committee on Police Training and Education (1992h). *Report on the evaluation adult learning in the workplace*. Toronto: The Ministry.

Ontario. Ministry of the Solicitor General. Strategic Planning Committee on Police Training and Education (1992i). *Report on evaluating learning systems*. Toronto: The Ministry.

Ontario. Ministry of the Solicitor General. Strategic Planning Committee on Police Training and Education (1992j). *Report on consultation with the police community on police specialties in the future*. Toronto: The Ministry.

Ontario. Ministry of the Solicitor General. Strategic Planning Committee on Police Training and Education (1992k). *Report on future policing issues in Ontario*. Toronto: The Ministry.

Ontario. Ministry of the Solicitor General. Strategic Planning Committee on Police Training and Education (1992l). *Report on relationship between higher education and police learning requirements*. Toronto: The Ministry.

ETHICAL CHALLENGE

Ethics involves an understanding of right action, character, integrity, and self-discipline. It is a topic that is complex and particularly relevant in the context of community policing. There is much interest in this area of study in many disciplines (e.g., medicine, law, politics), and police organizations in Canada are seized with this challenge. The following items are suggested because they offer an entry point for this topic that will lead to several other useful resources.

Delattre, Edwin J. (1994). *Character and cops: ethics in policing. 2nd ed.* Washington, D.C.: AEI Press.

Edwin Delattre is a professor of education and dean of the School of Education at Boston University, as well as a professor of philosophy. He has written an excellent study of ethics as it relates to policing. His approach is to range across the full range of human scholarship to find perspectives and insights that will help all of us make sense of the ethical dimension. He refers to Socrates, Plato, Aristotle, Kierkegaard, and John Stuart Mill in order to shed light on this difficult topic. For Delattre the question of character is absolutely critical to policing, and those who enter into this work should understand that it is a high calling that requires much of them. This is an excellent primer on the subject.

Delattre, Edwin J. (1998). "Leadership: a position? An activity? And . . . ?: ethics in policing." Reprint of a paper delivered at the *Leadership '97* conference sponsored by the Police Leadership Forum, November 6, 1997, in Sydney, Nova Scotia. Ottawa: The Police Leadership Forum. (Occasional papers collection).

This paper continues the line of thought that Delattre established in his earlier writings on ethics in policing. Here he presents a more personal perspective on the essential moral standards required for policing to a Canadian audience.

Mancuso, Maureen et al. (1998). *A question of ethics: Canadians speak out*. Toronto: Oxford University Press.

Maureen Mancuso (University of Guelph), Neil Nevitte (University of Toronto), Michael Atkinson (University of Saskatchewan), Ian Greene (York University), and André Blais (Université de Montréal) have collaborated in producing this book, which aims at capturing the flavour of Canadian opinion about ethics in the realm of politics. The authors have gathered data pertinent to our nation's sense of ethical behaviour in the political arena and have divided their survey results into four main categories of ethical concern:

- Conflict of interest;
- Gifts and gains;
- Patronage; and
- Lying.

In each of these categories, questions were posed to probe public opinion on acceptable behaviour among elected officials. Also, a series of hypothetical scenarios were described with "forced-choice" responses offering a limited range of options in making ethical determinations. A randomized listing of contacts allowed the survey to be administered to appropriate individuals in five regions: British Columbia, the West, Ontario, Quebec, and the Atlantic provinces. Three age groupings were agreed upon: the "Elders" (born 1945, or earlier), "Baby Boomers" (born 1946-65), and "Generation Xers" (born since 1966). This publication provides clear insight for many potential readers. It is a rendering of what Canadians think about right and wrong in political action. It is a glimpse into that formidable entity referred to as "public opinion."

SPECIAL CHALLENGE OF RACE RELATIONS

Ontario. Task Force on the Racial and Ethnic Implications of Police Hiring, Training, Promotion and Career Development (1980). *Policing in Ontario for the eighties: perceptions and reflections: report of the Task Force on the Racial and Ethnic Implications of Police Hiring, Training, Promotion and Career Development*. Toronto: Solicitor General of Ontario.

Ontario. Race Relations and Policing Task Force (1989). *The report of the Race Relations and Policing Task Force*. [Toronto]: The Task Force.

The fifty-seven recommendations of this task force set in place many of the changes that occurred in policing in Ontario in the early 1990s, including the creation of the Race Relations and Policing Unit, the Employment Equity regulation, the Special Investigations Unit, and other elements designed to address the problems of race relations and policing.

Ontario. Race Relations and Policing Task Force (1992). *The report of the Race Relations and Policing Task Force*. [Toronto]: The Task Force.

This is the final report of the task force and makes a further forty-two recommendations in order to move Ontario forward toward a model of "contemporary policing." This model recognizes and respects the diversity of peoples and their wide-ranging cultural and racial differences. It further recommends that the Race Relations and Policing Monitoring and Audit Board be assigned responsibility for evaluating the future development and implementation of employment equity plans in Ontario's police services.

Ontario. Ministry of the Solicitor General and Correctional Services. Police-Race Relations Monitoring Board (1996). *Good beginnings: a catalogue of race relations initiatives in Ontario*. Toronto: Police-Race Relations Monitoring Board.

Although the Police-Race Relations Monitoring Board was disbanded, this publication remains an informative and valuable catalogue of initiatives introduced by several police services across Ontario. It documents the many promising police-community partnerships that serve the overall principles of community policing.

CHAPTER EIGHT

The police must move in the direction of anticipating and forestalling crime. Eventually intervention will be useless. This state of affairs can come about in two ways: first, by constant surveillance, to the end that noxious intentions be known in advance and the police be able to act before the premeditated crime takes place; second, by the climate of social conformity . . . This goal presupposes the paternal surveillance of every citizen and, in addition, the closest possible tie-in with all other techniques—administrative, organizational, and psychological.

Jacques Ellul (1964). *The technological society.* New York : Vintage Books.

COMMUNITY POLICING MEETS TECHNOLOGY—CONFLICT OR COOPERATION?

LEARNING OBJECTIVES

1. Identify working definitions of *technology* and *technique.*
2. Apply the definitions of *technology* and *technique* to policing in Canada.
3. Explain the significance of technology to the birth of community policing in Canada.
4. Identify a *typology* of police technology.
5. List advantages of technology in the context of community policing.
6. List impediments to the introduction of community policing created by the deployment of technology.
7. Understand the role of the Internet in supporting, and undermining, the goals of community policing.
8. Identify the range of technological innovations that may threaten the principles of community policing.
9. Identify approaches to technology that support the advancement of community policing in Canada.

INTRODUCTION

This chapter will examine community policing, as an innovative mode of police organizational behaviour, in the context of technology. Because of the power of technology to dominate so much attention when thinking about modern organizations (Grant, 1969; Grant, 1986), it is important to understand the role that technology has played, or is likely to play, in the development of community policing in Canada.

There will be an attempt to provide a working definition of the term "technology" in order to focus this discussion. The definition chosen is not uncontroversial. However, it allows the reader to place some emphasis on the enormous transformative power of technology in the modern context. In addition, there will a further definition offered for "technique" and some consideration of its relevance to discussions about community policing. Also, this chapter will look at the range of technological elements that have been introduced into policing over the last few decades in order to establish a framework for looking at the complex impact of technology on community policing initiatives.

Issues with respect to the technology of weapons, surveillance, and control will be reviewed in an effort to see the persistent relevance of reactive methods in the realm of crime control and public safety. Their impact on community policing as a mode of police operation will be considered in some detail. Also, the impact of the Internet and other forms of computer technology will be reviewed and placed into the context of community policing as a "way of doing business." This discussion will lead to a brief consideration of the growth of computer crimes and how the criminal element has embraced the power and performance of new technologies to enhance its opportunities for criminal activity and victimization. Finally, there will be an acknowledgement of the role of new scientific methods, including DNA profiling and criminalistics, in the repertoire of police techniques.

It is important to introduce this chapter with a clear understanding that the use of technology inevitably involves fundamental choices that have substantive, perhaps irrevocable, implications for police organizations. For example, Bayley (1991) notes:

> Police departments will be choosing, whether they recognize it or not, between assigning personnel to crime prevention through intensive community interaction or to crime prevention through enhanced technology-assisted law enforcement. (p. 45)

In fact, there has been very little recognition of the accuracy of Bayley's observation. Many police organizations claim that they are able to bridge the gap between the openly interactive mode of community policing or crime prevention, and the technologically-enhanced vision of the earlier "professional" model of policing which tends toward the "watchman" style of policing, requiring minimal public engagement or input.

WHAT IS TECHNOLOGY?

Canada's foremost political philosopher, George Grant, spent a great deal of intellectual energy exploring the issue of modern technology, that is to say, technology that has come into being as a result of the modern project which has taken upon itself the challenge of overcoming human and non-human nature. For Grant, the word "technology" puts together the Greek word for "art" and the word for the "systematic study" of it, as the word "biology" puts together "bios" and "logos." As Grant further explains:

> In fact, the coining of the word 'technology' catches the novelty of that co-penetration of knowing and making. It also implies that we have brought the sciences and the arts into a new unity in our will to be masters of earth and beyond. (p. 12)

Technology is the force that drives modern society toward the conquest of nature for the benefit of mankind. However, as Grant goes on to explain, the problem with this new perspective on technology is that we are almost incapable of seeing the forest for the trees. If, as he suggests, we have a destiny that is tied up with technology understood as "progress," then every problem confronting human development is seen as an opportunity for further technological development (Grant, 1986). It is not surprising that science is seen as the key to solving the world's manifold problems. Another, more conventional, definition of technology is offered by Volti (1995):

> . . . a system based on the application of knowledge, manifested in physical objects and organizational forms, for the attainment of specific goals (p. 6)

Grant's definition of 'technology', and his understanding of its penetration of all facets of modern existence is derived, in part, from his reading of Jacques Ellul. As a controversial thinker, Ellul has made some clearly disturbing predictions about the impact of technology, and technique, on the life of modern man. Ellul's theories have been considered by Albritton (1995) in the context of community policing:

> . . . an attempt has been made to demonstrate the value of incorporating the much narrower focus on the internal characteristics of police reform like COP/POP [community-oriented policing/problem-oriented policing] into the macro-level understanding of policing in a 'Technological Society.' Ellul's definition and analysis of 'technique' provides such a holistic context, as well as, a powerful explanatory matrix in which to understand the nature and determinants of policing as a sociological and technical phenomenon. In this broader context, COP/POP can be understood not as a new or revolutionary strategy for policing, but as part of a 'progression' of technical innovations that has led to policing techniques that are now on the verge of penetrating the most intimate level of social existence. (pp. 32–33)

Albritton suggests that some versions of community policing are really efforts at enhancing police dominance of social life and that they merely accelerate and concentrate the control of the community by police agencies. While this may be argued by police leaders, it is certainly true that community policing entails, in most instances, a broadening of the police mandate, and anticipates a much deeper engagement of the police with the community.

Another useful term to understand in this context is "technique." This word has been applied more routinely in Europe as a substitute for "technology" which is, as Grant points out, a North American coinage. Ellul defines "technique" in the following clear manner:

> The term *technique*, as I use it, does not mean machines, technology, or this or that procedure for attaining an end. In our technological society, *technique* is the *totality of methods rationally arrived at and having absolute efficiency* (for a given stage of development) in *every* field of human activity. Its characteristics are new; the technique of the present has no common measure with that of the past. [emphasis in the original translation]

Ellul's understanding of "technique" allows for a clear perspective on the progression of methods offered up for the control of crime and the facilitation of police operations. It also makes it necessary to contemplate whether any given technique is contributing to the goals and objectives of public safety, or to the absolute efficiency of police organizations. If the professional model of policing, with its concentration on the power of technique to provide rapid response and crime control, has failed, it remains to be seen if police executives have sufficiently recognized the role of technology in that failure.

Typology of Police Technology

In order to frame this discussion of technology in the context of community policing, it is useful to have some shared understanding of the range of technologies (or techniques) applied by police organizations. Nogala (1995) offers a valuable "typology of police technology" that can guide any examination of the technological innovations pursued by police organizations:

- Surveillance & detection technologies;
- Identification technologies;
- Information processing technologies;
- Communication technologies;
- Organization & administration technologies;
- Intervention technologies; and
- Technologies for mobility.

Also, Nogala has provided an important listing of the various forces that drive technology in the realm of policing:

- Reinforcement of international police cooperation;
- Preventive orientation;
- Expanding techno-crimes;
- Continuing process of professionalization and specialization;
- Compulsion to rationalization;
- Technology as a weapon and the promise of technical solutions.

Here, the leading question may be: Which of these forces equally contributes to the "drive" toward community policing?

Technology and Police Professionalism

Science, which offers a prelude to technology, has been embraced in the modern world, and it has been encouraged to penetrate North American society to an exceptionally high degree. As observed by McCracken (1997):

> In the postwar period, science was a privileged form of discourse. It was the most credible "way of knowing" at our disposal. It was the great linchpin between progress on one side and technology on the other. It was the way progress was accomplished; it was the great fund of knowledge on which technology drew. It was the shrine of national aspiration, and North Americans deferred before it. Science was asked to specify the best arrangements for domestic living (medicine and home economics), the best way of organizing the world of work (time and motion studies), the best way of intervening with nature (hydroelectric projects), and the best way to address social problems (e.g., Kennedy's Camelot projects). It was regarded as the final authority on many topics. (p. 60)

Technology is conventionally accepted as the realm of experts. Police professionalism operates on an assumption that police specialists can develop a sufficient level of expertise to deal with the problems of crime and disorder. The important work of individuals like August Vollmer and O.W. Wilson was to build a "science" of policing that would gain a fully professional status. However, within an increasingly complex technological society, police professionals have had to turn to scientists and engineers in order to tap into their highly refined subject matter expertise, be it telecommunications, data processing, engineering, or the medical sciences. This challenge by technical experts for a place in the world of policing was foreshadowed by the report of the U.S. President's Commission on Law Enforcement and Administration of Justice in 1967. A portion of this report dealt with "science and technology," and its initial assumption is clearly that what modern policing requires is the greater enlistment of scientists and engineers in order to attack the technological problems of crime and social disorder. It is such a transparently self-serving assertion as to be humorous. However, the authors of this section of the report are clearly in earnest. It is also disturbing that they immediately equate the contributions of science and technology to the American war machine with the potential for advancement on the front of policing.

> More than two hundred thousand scientists and engineers are helping to solve military problems, but only a handful are helping to control the crimes that injure or frighten millions of Americans each year. (p. 245)

As Albritton (1995) indicates, policing had functioned fairly effectively within a closed system:

> . . . the highly centralized, bureaucratic, paramilitary, and professionally oriented model of police organization . . . remained a relatively stable example of technical efficiency well into the twentieth century. (p. 25)

It may be reasonable to question the veracity of the presumption that those hundreds of thousands of scientists and engineers had indeed "solved" the military problems referred to above. In any event, there is a presumption embedded in this statement that scientists, armed with their specialized expertise, could be turned to "control" crime. There is little thought given to the potential for community policing, community development, or problem-oriented policing as a means to the end of public safety and security. This raises the serious question as to whether or not the benefits of technology can ever be fully consistent with a community policing philosophy, rightly understood. Consider, for example, Ellul's observations about "experts":

> . . . experts are traditionally guardians of the objectivity of science, independent of opposing interests. They have a duty to speak the truth. But the public has now found out that experts, too, are part of the power game, that they belong to organizations, that they plead on behalf of those organizations. This is exactly in keeping with the technostructure as Galbraith analyzed it. Experts who belong to an organization always justify it, for it is from the organization that they receive their power and authority. Experts are thus an indispensable part of the technical system, for they are exponents of technological discourse. They tie everything together. (pp. 191–192)

There is a supreme confidence that the dynamism of science can be brought to bear on all problems of a human, and non-human, nature. This confidence is stated explicitly in the report of the U.S. President's Commission on Law Enforcement and Administration of Justice:

> Technology can indeed fill most reasonable requests and can thereby provide considerable help to law enforcement . . . Science can provide the capability, but the public as a whole must participate in the value discussion of whether or not the capability is worth its financial and social costs. (p. 246)

Of course, while the presentation above seems to imply that public debate will occur around the various "capabilities" offered by technology for law enforcement, in circumstances of this sort, the public inevitably turns to these self-same "experts" for guidance on questions of "value." Therefore, the role of the scientists is to develop the capability, to fashion its application, and to speak to the worth of that capability. It is unlikely that their interventions could be fully objective under such circumstances (Ellul, 1964, p. 435).

Speed and efficiency have become hallmarks of "professional" policing, in spite of research that undermines the value of the rapid response mentality. Manning (1992) in speaking to the issue of call response as being a key focus for police departments when considering new technology:

> The police have always viewed advanced communications, from the first telephone, to mobile two-way radios, hand-held radios, and computers, as a solution to the problem of responding equitably to public demand and crime control. Since 1967 the police have attempted to deal with rising demand by increasing their communicative capacity by means of computer-assisted dispatching. They sought, often with federal government assistance, to develop enhanced means of receiving and processing calls and to produce systematic, patterned, and considered response to these calls.
>
> This focus on communication was keynoted by the President's Commission's Report on Science and Technology (1967). This document, written primarily by computer scientists and engineers, claimed that information technology, namely computer-assisted dispatching, would reduce many of the vexing problems of policing, including crime control, strategic planning, evaluation and record-keeping. (p. 254)

Skolnick and Bayley (1986) comment on the difficulties that the rapid response mentality have caused for police organizations that are struggling to reorient the focus of their service, including the move to community policing:

> Police departments are also trying in various ways, though they don't like to admit it, to unplug the 911 emergency dispatch system selectively for patrol officers. By so doing they free themselves for community development and crime prevention activities of their own devising. In many cities the 911 system with its promise of emergency response has become a tyrannical burden. (p. 216)

In 1972, the Ontario Solicitor General announced the establishment of a task force to undertake a review of police administration, organization, and efficiency in the province. This task force made several recommendations for the improvement of policing in Ontario, including its organization and structure. However, the members of the task force were also alive to what they termed "the technological trap" which had separated police officers from the community (Ontario. Task Force on Policing in Ontario, 1974):

> The introduction of technology into the police role has tended to widen the gap between the citizen and his police force. Early stages of technological reform put police officers into automobiles and tied them to dispatch centres through radio. This had the effect of taking officers out of direct contact with people on the streets. Perhaps of equal importance, radio concentrated the officers' attention on incidents reported to police departments by telephone. While this improved the police ability to respond to citizens in need, it has also tended to shift their attention to cases needing immediate resolution and away from a continuity with a geographic area and its people. (p. 25)

It is interesting that only seven years after the scientists had counseled the U.S. President to accelerate the pace of technological innovation, and the placement of scientists in the service of policing, Canadian authorities are cautiously warning about the difficulties posed by technology. Because technological advancements have primarily been designed to benefit patrol operations, rather than the proactive approaches anticipated by community policing, systems for performance measurement have focused on efficiency in that application of those technologies. Again, the Ontario Task Force on Policing is instructive in this regard:

> Police forces, like other institutions in the post-war milieu, have sought performance measures which would help them gauge the efficiency with which they carry out their duties. Like other institutions, they have tended to seize on measures which are easily quantifiable—perhaps ignoring other, less tangible yardsticks of performance. Crime rates, clearance rates, warnings, arrest statistics, and other concrete variables have been introduced into police reporting schemes to assess efficiency. (p. 25)
>
> These technological reforms have reduced the personal contact...between the citizen and his police officer. They have made police work more impersonal, more incident-oriented and less concerned with the continuity of relationships with the community.
>
> In terms of the six functions we have identified for the police role, the introduction of technology in urban areas has tended to emphasize the 'response', 'crime solving' and 'law enforcement' functions to the detriment of 'referral', 'prevention' and 'public education'. (p. 26)

Manning describes some of the British police officers' experience with the controlling nature of technology, this time in the form of computer-assisted dispatch (CAD):

> Officers in a British police force, finding themselves responding to an increasing number of radio-transmitted calls, referred to themselves as "blue mice." They viewed themselves as acting in a frenetic, almost random manner, responding to unknown pressures and forces outside their control. (pp. 254–255)

Most significantly, Manning emphasizes the *organizational* impact of technology. As discussed earlier in Chapter Five, there is a substantial need for major transformation in police departments that are attempting to move to a community policing or problem-oriented model of service delivery. Equally, technology cannot be placed in the service of the police department without taking stock of the impact it will have on the entire organizational framework:

> Technology has the potential to alter a field of authority relations because it can increase or decrease choice, or autonomy, increase or decrease a span of control, or provide on-line supervisory information . . . A new technology brings with it new skill requirements for some workers, even while it may "de-skill" others. Any technological innovation introduces new uncertainties in the work flow and in role relationships. It may also stimulate responses designed either to restore the status quo or to modify it . . . In these ways, technology can alter social relations within an organization. (p. 255)

And:

> Police traditionally rely on verbal skills, persuasion, and violence to accomplish their ends. More sophisticated means of communicating are used primarily to manage officers and improve the administration of police work, rather than to enhance the quality of police-community relations. (p. 260)

Rossmo (1995) points out that police in the professional model were removed from the public by an extreme emphasis on technology.

> This overreliance on technology, bureaucracy and criminal law distanced the police from the community, handicapped their efforts and led to roadblocks that impeded improvement. One of the predictable outcomes of this professionalization was a shifting of social control responsibility from the citizenry to the police.
>
> It was implicitly assumed that a properly trained police force with sufficient resources and the latest technology should be able to control any problem and solve any crime. (p. 1)

This view of the transcendent power of technology, and the infinite capacity of the professional police force to conquer any problem or crime, has slowly disintegrated in the face of research and reality. However, this does not mean that technology can be ignored. Normandeau and Leighton (1990) point out that the future of policing in Canada will be tied to several significant technological developments, including the increased capacity of criminals to engage in illegal activity through the application and misuse of these very same technologies (Albanese, 1988). Because of the speed of innovation in many areas of technology, police departments will be compelled to make choices that may cause community policing initiatives to suffer due to a priority on introducing or reengineering one of the many technology "systems" in place within the department. Technology is often extremely expensive (e.g., computer systems, surveillance equipment, helicopters with infrared capabilities) and must be maintained in order to achieve appropriate levels of efficiency and effectiveness.

When community policing was introduced in Great Britain, its effectiveness was complicated by the heavy reliance on data being generated by dispatchers. There was little opportunity for officers to actually engage with community problems in meaningful ways because they were constantly being driven by isolated incidents. There were both organizational and individual impediments to the successful introduction of real, proactive, problem-solving policing. As Manning reports:

> Chatterton (1991) evaluated officers in a community policing scheme in five divisions in the north of England who had too little time to assess the information they were given (in the form of printouts of incidents made by dispatchers). Furthermore, they had little inclination to analyze

> it in detail and were given no rewards for doing so . . . This suggests that the police organization valorizes action over information, on the spot completion of tasks, and devalues paperwork other than that that can be completed with dispatch at the scene or shortly thereafter. (p. 266)

Frequently, new techniques, or technologies, have been introduced into policing without any transformation of the organizational structure or guiding purpose. In terms of community policing, this means that the fundamental reconstitution required in order to move in the appropriate directions is missing, and technology is used to bolster, even disguise, the lack of real change. Again, Manning is instructive in this area:

> The pattern of police deciding, the use and definitions of relevant information, and the constitution of technology sustain the discretion and autonomy of officers on the ground. Thus far, evidence suggests that officers use information technology in line with traditional practices. They reify the structure of authority, secrecy and mistrust extant in policing. While the amount of data possessed by police has increased, it is not clear that information has. Converting data into information requires a theory and objectives of gathering or storing any bit, and to guide the transformative process. (p. 270)

Police organizations tend to bolster the entrenched approaches to their core functions; there is a kind of gravitational pull to continue doing things in conventional ways. It is often perceived that community policing can be addressed on a superficial level, through programs and slogans that express an adherence to community integrity, problem-solving principles, and meaningful partnerships. However, if the technology that is sustaining and driving the organization's processes, functions, and systems is geared toward the "professional" model of policing, the goals and objectives of community policing will never be realized. With the centralizing impact of most technologies, it is not surprising that many information management systems in policing have been directed at controlling the officers in the activities. Manning (1992) speaks to the experience in this regard:

> It is particularly important to note that rationales for policing information systems were and are based on claims that they will increase efficiency, or effectiveness, or make the police more responsive. Yet, they have been used primarily to increase surveillance of patrol officers, have made no impact on strategic planning or management, and have not produced evidence of crime-solving potential.
>
> Furthermore, the entire technological edifice is based on the false and misleading assumption that policing is primarily about crime and crime control. As the research of the past twenty-five years has firmly established, policing is about differentially and situationally ordering the chaos of everyday life, providing a sense of neighborhood security, and responding to diverse human needs. (p. 271)

Hough (1990) has considered the impact of technology on policing, and asserts that the success of technology may serve to diminish, rather than increase, public scrutiny or accountability in police organizations. The more police departments are seen as having expertise in areas of law enforcement and public safety, the more they are guarded from outside influence.

> The technology, and the "crime-control" and "public service" models of policing on which it is premised, lend to police administration the appearance of a value-free, technical process; they

> support the image of professionalism. The dilemma is this: the appearance of professionalism protects any organisation from outside interference. If its functions are seen to require a highly specialist body of knowledge, then the competence of outsiders to assess its performance is limited. From this perspective, the models of policing on which the technology is premised need more exposure to criticism. (p. 355)

The desire to increase the engagement of the public in the project of policing, which characterizes community policing, argues in favour of "low-tech" approaches to policing functions. Buerger, Cohn, and Petrosino (1995) describe an experiment conducted by the police department in Minneapolis, Minnesota:

> The Hot Spots experiment hypothesized that an increase in the degree of formal public guardianship that was consistent over time—represented by the visibility of a police officer or officers, without regard to their specific activities—would be sufficient to change the ecological profile of a given place in a positive fashion, reducing the incidence of crime and disorder. (p. 239)

The mere presence of officers in areas identified as being problematic had a positive influence on public safety. These officers did not require any special equipment or sophisticated technology. They simply had to be on site to produce an effect. Further explorations in the area of crime "mapping" have demonstrated that police organizations can improve situations by letting the public in on their efforts at identifying "hot spots" in the community (Rich, 1998). McEwan and Taxman (1995) point to the need for information in support of community policing:

> . . . as police become more involved in community policing and problem-solving initiatives, information is going to be key. Maps provide police officers and communities with critical information that is easy to understand. The maps can be shared with the community to develop cohesive and collaborative efforts. Within the police department, the use of information, and maps particularly, can be critical in empowering the police to work on crime problems. The police officer, armed with easy-to-understand maps describing and analyzing the crime issues, can then become more responsible for developing proactive strategies. (p. 281)

Information gathered through a community-police partnership can be extremely effective in addressing potential crime problems. There have been several projects that have used the combined energies of the local community and the police to good effect. Block (1995) describes the "early warning system" approach:

> The Early Warning System [for Street Gang Violence in Chicago] project is founded on the premise that, since street-gang violence is spatially anchored and occurs as the culmination of escalating incidents of revenge and retaliation, information compiled by community and neighborhood organizations, as well as by law enforcement, could be used to develop an "early warning system" of neighborhoods in crisis . . . Continuing escalation would then be prevented by crisis intervention and dispute mediation, using both internal community influences and external police support. Such a program has shown success in two pilot projects in Chicago's Humboldt Park and in Philadelphia. . . but requires the strong support of neighborhood agencies, churches, community groups, and the police department. (p. 26–27)

A preference for the "low-tech" approach is perhaps characteristic of policing in Great Britain, if Hough's (1980a) observation continues to hold:

Judgment rather than technique, is at the heart of police administration. It is possible that recognition of this in part accounts for the skepticism of British police to management technology. (p. 354)

The need for better information and a deeper enlistment of the public in the overall approach to safety, security, and order is what matters more than the precise technologies applied. This observation has been clearly articulated by Cassels (1995):

> Information is the lifeblood of community policing, yet not enough systematic attention is given to information systems that support this model. Police managers can no longer separate information technology from organizational strategy. Organizational effectiveness will depend on practical technology applications that parallel the major shift to Community Policing . . . Police leaders must therefore change their emphasis from technology in the form of gadgets to technology that supports local neighborhood problem solving. (p. 113)

Canadian Police Technology and Research

The Canadian Police Research Centre (CPRC), a partnership involving the Canadian Association of Chiefs of Police, the Royal Canadian Mounted Police, and the National Research Council Canada, has been instrumental in providing leadership in the research, development, evaluation, and commercialization of police equipment and technology. A listing of the projects from the 1995-1996 annual report will demonstrate the breadth of the CPRC's activities:

Category A - Health and Safety—Protecting the Police in Hazardous Situations

- *Ampel probe*—for searching suspects, crime scenes, and exhibits while reducing the risk of biohazard, and avoiding the contamination of exhibits;
- *Blast suppressant foam containment system*—a sensitive project that has been successful in experimental trials and will likely be made available to first responders;
- *Bomb suit testing*—field experiments with the "next generation" of bomb disposal suits and helmets;
- *Soft body armour*—development of a Canadian standard for regular use soft body armour;
- *Miniature emergency response vehicle*—known as MERV, this is a remote-controlled, tracked robot equipped with an infrared pan camera and oleoresin capsicum (OC, or pepper) spray;
- *Remote wireless explosives disruptor initiator*—developed through a collaboration between the RCMP 'E' Division explosives disposal unit and the Canadian Bomb Data Centre, this is a light-weight, miniature transmitter/receiver system that can initiate explosive charges from a remote command post without the need for a ground line;
- *Review of OC spray*—study to review the current literature to identify possible risks associated with the operational use of oleoresin capsicum. A concurrent review is being undertaken by the U.S. National Institute of Justice; and

- *Smart-gun technology evaluation*—involves experiments aimed at developing a "personalized" weapon that can only be operated by an authorized user(s).

CATEGORY B—OPERATIONAL EFFECTIVENESS – FIGHTING CRIME, GATHERING INFORMATION, INTELLIGENCE AND EVIDENCE

- *Explosive resistant vehicle*—use of prototype vehicles to determine the performance of explosive resistant materials;
- *Forensic entomology*—involves the gathering of data relating to insect succession on carcasses in shallow graves, including research in various climatic regions of Canada;
- *Geographic profiling*—based on computer software specific to the Vancouver Police Department, a geographic profiling product, known as "Rigel," has been developed for sale to other police agencies worldwide. This technology was invented by Dr. Kim Rossmo, a detective with the Vancouver Police. The system assigns statistical probabilities that function to predict the most likely location of an offender's residence. This technology has been purchased by the OPP, and there is currently interest expressed by departments in the U.S., Japan, Europe, and South Africa;
- *Public Safety Network* (formerly known as CPRCNet)—this is an umbrella organization that provides Internet capabilities that focus on the provision of services for clients in law enforcement and community-based policing organizations. The network is supported by the Canadian Association of Chiefs of Police, the Canadian Police College, the Canadian Police Research Centre, and several other organizations;
- *Shift scheduling software evaluation*—involves an operational evaluation of software for shift scheduling of police officers, called "Watch Commander," developed in British Columbia. Allows the user to interactively modify days off and shift rotations in order to provide better coverage; and
- *Use of force training simulators evaluation*—under the auspices of a working group established at the request of the Canadian Association of Chiefs of Police, information is being collected and analyzed on commercially available systems, including a study of the pedagogical bases for the use of simulator training.

CATEGORY C—PROTECTING THE PUBLIC—TRAFFIC, CUSTODY, CRIME PREVENTION

- *CACP website*—established at the request of the Internet Subcommittee of the Canadian Association of Chiefs of Police Informatics Committee;
- *Evaluation of vehicle stopping device (ROADSPIKE)*—involves a retractable spiked barrier strip that holds hollow spikes that police can deploy in advance of a fleeing vehicle to safely deflate its tires. This technology was originally developed through a U.S. National Institute of Justice project conducted by the Idaho National Engineering Laboratory and is commercially available. The product evaluation is being carried out through the British Columbia Justice Institute as one component of a review of stopping devices and high speed pursuit strategies; and

- *Less than lethal technology* (ring airfoil projectile)—this technology was devised more than twenty years ago by the U.S. military, and involves an aerodynamically designed flat flight trajectory that has a high spin velocity but low muzzle velocity, meaning that it is less than lethal at the muzzle. The U.S. National Institute of Justice is studying its application for civil law enforcement and the CPRC's Technology Development Advisory Council in Ottawa is participating in demonstrations of this technology as a delivery system for oleoresin capsicum (OC) spray or CS gas.

DETECTIVES, INVESTIGATIONS, AND THE SHARING OF INFORMATION

Police detectives were often considered a special elite, and frequently, the best route to promotion was through successful work in this organizational "silo." Often the highest office was held by a senior officer who had shown great talent among the detective ranks, not from the administrative or field levels of the organization. Detective work has frequently been studied in terms of its lack of effectiveness and the disproportionate resources required for little gain (Greenwood and Petersilia, 1975). The detectives have a heightened sense of proprietary interest over their sources of information. However, what Manning suggests for police organizations generally may be extended to the detective branch:

> Unlike the Weberian model of bureaucracy that features centralized, formatted, and accessible records and files, policing remains informationally highly decentralized . . . Information clusters into a few richly textured and coherent islands isolated from other islands of information. Only under certain specified conditions, usually when a crisis draws together police resources, does police information become accessible, understood, generally shared, reproducible knowledge held collectively in organization records. (p. 263)

The guarded use of information in police organizations, especially among detectives, is noted by Moland (1998):

> In the past, for detectives to become successful and remain so, they had to acquire and develop reliable information sources. The best detectives sustained themselves on the energy of their exclusive information resources—personal contacts, confidential informants, field notes, historical files and other data sets that remained carefully guarded secrets . . . The problem with this system is that many detectives developed institutional expertise, knowledge, or memories of particular types of investigations, but often failed to share, transfer, or document their knowledge. (p. 51)

In Canada, most recently, the systemic failure of police investigators to fully and functionally cooperate has been profiled through the review of the Bernardo investigation, conducted on behalf of the Ontario Solicitor General and Minister of Correctional Services by Mr. Justice Archie Campbell (Ontario. Review of the Bernardo Investigation, 1996). In essence, Mr. Justice Campbell's mandate was to review the Green Ribbon Task Force and the Metropolitan Toronto Police Service investigations into specific deaths and sexual assaults and report on issues and recommended policies and procedures. What this review revealed has generated considerable activity in the Ontario law enforcement community around change.

While Campbell avoids assigning individual blame with regard to these investigations, he is clearly disturbed by the system in which police investigators operate:

> The Bernardo case, like every similar investigation, had its share of human error. But this is not a story of human error or lack of dedication or investigative skill. It is a story of systemic failure . . .
>
> What is needed is a system of case management for major and interjurisdictional serial predator investigations, a system that corrects the defects demonstrated by this and so many similar cases. A case management system is needed that is based on cooperation, rather than rivalry, among law enforcement agencies . . . There were times during the separate investigations of the Scarborough rapes and the St. Catharines rapes and murders that the different police forces might as well have been operating in different countries . . . Because of the systemic weaknesses and the inability of the different law enforcement agencies to pool their information and co-operate effectively, Bernardo fell through the cracks. (pp. 1–2)

In order to utilize the power of computers and the information technology available, there would clearly need to be a more open approach to the generation of information and the sharing of processed information (i.e., knowledge) within the department. Mr. Justice Campbell turned to an important study conducted of the Rand Corporation by Greenwood and Petersilia in 1975 that looked at American police investigations, and reiterated some of the study's key insights in his final report:

- Most cases are solved by means of information spontaneously provided by a source other than those developed by the investigator;
- A small number of serious cases requires special investigative skills and experience;
- Computerized information systems can greatly increase the effectiveness of criminal investigations;
- Most police departments collect more evidence than they can process productively and a greater proportion of police resources should be devoted to processing and analysis of information. (pp. 271–272)

Rengert (1995) observes that police have been doing "spatial analysis" for many years. However, with the advent of the microcomputer, there are now much easier and more reliable ways to capture the expertise they have developed and transmit that knowledge to other officers. Rossmo (1995) has provided substantial background on the use of "environmental criminology" to assist in the investigation of serial and violent criminals. His examination of geographic considerations has documented several instances in which careful analysis of the setting of a crime has been valuable for police investigators. Geographic analysis has been used by the Indian Police Service, the LAPD in the search for the Hillside Strangler, and by Her Majesty's Inspectorate of Constabulary for tracking the Yorkshire Ripper. In Fort Worth, Texas, investigators developed an approach they termed "geoforensic analysis" which is based on the hypothesis that serial murder or rape tends to form place-time patterns distinct from "normal" crimes. As Rossmo (1995) explains, these approaches rely heavily on computer technology and fairly sophisticated techniques:

> . . . combining centrographic principles and journey to crime research in a manner informed by environmental criminological theory can produce a viable method for predicting the location

> of offender residence from crime site coordinates. One such effort is criminal geographic targeting (CGT), a computerized geographic profiling used in police investigations of complex serial crimes. (pp. 223–224)

Computer technology has driven a great deal of the innovation in investigative techniques over the last several decades. However, if technology is going to be placed at the disposal of community policing initiatives, there must first be a profound recognition of the mindset of police criminal investigators, particularly their institutionalized reluctance to share information, even with other investigators.

As noted by McEwen and Taxman (1995):

> Three trends have spurred the growth and interest in automated mapping in recent years. One trend is related to the microcomputer revolution over the last ten years, in which the capabilities of microcomputers have improved while costs have simultaneously decreased. Desktop microcomputers now have the computing and storage power of large mainframe computers that once filled entire . . . rooms. Coupled with these hardware improvements, the second trend has been improved software mapping systems . . . These packages facilitate automated mapping by providing a set of menus of various tasks involved in making a map. The combination of faster microcomputers and mapping packages has provided police departments with new capabilities for compiling and analyzing data. Improved computer resources are driving the final trend of improved geographic base files. (p. 261)

> Automated mapping systems offer potential for having major impacts on the strategies and tactics of police departments. Most importantly, the maps provide an approach to examine issues, information and problems with existing information. Automated mapping provides the mechanisms to train staff in how to use information differently in police departments, especially the use of information to develop effective police responses. In most cases, these responses can be in the form of proactive policing or mobilizing the community to address crime related issues. (p. 280)

> . . . as police become more involved in community policing and problem-solving initiatives, information is going to be key. Maps provide police officers and communities with critical information that is easy to understand. The maps can be shared with the community to develop cohesive and collaborative efforts. Within the police department, the use of information, and maps particularly, can be critical in empowering the police to work on crime problems. The police officer, armed with easy-to-understand maps describing and analyzing the crime issues, can then become more responsible for developing proactive strategies. (p. 281)

The management of information is a critical issue for police administrators, and this is true in the context of community policing as well as in more traditional modes of law enforcement. The difficulty relates to the introduction of systems that facilitate, rather than hinder, the dissemination of information to the public. Police, as has already been established, have a natural reluctance to share information even within their own profession (Ontario. Bernardo Investigation Review, 1996), unless there are overwhelmingly compelling reasons to do so. There continues to be optimism that the new forms of crime analysis, supported by computer technology, statistical methods, and mapping, can generate information that will be valuable to the cause of community policing. As Reuland (1997) observes:

> The need for efficient and effective management of information is further heightened in those departments that adopt a community policing approach. This approach involves developing partnerships with the community to solve recurring problems. To engage the community fully

> and resolve these problems successfully, police departments and community members require access to large amounts of information about the problems they are attempting to solve. Experience has demonstrated how a well-organized crime analysis unit coupled with appropriate technologies can assist a department in providing such information in support of community policing. (p. 1)

Crime analysis follows a fairly predictable pattern, in general terms. In essence, it consists of five key components:

- Data collection.
- Data collation.
- Data analysis.
- Dissemination of the results of analysis.
- Feedback.

In order to fully understand the components of an effective crime analysis database, Simmons and Kenney (1997) have prepared a model that is made up of four areas:

- Domains;
- Categories;
- Data fields; and
- Values.

Table 8-1, below, provides the range of elements that could be incorporated into a functional crime analysis database:

TABLE 8-1

CRIME ANALYSIS DATABASES—BASIC STRUCTURE

Events	Modus Operandi	Persons
Identifiers	*Identifiers*	*Identifiers*
Geography	*Crimes Against Property*	*History*
Temporal	*Crimes Against Persons*	*General Descriptors*
Target	*General*	*Specific Descriptors*
General		
Identifiers	*Identifiers*	*Identifiers*
Report Number	Report Number	Report Number
* Offense Type/ Classification	*Crime Against Property*	* Person (Victim, Suspect, etc.)
	* Point of Entry	*History*
Geography	* Method of Entry	Reason for Contact
Street Address	* Suspect's Actions	Date of Contact
Large Area (District Zone)	* Victim's Location	Address of Contact
	* Elements	Area of Contact (Large)
Medium Area (Sector/Beat)	* Tools Used	Area of Contact (Medium)
	Crime Against Person	Area of Contact (Small)
Small Area (Grid/ Reporting Area	* Suspect's Actions	*General Descriptors*
	* Threats or Force Used	Name

Temporal	* Victim Forced to do	Alias
Beginning Date	* Pretext/Offered	Nickname
Beginning Time	* Suspect's Characteristics	DOB
Ending Date	* Victim Movement	SSN#
Ending Time	* Victim Location	ID Type
* Shift	* Victim Type	Home Address
General	***General***	Home Phone
* Case Status	* Method of Escape	Place of Employment
Prints		Work Phone
* Premise/Location Type		***Specific Descriptors***
Premise/Location Name		Race
Gang Related		Sex
Drugs Used		Age
Alcohol Used		Height
Domestic Related		Weight
Reporting Officer		* Build
Narrative		* Complexion
		* Facial Hair
		* Speech
		* Hair Length
		* Hair Style
		* Hair Colour
		* Demeanor
		* Scar
		* Tattoo Type
		* Tattoo Location
		* Clothing
		* Weapon Type
		* Weapon Action
		* Weapon Feature
		* Weapon Colour

Vehicles	Property	Calls For Service
Identifiers	***Identifiers***	***Identifiers***
Descriptors	***Temporal***	***Temporal***
Geography	***Descriptors***	***Geography***
Temporal		***General***
		Descriptors
Identifiers	***Identifiers***	***Identifiers***
Report Number	Report Number	Report Number
* Vehicle Classification	Property Class (Stolen, Recovered)	***Temporal***
Descriptors		Date of Incident

Make	***Temporal***	Time of Incident
Model	Stolen Date	Date of Call
Body Style	Recovery Date	Time Call Received
Year	***Descriptors***	Time Unit Arrived
Colour	Brand	Time Unit Cleared
Vehicle Features	Model	***Geography***
Tag Number	Serial Number	Address of Incident
State of Tag	Size/Caliber/Gauge	Large Area Boundary
VIN Number	Colour	Medium Area Boundary
Geography	* Property Type	Small Area Boundary
Recovery Address		***General***
Recovery Area (Large)		Complainant Name
Recovery Area (Medium)		Complainant Address
Recovery Area (Small)		Complainant Telephone
		Race
		Sex
		Age
		First Unit
		Second Unit
		First Backup Unit
		Second Backup Unit
		Family Violence Y/N
		Business Name
		Descriptors
		* Nature/Type of Call
		* Priority of Call
		* Disposition of Call

* Denotes data elements suitable for forced-choice checklists.
Source: Simmons, John and Dennis J. Kenney (1997). "The stages of crime analysis." In Reuland, Melissa Miller (ed.) *Information management and crime analysis: practitioners' recipes for success*. Washington, D.C.: Police Executive Research Forum, pp. 17–18. Used with permission.

WEAPONS, SURVEILLANCE, AND CONTROL

Meyer (1992) makes a clear point when considering the contribution of technology to police use of force tactics:

> If we have the technology to put a man on the moon and return him safely to Earth, why don't we have the technology to put a man on the ground and take him safely to jail? (p. 4)

Geller and Scott (1992) cite the work of Clede and Parsons (1987) who have researched the literature on non-lethal weapons and have proposed the following repertoire of control techniques:

- *Body language*—includes stance and eye contact;
- *Oral control*—includes a commanding voice and the use of clear, direct instructions;

- *Unarmed restraints*—also known as the "come-along" holds (e.g., pressure points, fingerlock, wrist lock);
- *Neck restraints*—includes the carotid restraint, lateral vascular neck restraint, bar arm holds;
- *Short sticks and batons*—covers such products as the Nunchaku and Kubotan martial arts sticks as well as longer batons and break-resistant flashlights;
- *Handcuffing techniques*;
- *Knife-fighting techniques*;
- *Mechanical devices*—including capture nets, physical impact weapons that launch non-lethal projectiles such as rubber rounds, bean bags, and stun grenades, pneumatic piston impact weapons;
- *Electronic devices*—includes stun guns (e.g., the TASER); and
- *Chemical devices*—includes products like Mace (or other CS and CN gases), tear gas, smoke, and oleoresin capsicum (commonly referred to as "pepper spray").

Connor (1991) specifies 16 levels of police force that represent a range of applications that follow a path of escalating resistance:

No Force—
- Controlled confrontation
- Body language
- Verbal persuasion

Ordinary Force—
- Contact controls
- Joint restraints
- Weapon-assisted leverage techniques
- Nerve center controls
- Weapon-assisted pain compliance techniques
- Chemical irritants
- Electrical devices
- Intimate impact devices

Extraordinary Force—
- Extended impact weapons
- Weaponless techniques with debilitating potential
- Weapon techniques with debilitating potential
- Service firearm
- Supplemental firearm

The issue of police shootings and officer safety may seem removed from community policing; however, there is a connection. Geller and Scott (1992) in their comprehensive study of police deadly force make the following observations:

> It is possible that the enhancement and qualitative improvement of officer discretion implied by community policing and problem-oriented policing strategies may have a long-term bearing

> on the nature and prevalence of officer-involved shootings. Some police executives who have been leaders in the movement toward these new strategic thrusts . . . believe these methods will pay fringe benefits in greater officer safety and fewer injuries to police adversaries. Some proponents of community policing believe these benefits have already been realized in some locales . . . Hard data as yet are hard to find. (p. 415)
>
> Another illustration of the need for more 'holistic' thinking about officer safety, violence reduction, and community policing comes from the Christopher Commission's report on the LAPD . . . A centerpiece of its sweeping recommendations for reforming abuse-of-force problems in the LAPD, announced after 100 days of deliberations, was adoption of community policing as the dominant modus operandi of the department. Another suggestion, intended to avert racial discrimination in assignments practices and to break up cliques of officers who were bad influences on one another, was to rotate periodically officers from one unit and geographical area to another . . . If members of the Christopher Commission were aware of the tension between urging implementation of community policing and recommending stepped up rotation of beat officers, they showed no indication of it. A cornerstone of community policing, of course, is leaving officers in their units and beats for long periods so that they get to know and trust one another . . . Officers can be forgiven if they are left confused, concerned, and even resentful by recommendations that appear to be sending mixed messages. (pp. 415–416)

Officers are constantly placed in the difficult position of being trained for the most extreme levels of violence and aggression while simultaneously being schooled in the need to demonstrate extreme constraint in the application of force. Community policing would seem to support minimal amounts of force. Indeed, community policing seeks to entirely avoid the need for force. However, this positive goal is not realistic and it would be fatuous to expect that any approach to community policing could achieve complete success in this regard. The sheer diversity of options available to officers in protecting themselves and the public clearly demonstrates the problems faced by police organizations when attempting to strike the appropriate balance between the interests of community policing and the potential for violence endemic to law enforcement. The human and financial capital invested in the research, development, proficiency training, and evaluation of lethal and less-than-lethal force options indicates that it is central to policing in Canada. Geller and Scott (1992) make explicit what common sense would suggest with regard to the front-line officer's preferences:

> Frequently, when officers sense any conflict between the requirements of community policing and officer safety, they quite understandably resolve all doubts in favor of officer protection. But sometimes the intuitive reaction is not the right one, not the one presenting the greatest long-term benefits in terms of officer welfare. (p. 415)

Where community policing, as a mindset for effective policing, has significantly contributed to the evolution of police use of force policies is in the area of officer training. By making community development, or community wellness, a central focus of all officer activities, there has been a much clearer focus on improving the quality of learning around police use of force. The growth of a comprehensive approach that places an emphasis on problem-solving and decision-making allows for better programs geared to officer safety. Geller and Scott (1992) underline this development:

> The empirical literature has recommended training in four basic areas: (1) policy interpretation; (2) human relations and "cultural awareness" skills; (3) conflict management techniques (training in physical, verbal, and judgmental "violence-reduction" tactics); and (4) tactics for the proficient use of weapons when shooting is required. (p. 297)

By linking police training for the use of force with more fundamental decision-making skills that are based on the principles and practice of community policing, it is possible to broaden the officer's capacity to arrive at a successful resolution of potentially violent situations. If front-line officers are provided with the support they need to develop conflict resolution skills, they are in a better position to avoid the application of force. Geller and Scott (1992) support this approach:

> Regardless of the level of specificity contained in different decision models, . . . virtually all analysts who have considered the topic of potentially violent police encounters concur that through "early" decisions concerning how, when, and whether to engage a suspect in a confrontation, police often can influence the eventual outcome of confrontations . . . The tactics employed by highly skilled hostage-barricade team members in containing, controlling, and resolving these dramatic events is perhaps the most obvious illustration of how police need not simply passively accept all situations as they present themselves but can shape events and outcomes in desirable directions . . . (pp. 328–329)

LEADER IN COMMUNITY POLICING

Bill Closs, Chief of Police, Kingston Police

What is your understanding of "community policing" and its importance to the evolution of policing in Canada?

I have seen and heard so many definitions of community policing, have been told or read many times how it is applied in various communities right across Canada and the United States, that it really doesn't matter how anybody in the community, be it police officer, politician, or academic defines it. With everybody putting their own "spin" on community policing, I am of the belief that it is a philosophy, or a way of thinking about police service delivery, that enhances the community's quality of life. The one hundred definitions I have heard create confusion but are probably all correct because they usually contain buzz words such as prevention, problem solving, and partnerships. What is most important for police agencies and police leaders to recognize is that individual communities and neighbourhoods play the key role in determining what it is and how it is applied. Community policing means one thing to the elderly, another to the downtown business association, something else to people living on the streets, in the suburbs, or in low income housing complexes. Whatever community policing may be to a village of 1000 may be different to a city of 100 000. Whatever community policing might mean to a university

Bill Closs LEADER IN COMMUNITY POLICING continued

may be different from what it means to an elementary school. Whatever it may mean to an urban city core may be different than what it means to our rural constituents. There are constants. Community policing always involves community participation, community involvement, community consensus, community decision-making, and community priority setting. It is the single most important driving force for moving police organizations away from the mentality of the "thin blue line," "us against them," in the so-called "war against crime." It requires police leaders to recognize that the authority of rank must be tempered with the empowerment of the officer and citizen on the street.

Outline your own contribution to the development of "community policing" in Canada.

As a Chief Superintendent with the Ontario Provincial Police, I was a founding member of a small group of committed volunteers who organized a 1994 Community Policing Conference which developed into a professional association, known as the Community Policing Advisory Council. Today, as Kingston's Police Chief, my focus is on creating internal and external processes that empower street officers and members of our community to influence policing and the quality of life in our city. This is being accomplished in various ways such as: community development training for police officers that enhances their conflict resolution and problem solving skills; the Chief's Advisory Council that brings together various people and agencies throughout the city in a consultative process; empowering street officers through the creation of neighbourhood teams and team leaders across police platoon structures; formally entrenching in the police organizational structure community volunteers who patrol our city streets; initiatives requiring all new police recruits to spend time working with community groups, such as the Salvation Army, detoxification centres, and shelters for the homeless.

What is your view of the future of "community policing" in Canada?

Evolution takes time, but community policing will continue to make its way into police organizations and communities, improving the way we do business together. It will not be easy—we are talking about changing one hundred years of the "police way" of doing things, of the police way of thinking, of the police hierarchy. If you are a community person wanting to participate in a meaningful way, you may view its progress as moving at a snail's pace. If you are a street officer committed to the philosophy, waiting for the empowerment and flexibility, the organization is moving too slowly. On the other hand, some police supervisors, managers, and leaders may feel threatened by community policing, hoping

Bill Closs LEADER IN COMMUNITY POLICING continued

to see progress and change creep along as slowly as possible. In the end, the people and their community will move forward, while some police managers will be shoved aside.

What other comments would you like to make on this topic?

The community has often been told that the police work in partnership with them. What does "partnership" mean? The police will participate, or agree, as long as the respective police leader/manager says "OK," as long as it's covered in police policy and procedures. Is the partnership 80/20, 60/40, or 50/50?

I wonder where "labour" fits into the community policing equation. Police associations/unions have a shared responsibility to look after their officers, make sure they've got the best of equipment, clothing, and vehicles. But the community pays for it all! Is there a role for police associations in community policing? You bet there is . . . Labour leaders should respond by encouraging their membership to be flexible and cooperative with the community.

Our citizens buy into community policing because they are involved, taking action, sharing in the responsibility, taking ownership. Patrol officers on the street accept the philosophy because it allows them to become closer with the community, empowered with flexibility and influence. If only police managers could leave behind their "baggage" of twenty-five years in a policing career where the norm was taking orders in a paramilitary structure. Finally it's their turn to be the boss. Just when these managers get the "rank," the "power," the "control," along comes community policing asking them to share it all with the people.

The one major impediment to the advancement of the community's role in policing is police managers holding on to their fantasy of power and control. True leaders understand their power comes from the people, not from the braid on the uniform.

THE INTERNET, KNOWLEDGE, AND COMMUNITY INTERESTS

Manning (1998) speaking about the Internet underscores its potential as an excellent source of information:

> . . . the Internet can serve as a convenient community resource. Web sites on the Internet can easily provide crime statistics (eventually real-time crime statistics and interactive maps), crime prevention information, links to other criminal justice resources, and contact information for any unit or even individual within an organization. (p. 99)

Davenport and Prusak (1998), who have written extensively on the creation of knowledge within organizations, see the merits of the Internet for any type of corporation (public or private) that needs to collect, analyze, disseminate, and evaluate large amounts of information. The goal of these activities is to create real knowledge that can function to increase the working capacity of all individuals within the organization. They note the successes of companies such as Hewlett-Packard that have highly effective web-based Intranets. They have considerable praise for the Internet and provide an insight into the ways in which it can be turned to good effect for community policing:

> The best example of a broad knowledge repository is the Internet. As a source of outside knowledge, the Internet can overcome some of the disadvantages of the locales and asymmetry of knowledge, since a subject search will return results from the whole system. Localness is not an issue because it does not matter to the user where the material resides. The hypertext that has made the World Wide Web such a success allows related content to be linked regardless of its physical location. (p. 130)

Unfortunately, everyone has access to the Internet, and this fact has given rise to the challenge of controlling the illegal and criminal uses of this medium. Hate crimes, fraud, pornography are but three of the offences that have been catapulted into new realms of activity due to the incredible reach and the anonymity of the World Wide Web. There has never been such an explosion of hate literature, nor the capacity for like-minded individuals to feed off each other's support through the Internet. Examples of this have been documented by the Anti-Defamation League (1998), and it poses a real threat to community policing when it becomes impossible to control what seeps into your "community" through such a channel.

Furthermore, the Internet has not reached a stage of perfection in the quality of its content, nor is it likely to in the near future, given its ungovernable nature. As noted by Davenport and Prusak (1998):

> Although these systems partly solve the problem of locating knowledge, they often compound the problem of judging the knowledge that is being provided. As anyone who has carried out an Internet search knows, the overwhelming majority of "hits" supplied by search engines are irrelevant or worthless. A tremendous amount of time can be wasted sorting through trash to find a few treasures. (p. 131)

But, again, the technologies that are brought to bear on the challenges of modern policing cannot be applied piecemeal, or without some serious contemplation of their impact on the overall direction and relational dynamics of the organization. Community policing sets much stock in the partnership between the public and the police. Technologies that hasten the deepening of this partnership may be seen as good. However, it has been demonstrated that police organizations are resistant to change, and one of the points of resistance is the sharing of information, beyond the most pasteurized or mundane, with members of the public.

DNA PROFILING, VICLAS, AND CRIMINALISTICS

Several scientific refinements have been made over the last several decades that have had a real impact on the way in which police business is conducted in Canada. There are now applied techniques that appear to make the scientist or engineer virtually indispensable to the policing enterprise. While police officers frequently develop the skills to apply these new techniques further enhancing the quality of policing in Canada, it is more often the case that police departments develop a substantial reliance on the "experts" who have introduced these new technologies and, as a result, become closely tied with those organizations, institutions, and agencies that are able to sustain the research and development necessary for their perpetuation. Whether or not this constitutes a threat to the capacity of a police organization to vigorously pursue a vision of community policing remains to be determined. However, it rests with Canadian police leaders to ensure that the demands of community policing are not overwhelmed by technologies that are inaccessible to the public, or beyond the reach of public accountability mechanisms.

This section will look at three areas in which sophisticated techniques or technologies have been directly applied to the "business" of policing:

- DNA profiling;
- ViCLAS; and
- Criminalistics.

DNA Profiling

With the discovery of Deoxyribonucleic Acid (DNA), called the building block of life, there has been a growth in the technologies devoted to the identification of possible suspects (Solicitor General Canada, 1996; Saferstein, 1998). This represents the application of scientific method to the investigation of crime, and has been highly publicized throughout North America and Europe in many dramatic cases. In Canada, there has been substantial support for the use of DNA profiling, and it has been advanced through legislation and other government-sponsored initiatives.

ViCLAS

The Violent Crime Linkage Analysis System (ViCLAS) was "... designed to capture, collate, and compare predatory/sexual crimes of violence through the analysis of victimology, suspect description, modus operandi, forensic and behavioural data." (Ontario. Bernardo Investigation Review, 1996). This system has been operated through the Behavioural Sciences Section of the OPP since December 1993, and has been recognized throughout the world as the most effective automated system of its type currently available. As a result of its promise for continuing success, Mr. Justice Campbell recommended that there be mandatory ViCLAS reporting by all Ontario police services of crimes that fall under the submission criteria established. This recommendation was undertaken as part of the recent amendments to the *Police Services Act*. Furthermore, several other jurisdictions in the U.S., Europe, and Australia are considering the introduction of ViCLAS as an automated investigation tool.

Criminalistics

Criminalistics is composed of many elements drawn from what may be called forensic science. Saferstein (1998) indicates that forensic science:

> . . . is the application of science to those criminal and civil laws that are enforced by police agencies in a criminal justice system. (p. 2)

In order to see the range of disciplines that fall under the "forensic" category, Saferstein (1998) includes the following:

- Forensic pathology;
- Forensic anthropology;
- Forensic entymology;
- Forensic psychiatry;
- Forensic odontology; and
- Forensic engineering.

Furthermore, it may be useful to consider the extensive components that make up the field commonly referred to as criminalistics. Clearly there is an enormous investment in the pursuit of scientific and technical expertise and knowledge for the purposes of modern policing. When community policing is pursued as a vision for the delivery of service, it must be able to somehow acknowledge and accommodate the massive expenditures made in the following areas:

- *Organic analysis*—including analytical techniques, chromatography, spectrophotometry;
- *Inorganic analysis*—including the emission spectrum of elements, atomic absorption spectrophotometry, neutron activation analysis, x-ray diffraction;
- *Hair, fibres, and paint analysis*;
- *Drugs*—including the identification, collection, and presentation of drug evidence;
- *Forensic toxicology*—including the toxicology of alcohol;
- *Forensic aspects of arson and explosion investigations*—including the chemistry of fire, analysis of flammable residues, and types of explosives;
- *Forensic serology*—including the characteristics of bloodstains, stain patterns of blood, forensic characterization of semen;
- *DNA*—including replication of DNA and DNA typing;
- *Fingerprints*—including classification of fingerprints, automated fingerprint identification systems, digital imaging for fingerprint enhancement;
- *Firearms, toolmarks, and other impressions*—including bullet comparisons, automated firearm search systems, gunpowder residues, collection and preservation of firearm evidence; and
- *Document and voice examination*—including handwriting comparisons, typewriting comparisons, and voice examination.

SUMMARY

Can a vision of the future of policing that places the highest priority on community partnerships, the "co-production" of social order, and other integrated approaches that constitute the principles of community policing be pursued within the context of a technological society? There is no easy answer to this question. However, it is warranted to consider the reality of the dominance of technique in modern society. While it may not be apparent that the progression is to an oppressive style of policing, the words of Ellul (1964) are sobering, to say the least:

> At a still more advanced stage of police technique, even . . . diffuse terror gradually dissipates. The police exist only to protect "good citizens." They no longer carry out raids and there is nothing mysterious about them; therefore they are not felt to be oppressive. Police work has become "scientific." Their files contain dossiers of every citizen. The police are in a position to lay hands on anyone "wanted" at any moment, and this obviates to a great degree the necessity of doing so. No one can evade the police or disappear. But, then, no one wants to. An electronic dossier is not particularly fearsome. (p. 413)

There is no question that some technological developments have made crime control easier and offered solutions to the problems of social order. DNA profiling has led to several convictions (the O.J. Simpson case notwithstanding). The ViCLAS program pioneered by the OPP has been of interest and importance to policing, and has received the solid support of several authorities (Ontario. Bernardo Investigation Review, 1996). The research into non-lethal force technologies and techniques is valuable for the officers they may protect and the citizens that may be shielded from more extreme levels of force. Advances in information technology offer some promising options for managing the prodigious amounts of data that police organizations are seeking to transform into useful information or knowledge (Davenport and Prusak, 1998).

The advances that continuously burst forth from the technological enterprises that feed law enforcement have varying degrees of absolute value. This chapter has attempted to examine the challenge of technology in the context of community policing. It is clear that technology cannot be seen as a "holy grail" for the problems of policing in Canada. However, it is to be hoped that technology does not reveal itself as a Pandora's box, producing more problems than it solves.

QUESTIONS FOR CONSIDERATION

1. Do the working definitions of technology and technique cover the range of methods and materials applied by modern policing?
2. What are the concerns surrounding technology that led police organizations toward the development of community policing?
3. How can technology be applied for the benefit of community policing?
4. What are the elements of the typology of police technology presented in the chapter?
5. How has the Internet contributed to the development of community policing?
6. How has the Internet complicated policing in Canada, and elsewhere?
7. Are there ways in which weapons and surveillance research can contribute to community policing in Canada?
8. What are the benefits of crime mapping, and other techniques of crime analysis, from a community policing perspective?

9. What are some ways in which technology can be used to generate knowledge about crime patterns and public order issues for the benefit of the community?

REFERENCES

Albanese, Jay S. (1988). "Tomorrow's thieves." *The Futurist*, Vol. 6, no. 5, pp. 24–28.

Albritton, James S. (1995). "The technique of community-oriented policing: an alternative explanation." In Kratcoski, Peter C. and Duane Dukes (eds.). *Issues in community policing*. Highland Heights, Ky.; Cincinnati, Ohio : Academy of Criminal Justice Sciences, Northern Kentucky University; Anderson Publishing Co.

Alpert, Geoffrey P. and Roger G. Dunham (1995). *Police use of deadly force: statistical analysis of the Metro-Dade police department*. Washington, D.C.: Police Executive Research Forum.

Anti-Defamation League (1998). *Explosion of hate: the growing danger of the National Alliance*. New York: Anti-Defamation League.

Barley, S.R. (1986). "Technology as an occasion for structuring." *Administrative Science Quarterly*, Vol. 31, pp. 78–108.

Barley, S.R. (1988). "Technology power and the social organization of work." In Bardach, S. (ed.), *Research in the sociology of organizations: vol. 6*. Greenwich, Conn.: JAI Press.

Bayley, David (1991). "Strategy." In Chacko, James and Stephen E. Nancoo (eds.). *Community policing in Canada*. Toronto: Canadian Scholars' Press Inc.

Block, Carolyn Rebecca (1995). "STAC hot-spot areas: a statistical tool for law enforcement decisions." In Block, Carolyn Rebecca, Margaret Dabdoub and Suzanne Fregly (eds.) (1995). *Crime analysis through computer mapping*. Washington, D.C.: Police Executive Research Forum.

Block, Carolyn Rebecca, Margaret Dabdoub and Suzanne Fregly (eds.) (1995). *Crime analysis through computer mapping*. Washington, D.C.: Police Executive Research Forum.

Block, Richard (1995). "Spatial analysis in the evaluation of the 'CAPS' community policing program in Chicago." In Block, Carolyn Rebecca, Margaret Dabdoub and Suzanne Fregly (eds.) (1995). *Crime analysis through computer mapping*. Washington, D.C.: Police Executive Research Forum.

Buerger, Michael E. and Ellen G. Cohn, and Anthony J. Petrosino (1995). "Defining the 'hot spots of crime': operationalizing theoretical concepts for field research." In Eck, John E. and David Weisburd (eds.). *Crime and place*. Monsey, N.Y.; Washington, D.C.: Criminal Justice Press and the Police Executive Research Forum. (Crime prevention studies; no. 4).

Canadian Police Research Centre (1996). *Annual report 1995-1996*. Ottawa: Minister of Supply and Services Canada.

Canadian Police Research Centre (1998). *Canadian Police Research Centre 1997-1998 annual report*. Ottawa : Minister of Supply and Services Canada.

Cassels, David (1995). "The role of technology on crime and policing." Paper presented in the Management challenges in 21st century policing conference proceedings.

Chaiken, J.M. (1975). *Patrol allocation methodology for police departments*. Santa Monica: RAND.

Chaiken, J. M. et al. (1975). *Criminal justice models: an overview*. Santa Monica, Calif.: Rand.

Chatterton, M. (1991). "Organizational constraints on the use of information and information technology in problem-focused area policing." Unpublished paper presented to the British Criminology Conference, York, Great Britain.

Clede, Bill and Keven Parsons (1987). *Police nonlethal force manual: your choices this side of deadly*. Harrisburg, Penn.: Stackpole Books.

Connor, Gregory J. (1990). "Use of force continuum: phase II." *Law and Order* (March), pp. 30–32.

Davenport, Thomas H. and Laurence Prusak (1998). *Working knowledge: how organizations manage what they know*. Boston: Harvard Business School Press.

Drummy, Patrick (1997). "Public access to police information." In Reuland, Melissa Miller (ed.) *Information management and crime analysis*. Washington, D.C.: Police Executive Research Forum.

Eck, John E. and David Weisburd (eds.) (1995). *Crime and place*. Monsey, N.Y.; Washington, D.C.: Criminal Justice Press and the Police Executive Research Forum. (Crime prevention studies; no. 4)

Ellul, Jacques (1964). *The technological society*. Translated from the French by John Wilkinson. New York: Vintage Books.

Ellul, Jacques (1990). *The technological bluff*. Translated by Geoffrey W. Bromiley. Grand Rapids, Mich.: William B. Eerdmans Publishing Co.

Fraser, Craig B. et al. (1998). *Challenge to change: the 21st century policing project*. Washington, D.C.: Police Executive Research Forum.

Geller, William A. and Michael S. Scott (1992). *Deadly force: what we know: a practitioner's desk reference on police-involved shootings*. Washington, D.C.: Police Executive Research Forum

Geller, William A. and Guy Swanger (1995). *Managing innovation in policing: the untapped potential of the middle manager*. Washington, D.C.: Police Executive Research Forum.

Grant, George (1969). *Technology and empire: perspectives on North America*. Toronto: House of Anansi.

Grant, George (1986). "Thinking about technology." In Grant, George, *Technology and justice*. Toronto: Anansi.

Greenwood, P.W. and J. Petersilia (1975). *The criminal investigation process volume 1: summary and policy implications*. Santa Monica, Calif.: RAND Corporation.

Harries, Keith (1990). *Geographic factors in policing*. Washington, D.C.: Police Executive Research Forum.

Hicks, Douglas and Jim Wilson (1995). "Geographic analysis to support community policing: an information foundation for the future." In Block, Carolyn Rebecca, Margaret Dabdoub and Suzanne Fregly (eds.) (1995). *Crime analysis through computer mapping*. Washington, D.C. : Police Executive Research Forum.

Hoover, Larry T. (ed.) (1992) *Police management: issues & perspectives*. Washington, D.C.: Police Executive Research Forum.

Hough, J.M. (1980a). "Managing with less technology: the impact of information technology in police management." *British Journal of Criminology*, Vol. 20, no. 4 (October), pp. 344–357.

Hough, J.M. (1980b). *Uniformed police work and management technology*. London: Her Majesty's Stationery Office. (Home Office Research and Planning Unit paper; no. 1).

LaVigne, Nancy G. and Julie Wartell (eds.) (1998). *Crime mapping case studies: successes in the field*. Washington, D.C.: Police Executive Research Forum.

Lyons, Thomas, Gail O. Burton and Richard C. Sonnichsen (1989). "Law enforcement views on future policing issues." In Loree, Donald J. (ed.). *Future issues in policing: symposium proceedings*. Ottawa: Canadian Police College.

Manning, Walt W. (1997). "Criminal justice technology: planning for the future." In Reuland, Melissa Miller (ed.) *Information management and crime analysis*. Washington, D.C.: Police Executive Research Forum.

McCracken, Grant (1997). *Plenitude: volume 1 of the series: culture by commotion*. Toronto: Periph.: Fluide.

McEwen, J. Thomas and Faye S. Taxman (1995). "Applications of computer mapping to police operations." In Eck, John E. and David Weisburd (eds.). *Crime and place*. Monsey, N.Y.; Washington, D.C.: Criminal Justice Press and the Police Executive Research Forum. (Crime prevention studies; no. 4).

Meyer, Greg (1992). "The sorry state of police tactics: what to do?" *Crime Control Digest*, Vol. 1 (June), pp. 3–5.

Moland, Robert (1997). "Investigative technologies." In Reuland, Melissa Miller (ed.) *Information management and crime analysis*. Washington, D.C.: Police Executive Research Forum.

Nogala, Detlef (1995). "The future role of technology in policing." In Brodeur, Jean-Paul (ed.). *Comparisons in policing: an international perspective*. Aldershot : Avebury.

Normandeau, André and Barry Leighton (1990). *A vision of the future of policing in Canada: police-challenge 2000: background document*. Ottawa: Solicitor General of Canada, Police and Security Branch.

Ontario. Bernardo Investigation Review (1996). *Bernardo investigation review: report of Mr. Justice Archie Campbell*. Toronto: The Bernardo Investigation Review.

Ontario. Strategic Planning Committee on Police Training and Education (1992). *Report on future policing issues for Ontario*. Toronto: Ministry of the Solicitor General.

Ontario. Task Force on Policing in Ontario (1974). *The Task Force on Policing in Ontario: report to the Solicitor General*. Toronto: The Task Force.

Peak, Kenneth J. and Ronald W. Glensor (1996). *Community policing and problem solving: strategies and practices*. Upper Saddle River, N.J.: Prentice Hall.

Reiner, Robert (1995). "Community policing in England and Wales." In Brodeur, Jean-Paul (ed.). *Comparisons in policing: an international perspective*. Aldershot: Avebury.

Rengert, George F. (1995a). "More than just a pretty map: how can spatial analysis support police decisions?" In Block, Carolyn Rebecca, Margaret Dabdoub and Suzanne Fregly (eds.) (1995). *Crime analysis through computer mapping*. Washington, D.C.: Police Executive Research Forum.

Rengert, George F. (1995b). "Comparing cognitive hot spots to crime hot spots." In Block, Carolyn Rebecca, Margaret Dabdoub and Suzanne Fregly (eds.) (1995). *Crime analysis through computer mapping*. Washington, D.C.: Police Executive Research Forum.

Reuland, Melissa Miller (ed.) (1997). *Information management and crime analysis*. Washington, D.C.: Police Executive Research Forum.

Rewers, Ronald F. and Lynn A. Green (1995). "The Chicago area four geoarchive: an information foundation for community policing." In Block, Carolyn Rebecca, Margaret Dabdoub and Suzanne Fregly (eds.) (1995). *Crime analysis through computer mapping*. Washington, D.C.: Police Executive Research Forum.

Rich, Thomas (1998). "Crime mapping by community organizations: initial successes in Hartford's Blue Hills neighborhood. In Lavigne, Nancy and Julie Wartell (eds.) *Crime mapping case studies: successes in the field*. Washington, D.C.: Police Executive Research Forum.

Robbins, Stephen R. and Nancy Langton (1999). *Organizational behaviour: concepts, controversies, applications. Canadian edition.* Scarborough, Ont.: Prentice Hall Canada.

Rosoff, Stephen M, Henry N. Pontell and Robert Tillman (1998). *Profit without honor: white-collar crime and the looting of America*. Upper Saddle River, N.J.: Prentice Hall.

Rossmo, D. Kim (1995a). "Place, space, and police investigations: hunting serial violent criminals." In Eck, John E. and David Weisburd (eds.). *Crime and place*. Monsey, N.Y.; Washington, D.C.: Criminal Justice Press and the Police Executive Research Forum. (Crime prevention studies; no. 4).

Rossmo, D. Kim (1995b). "Strategic crime patterning: problem-oriented policing and displacement." In Block, Carolyn Rebecca, Margaret Dabdoub and Suzanne Fregly (eds.) (1995). *Crime analysis through computer mapping*. Washington, D.C.: Police Executive Research Forum.

Saferstein, Richard (1998). *Criminalistics: an introduction to forensic science. 6th ed.* Upper Saddle River, N.J.: Prentice Hall.

Simmons, John and Dennis J. Kenney (1997). "The stages of crime analysis." In Reuland, Melissa Miller (ed.) *Information management and crime analysis*. Washington, D.C.: Police Executive Research Forum.

Skolnick, Jerome H. and David H. Bayley (1986). *The new blue line: police innovation in six American cities*. New York: The Free Press.

Solicitor General Canada (1996). *Establishing a national DNA data bank: a consultation document*. Ottawa: Minister of Supply and Services Canada.

Sparrow, M. (1993). "Information systems and the development of policing." Washington, D.C.: National Institute of Justice and Program in Criminal Justice Policy and Management, John F. Kennedy School of Government.(Perspective on policing; no. 16)

Thibault, Edward A., Lawrence M. Lynch, R. Bruce McBride (1998). *Proactive police management. 4th ed.* Upper Saddle River, N.J.: Prentice Hall.

U.S. President's Commission on Law Enforcement and Administration of Justice (1967). *The challenge of crime in a free society: a report by the President's Commission on Law Enforcement and Administration of Justice*. Washington, D.C.: U.S. Government Printing Office.

Volti, Rudi (1995). *Society and technological change. 3rd ed.* New York: St. Martin's Press

TOWARD MORE LEARNING

One valuable approach to learning more about the uses of technology in policing is to check the web sites for any number of Canadian police organizations to see how they are using the Internet as a medium for advancing their commitment to the principles of community policing. For a deeper understanding of the types of technologies and techniques currently under review and consideration in Canada, it is suggested that the website for the Canadian Police Research Centre be consulted.

Canadian Police Research Centre
Website: http://www.cprc.org

The National Criminal Justice Research Service (NCJRS), sponsored by the U.S. National Institute of Justice provides a substantial amount of funding to various research projects that will develop new technologies and applications for police organizations. The research conducted under the auspices of the NIJ often pertains to the application of community policing, and it frequently supports projects such as those discussed in this chapter that have the potential to engage the public more deeply in the work of policing.

National Criminal Justice Research Service
Website: http://www.ncjrs.org

Also of interest for those studying new technological developments in the area of policing is the Office of International Criminal Justice. This agency maintains a website that has several components, including the following publications:

Crime and Justice International;
CJ Technology News;
The Keeper's Choice; and
Criminal Organizations

The website covers press releases, reports, and news articles originating from several criminal justice sources. Users can browse the Criminal Justice Internet Applications Online Handbook.

Website: http://www.acsp.uic.edu/

On an international front, the European Union offers an interesting example of transnational cooperation among law enforcement agencies, frequently involving technological innovations that support such cooperation. Formally inaugurated in 1993, the European Union (EU) is currently comprised of: Austria, Belgium, Denmark, Finland, France, Germany, Greece, Ireland, Italy, Luxembourg, the Netherlands, Portugal, Spain, Sweden, and the United Kingdom. The following websites may be interest for further study:

Website: http://europa.eu.int/index.htm—a comprehensive, searchable site that provides a gateway to details on all aspects of the EU;

Website: http://europa.eu.int/abc/doc/off/bull/en/welcome.htm—the Bulletin of the European Union, which constitutes the official report on the activities of European Community institutions; and

Website: http://europa.eu.int/comm/sg/scadplus/scad_en.htm—this site contains EU criminal justice policies and actions. Click on "European Union Policies" then "Area of Freedom, Security and Justice." This area contains several items relating to the EUROPOL Convention.

CHAPTER NINE

In Moulmein, in lower Burma, I was hated by large numbers of people—the only time in my life that I have been important enough for this to happen to me. I was sub-divisional police officer of the town, and in an aimless, petty kind of way anti-European feeling was very bitter . . . As a police officer I was an obvious target and was baited whenever it seemed safe to do so.

George Orwell (1962). "Shooting an elephant." In *Inside the whale.* London : Penguin Books, p. 91.

"FRIENDLY FIRE"—QUESTIONING THE ASSUMPTIONS AND ASSERTIONS OF COMMUNITY POLICING

LEARNING OBJECTIVES

1. Identify several critiques of the philosophy and practice of community policing.
2. List specific weaknesses found in the theory and practice of community policing.
3. Develop a broader understanding of the complexity of the community policing issue in Canada.

INTRODUCTION

This chapter will consider some of the criticisms that have been levelled at community policing, both as a philosophy and as a practice. The majority of those who question the assumptions or assertions associated with community policing do so out of a genuine interest in improving the state of policing. While these criticisms derive from different intellectual and ideological sources, they are typically aimed at the betterment of public safety and security. Some of the arguments are focused more on the application of community policing rather than attacking the soundness of community policing as a viable theory of modern policing. In any event, if community policing is to survive as an important approach to the challenge of policing in Canada, both theoretical and practical concerns will need to be addressed in a direct and intelligible manner (Weatheritt, 1985).

There will be an examination of the following groups and their contributions to the debate over community policing:

- Police practitioners (including police associations);
- Police governing authorities;
- Academic community;

- Community groups; and
- Municipal governments.

Each of these groups has a distinctive perspective on the validity and viability of community policing, and each perspective should be assessed in order to arrive at a deeper appreciation of the potential of community policing as a functional philosophy for Canadian police organizations. Of course, there may be overlap in the perspectives advanced by each of these groups, and none of the groups identified subscribes to a monolithic viewpoint on this issue of community policing. What this chapter hopes to encourage is the careful scrutiny of community policing as a means of policing that permits open dialogue and discussion, moving toward systematic improvement in the delivery of services to the public.

POLICE PRACTITIONERS

All those who actually deliver police services may be considered "police practitioners." Therefore, this is a fairly broad category that, for the purposes of this chapter, will include the following key groups:

- Police executives and senior managers;
- Police associations;
- Police middle managers; and
- Front-line officers.

The concerns raised by these groups with respect to community policing may be distinct from one another in substance. However, they all issue from essentially the same source: one based on direct experience with the actual, day-to-day, operational delivery of policing services. Certainly the chief of police is further removed from the immediate contingencies of patrol than the front-line constable, but both individuals carry the organization's responsibility for the delivery of policing services to the community in a real and tangible way. While it is often difficult to differentiate between overall *organizational* policy and *operational* policy in a police department (Murphy and Muir, 1985), it may simplify the presentation in this chapter to suggest that the concerns raised in this section pertain to those originating with individuals who function to offer policing services to the public. They are the individuals who wear the department's uniform, or directly draw their livelihood from the police department.

One pre-eminent critic of much of the conventional wisdom surrounding community policing is Chris Braiden (see "Leader in Community Policing" profile, this chapter). Having had an opportunity to consider Canadian policing as an enlightened observer while seconded to the federal Solicitor General's office, Braiden was able to view developments in policing with a higher degree of knowledgeable objectivity than most serving officers enjoy. As a result of his careful study, he has arrived at some valuable insights that warrant the attention of anyone looking to embark upon or to understand the nature of community policing in Canada.

Braiden has made the point that community policing is, in fact, nothing new, and draws upon a deep well of experience that has been available to police officers for many decades. As indicated in Chapter 1, there are roots of community policing that are easily seen through a consideration of the available historical record. As Braiden (1990) indicates:

> To suggest that community policing is new is to suggest that up until now, we have been policing someone else. When one goes fishing, there is no need to say "fishing for fish." Fishing says it all. So does policing. The adjective "community" is redundant; so are the arguments for and against community policing. For me, that community policing is the cause is a settled issue; what the problems are and how they should be attacked in a given community is another matter. (p. 107)

Of course, there are layers of complexity that Braiden has chosen to leave undisturbed in his presentation on community policing. Policing as a human activity is not as simple as fishing. Indeed, fishing itself may be seen as substantially complicated when one considers the multitude of types of fish, or the wide range of techniques and tools used to achieve successful results in fishing. Policing has always been directed towards the "community" in some fashion. However, it has never been a simple matter to define that community, or to state that what the police have offered the community has always and everywhere been the same. Braiden's point is well taken, if he intends to tell his readers that policing has always been embedded in communities and that the purpose of policing is to support communities in a significant way. The argument about community policing, however, has been joined, and it is important that people who deliver policing services do not forget, as seems to have occurred during the period of "professional" policing, that the community is the focus of activity, not some nebulous goal of efficiency. Community policing also implies that the community is part of the policing equation, and not simply a passive recipient of the service. The concept of partnership is central to this type of policing model, and constitutes part of what differentiates community policing from the "old school" of policing.

Another point made by Braiden (1990) relates to the earlier investments in crime prevention, which can be seen as a precursor to community policing. When crime prevention became popular, there were many police departments that introduced programs under this umbrella. The results of such programs were not really targeted to the problems that community policing is designed to address. As Braiden observes:

> Over the past two decades, there has been a great proliferation of crime prevention units in police organizations. These units are usually staffed by people who, soon after becoming police officers, decide they don't really like arresting drunks but do like the pay and the permanency so they stay, but dodge the bullet. They become crime prevention specialists, usually attired in suits with a briefcase who specialize in giving lectures, invariably to the converted, about crimes they will probably never encounter and who are urged to call the cops with their slightest worries. Unfortunately, these people are noticeable by their absence in the neighborhoods most in need of their services. (p. 98)

Clearly, Braiden is critical of the *appearance* of action on the part of police departments and individual officers, when real problems remain to be addressed in real neighbourhoods. Certainly, the idea of a "program" of crime prevention, with its phalanx of pamphlets, does not conform to the underlying principles of a philosophy of community policing that requires a fundamental re-examination of the whole structure of the police organization, and an entirely new way of attacking the problems of crime and disorder in conjunction with community members. Braiden's exposure to the recent history of policing in Canada has resulted in words of warning for those organizations that only intend to dabble in the realm of community policing (1990):

> Many organizations claim to be "into" community policing. From what I have seen, I'm skeptical. In most cases, copious policy is written and lectures given by bosses to Grunts but little structural or intellectual change at the management level results . . . It is imperative that organization restructuring occurs <u>first</u> to accommodate community policing. If it is superimposed on convention, convention will win out and we will have a repeat of the team policing debacle of the 1970s. (pp. 108–109, emphasis in original)

Police organizations have, out of necessity in many instances, become extremely adept at projecting a positive image. Part of this arises out of the constant media scrutiny that police departments are subjected to, as well as their legitimate desire to manage (even manipulate) public impressions (Manning, 1985).

In correspondence received from a chief of police in charge of a small department in Ontario, the following expression of uncertainty about community policing is enlightening:

> Thank you for your consideration of this small Police Service when compiling data to complete the global and monumental task of outlining "Foundations of Community Policing." Unfortunately, all I can offer you by way of assistance has been gained via street experience, intuition and a total commitment to maintaining a peaceful way of life for the Village I serve.
>
> I am confused by the term "Community Policing," having been a Municipal police officer in three other departments over the past thirty years and fulfilled then what that term seems to mean today. My current . . . Police Service has no formal plan or rules to go by when "Community Policing." However, I feel that what we do in the name of community "well-being" is successful.

This chief's point of view is difficult to argue with, given the clearly stated commitment to the "well-being" of his community and his conscious blending of "street experience" and "intuition." There is much to argue in favour of the rural roots of community policing, and Murphy (1985) sees the seeds of this model in Canadian policing practice. When police services have never become isolated from their public, the "discovery" of community is much like finding something that was never lost.

Ex-commissioner of the RCMP, Norman Inkster (1993) has noted that there was some doubt among police practitioners about the whole proactive approach to policing:

> I recognize there are those concerned individuals who harbour legitimate concerns that preventive policing is a trendy concept only recently in vogue, one perhaps antithetical to our own traditions, and therefore subject to deep suspicion. (p. 236)

Bennett (1994) refers to Reiner's (1991) extensive survey of chief constables in Great Britain in order to indicate that support for community policing is not absolute:

> When asked specifically about community policing, 55% declared some support for the notion, whereas 45% thought that it was a meaningless expression. A common view of chief constables was that community policing was no more than traditional policing or "old wine in new bottles." (p. 231)

THE POLICE MIDDLE MANAGER

While several observers have diagnosed significant potential difficulty at the middle management layer of police organizations (Sadd and Grinc, 1994; Zhao, Thurman, and Lovrich, 1995), others have looked upon this level as a source of support for innovation (Geller and Swanger, 1995). Trojanowicz and Bucqueroux (1990) note the dilemma of the police middle manager in the context of community policing by way of reference to Robert Lunney (see "Leader in Community Policing" profile, Chapter Two):

> According to Robert Lunney, who has been called the father of Community Policing in Canada, middle managers fall into three groups: those who will support the new concept's goals and orientation immediately, those who will become converts over time, and those who will never embrace the philosophy. (p. 351)

Roberg (1994) points to the central importance of the police middle manager in moving toward community policing:

> Successful implementation of community policing will depend on how well mid-level police managers, most of whom have based their careers on traditional policing, accept such changes. It has been well documented that the team policing experiments failed, to a large extent, because of mid-level managers' reluctance to accept the proposed changes in philosophy and operating style. (p. 252)

Finally, police leaders need to prepare their middle managers with a convincing and compelling role to play in the new world of community policing. As highlighted by Mastrofski (1998):

> As top police executives attempt to implement the many reforms of community policing, they often meet resistance from middle managers who achieved their status by pursuing another model of police organization, or who are threatened by the uncertainty of changing the status quo, or who have superiors who cannot conceive or articulate a useful role middle managers might play. (p. 173)

At one extreme in the continuum of organizational changes that may be seen to have been begun with the interest in community policing is the whole notion of private policing (Shearing and Stenning, 1981; Shearing, 1982; Shearing and Stenning, 1987; Stenning, 1989). If community policing calls forth a more extensive engagement of the community to address crime and disorder problems, it also implies that the whole enterprise of policing may be done in ways that are different from the norm; that is public policing. By widening the net to include a range of approaches that are not strictly limited to conventional public police officers, there are more opportunities for innovation. As noted by Johnston (1998) with regard to recent experience in Great Britain:

> Not all police responses . . . [to the sectoral restructuring of the police] have, however, been defensive. At least two British chief constables have proposed that public police organizations be permitted to establish their own security companies in order to compete at a local level with the private sector. Such companies . . . might be set up as trusts, all revenue being plowed back into public police authority budgets. One proponent of this view . . . notes that its successful implementation would require the Home Office to make various policy changes. First, the

> police would need the authority to deploy patrols consisting of police auxiliaries—uniformed personnel with less training, fewer powers, and lower pay than regular officers. In addition, Blair suggests, consideration should be given to allowing police authorities the right to hire private security firms to work under police direction. Second, police authorities should be allowed to generate income from the many services that are, currently, offered free to the public, such as crime prevention advice. (p. 199)

POLICE ASSOCIATIONS

Police associations in Canada represent the collective bargaining interests of rank and file members of police departments. In that capacity, they have a legitimate and significant role in policing. Generally, police associations do not speak ill of the philosophy of community policing. However, the mechanics of its implementation frequently bring forward practical problems that cause difficulties for their membership. For example, the notion of police departments being encouraged, or even allowed, to engage a class of uniformed employee other than a fully-sworn, fully-trained police officer is subject to intense debate in Canada. Police associations are on record as being fundamentally opposed to this whole concept, and are critical of any recommendations that would facilitate the kind of broadening of the scope of auxiliary activities without very strict guidelines in place to control those activities. In Ontario, the collective view of the police associations is that auxiliary personnel should only be used to support and assist regular officers, not to augment their activities. As noted by the Police Association of Ontario (1996):

> In an environment supported with appropriate regulatory control, standards, and enforcement, a role could exist for trained auxiliary persons who can provide assistance to members of police services in extraordinary circumstances. The use of such persons must be restricted to special events, major emergencies or disasters, for which the members of the police force are not sufficiently numerous to deal with, provided that such functions do not require or involve:
>
> 1. The authority of a police officer;
> 2. The member to be armed with a firearm or weapon;
> 3. Emergency response;
> 4. Investigation, surveillance or forensic identification;
> 5. Law enforcement, arrest / laying of charges or provincial offence notices;
> 6. Preserving the peace or public order;
> 7. Performance of general police duties; or
> 8. The operation of police equipment or vehicles. (pp. 24–25)

The position taken in this instance clearly indicates that there is little acceptance for the idea, fundamental to community policing, that the public police are not the sole providers of safety, security, and order in the community. The police association presentation provides little hope that an entirely new category of officer, as referred to by Johnston, above, could come into being to extend a range of services currently offered by a police department in Canada. While issues of selection, training, uniform, equipment, and supervision are of considerable importance to the possible deployment of auxiliary members, it would appear from the stated position of the Police Association of Ontario, that a meaningful extension of public safety responsibilities cannot win favour with police unions at this time. The idea that police departments should "staff up" for all eventualities that relate to police services is entirely contrary to the relevant research on police effectiveness in areas such as rapid response, criminal investigations, or routine patrol.

The opposition of police associations to auxiliary members in policing is even further intensified in the case of volunteers. Police associations view such individuals as unpaid personnel who are not welcome in the world of professional policing. For example, on this issue, the Police Association of Ontario (1996) has stated that it is:

> . . . unequivocally opposed to the intrusion of unpaid volunteers in the workplace. Volunteers should not be involved in any activity which would otherwise constitute employment or work for a member of a police service.
>
> Opportunities exist for volunteers to contribute to community safety, through programs operated outside of the police service. These programs may include:
>
> 1. Providing comfort to victims of crime;
> 2. Assistance in searches for missing persons;
> 3. Neighborhood watch programs, but not neighborhood patrols;
> 4. Block parents; or
> 5. Apartment watch.
>
> Community policing is not effectively promoted or achieved through the use of volunteer patrols, "cop-fare" or "weekend warriors." These programs only serve to further deplete police resources and genuine interaction with the community at large, and create a false sense of security within the community. (p. 26)

The substance of this opposition to the use of auxiliary, or less than fully-sworn, members (with varying degrees of formal responsibility), or unpaid volunteers underlines a significant impediment to police and community leaders who wish to see a redefinition of the role responsibilities for public safety and security. Police associations, speaking essentially on behalf of their paid members, are looking upon policing as their workplace, and they are seeking to ensure that no incursions are made on their domain of employment. Unfortunately, public safety, community well-being, and individual security are not matters that are exclusively within the scope of a police association's collective bargaining powers. There will likely be much more discussion and discomfort around these issues as communities and police leaders grapple with the imperatives of community policing and public service.

Unfortunately, the challenge of policing in Canada relates to the inability of the traditional or "professional" approaches to address the problems of crime and disorder. It is not possible to continue adding police officers to the authorized strength of individual departments and expect that we will arrive at safer communities. It has been argued, in an American context, that in order to arrive at levels of police to violent crimes enjoyed in the 1960s, it would be necessary to add more than **five million** police officers (Walinsky, 1995). In this vein, Murphy (1993) insists that:

> [w]ithout a broader, more flexible policing perspective, we run the risk of having expensive, well equipped, resource rich police departments in communities bereft of the basic social and economic structures and services it takes to maintain community peace and order. (p. 60)

Stenning (1996) has offered an interesting perspective on an early section of the Ontario *Police Services Act*, dealing with the responsibilities of a municipal police services board:

> 31 (1) A board is responsible for the provision of police services and for law enforcement and crime prevention in the municipality.

Stenning's view is that this section could be considered as an opportunity for police services boards to use alternative approaches to public safety and gave them authority to make certain determinations about how best to engage in the delivery of three, distinct categories of activity (i.e., police services, law enforcement, and crime prevention). No definition of "police services" was offered in this version of the *Police Services Act*, and Stenning detected a lack of legislative clarity in this specific context. With the enactment of Bill 105, this section was amended to focus on "police services," which are defined early in the legislation:

> 31 (1) A board is responsible for the provision of adequate and effective police services in the municipality.
>
> 4 (2) Adequate and effective police services must include, at a minimum, all of the following police services:
>
> 1. Crime prevention.
> 2. Law enforcement.
> 3. Assistance to victims of crime.
> 4. Public order maintenance.
> 5. Emergency response.

In response to organized opposition to certain areas of police reform originating with police associations, it would be possible to circumscribe existing police departments with a minimal version of their previous responsibilities. Murphy (1993) envisions a minimal version of public policing which:

> . . . suggests that we see police exclusively as a reactive, crime focused service and that community policing functions such as general peace keeping, order maintenance, patrol, mediation and information services should be provided either by empowered civilian agencies or by private security agencies. (p. 61)

As further pointed out by Murphy (1993), there are challenges to redistributing the responsibilities for policing:

> The negotiation of the breakdown of policing responsibility among different organizations is not likely to be easy. At the moment, to the extent that such a process exists at all, it is not a very public process. However, it does appear that there are certain functions now carried out by sworn police officers where special police powers are not necessary for the task. (p. 102)

The difficulties that can arise when community policing is viewed as a threat to the collective bargaining of association members is expressed by Bradburn (1995):

> As police services advance their efforts toward community policing, collective agreement provisions may be found to impede their ability to respond to community needs. For example, if solving a community problem required changing a rigid shift rotation found in a collective agreement to one permitting a great deal of flexibility, management bargaining proposals to affect this change may be resisted by the police association due to the impact this would have on the personal lives of its membership. Balancing community needs and members' needs may become a challenge for police associations and managers alike as community consultation becomes a bigger part of management processes. (pp. 268–269)

Staying with the theme of collective bargaining, it has been observed that police associations have become extremely efficient in their efforts on behalf of their members, and that this very efficiency may well pose a threat to effective community policing. Murphy (1993) notes:

> Sophisticated union bargaining strategies, backed by aggressive public pressure campaigns, work-to-rule or strike actions have in many cases helped enshrine rules governing work, organization, discipline and work assignment that undermine effective police management.
>
> Examples of demonstrable inefficient police resource practices that are beyond management control are: the 4 day week, shift scheduling, 12 hour shifts, 2 person cars, some promotional procedures, and work assignment rules. In short, 'who works where, when and how'—essential management decisions—are often determined by police association rules and contract requirements . . . As a result, union/association influence and control over police work has made it difficult for police managers to engage in the reforms required to produce truly effective and efficient policing. (p. 49)

Police associations, having worked methodically and aggressively on behalf of their members' interests, demonstrate a degree of unwillingness to embrace new, more flexible working arrangements that might be advantageous from a community safety point of view. As Skogan and Hartnett (1997) have determined from their intensive examination of a community policing project in Chicago, Illinois:

> Left to their own devices, officers will easily fall back on the usual tools of their trade, including issuing warnings and making arrests. But many of the quality-of-life issues that plague urban neighborhoods cannot be solved by traditional police methods, nor by departments that act on their own. (p. 164)

This presents a problem for police leaders who want to venture into the wider world implied by community policing, with its expectations of meaningful community partnerships. If the department only uses the conventional tools of the trade, it will not be able to mobilize or "catalyze" the kind of alternative approaches suggested within the problem-oriented format, or through neighbourhood policing initiatives.

POLICE GOVERNING AUTHORITIES

The role of police commissions, police services boards, or some other form of civilian governing authority, is intended to fulfil the role of accountability for police services in Canada (Stenning, 1981; Hann et al., 1985). They are, in effect, the board of directors for the police service, and are meant to provide overall direction and guidance to the department, as well as being a touchstone for the larger community. As Saldanha (1995) suggests:

> A Board, whether in the public, not-for-profit or private sector is the singular, key component in determining the effectiveness, accountability, and responsiveness of an organization to its stakeholder group or groups; that in a modern society, the role of a local police board, intimately connected to the constituents it serves, is far more relevant to the community than a provincial or state oversight body; that the ability of a police board to fulfil its role is fundamentally linked to the proposition that the most important function of a police service is to be responsive to the needs of the community . . . (p. 40)

Saldanha (1995) goes on to assert that:

> [t]he role of a police board, in the absence of private sector motivation, is to be the voice of the community in ensuring this process of integration is a priority and becomes pivotal in the core of the police service. It accomplishes this task through the establishment of policy, the setting of police priorities and by selecting the right individual to put its vision into action. (p. 43)

Police boards, across Canada, are akin to the community consultative committees in place in Great Britain. Stenning (1995) has identified some important concerns with respect to the selection of board members, their training and preparation for serving on such a special purpose body, and the resources that are committed in order to allow the board to fully function in an independent manner. Generally, police governing authorities are looking for innovative ways of establishing and sustaining more effective and efficient police services. They frequently are engaged in a consideration of alternative approaches that would make the service more sensitive to changing community needs. This often involves a serious examination of better ways to use civilian members, auxiliaries, volunteers, and other municipal, private, or non-profit personnel. However, because many police governing authorities do not have the resources to independently pursue these objectives, they are compelled to rely on police departments in which strong support for these alternative measures is absent (Stenning, 1996; Eng, 1998). In Canada, police boards are frequently in a position similar to that described by Morgan, and referred to by Weatheritt (1985):

> Morgan says police community consultative committees lack professional competence because they are ignorant. They are totally reliant on the police for information that is often unavailable or is not forthcoming or is aimed primarily at educating them into the police view of things. (p. 172)

This is not to suggest that board members do not bring superior skills and abilities to their role as representatives of the community, or as civilian governors of a significant public service. Their knowledge of police administration and function will very likely be minimal and, therefore, the whole range of issues around selection, training, development, and resourcing becomes critical to the success of police boards, particularly in the context of community policing.

ACADEMIC COMMUNITY

The academic world has shown a continuous, but not exhaustive, interest in matters pertaining to community policing. However, there is a growing body of literature that is taking stock of the efforts of police organizations to realize the promise of community policing, particularly in North America. Within the disciplines of criminology, sociology, and political science there has been an uneasy engagement with police organizations over the past several decades. What has been occurring in recent years is a more willing partnership between the academic community and police departments. There has been a growing awareness of the value of shared purpose. This has been demonstrated in the area of community policing, with its focus on substantive organizational change, the need for community involvement through surveys and other forms of consultation, and the educational challenge that relates to an evolution of new modes of operation. All these features of community policing have heightened awareness of the role that academic researchers can play in the realm of policing.

The value of the academic observer of community policing lies, in part, in their capacity to offer a critique of this approach in its many forms. Albritton (1995) alerts the leaders in community policing to the dangers of being too complacent in their support for this approach:

> . . . for the "innovators" the onus is on the disbeliever to disprove the validity of the new paradigm. The innovators are committed 'only' to the technical imperative of change itself—i.e., the old or traditional that must be discarded in the face of the new and commonsensical that must be immediately embraced. This mentality represents nonetheless, a primary weakness in the general COP/POP [community-oriented policing/problem-oriented policing] orientation. The reluctance of proponents to subject their most cherished assumptions to honest scrutiny and critical evaluation actually defies the logic upon which their original demands for change in policing were purportedly based. (p. 31)

One key theme that flows through the academic literature on community policing relates to the absolute requirement for fundamental organizational change (see Chapter 4). As McLaughlin and Donahue (1995) observe:

> If police are going to be co-producers of public safety with the community, it would seem axiomatic that community policing cannot thrive in an autocratic or otherwise rigidly hierarchical environment. But the fact that police departments are precisely this way, and are not likely to change in any substantive fashion, presents undoubtedly the greatest impediment to its potential usefulness as a policing strategy. (p. 126)

Kelling and Moore (1985) lament the persistence of the image of police as being the "thin blue line" between communities and crime. This perception, both internal and external, has been an obstacle to community policing:

> The metaphor of the thin blue line reinforced the need to create isolated independence and autonomy in terms that were acceptable to the public. The patrol car became the symbol of policing during the 1930s and 1940s; when equipped with a radio it was at the limits of technology. It represented mobility, power, conspicuous presence, control of officers and professional distance from citizens. (p. 15)

This view has also supported the ongoing efforts of police associations to circumvent, or severely circumscribe, any efforts at broadening the role of "special constables," auxiliaries, and volunteers in Canadian policing.

There is some considerable emphasis on the importance of training for police officers in moving to a community policing model, representing as it does, a transformed attitude to the "business" of policing (see also Chapter Seven, above). Manning (1985) speaks directly to the challenge of training:

> The issue of 'training' of community police officers has been almost completely overlooked. The actual duties and obligations of community patrol officers, with the exception of the Flint research, is not specified. What do these people do for the hours they are meant to be 'patrolling'? There is little mention on evaluation reports of such matters as training, rewards, motivation, promotion opportunities, and the like. Unless these basic organizational matters are studied and structural changes made to reflect these realities, the ideational nature of community policing will remain. (p. 34)

The evidence of several Canadian police organizations (notably Halifax, Halton Region, Edmonton, and Calgary) would suggest that the training, or learning, dimension has not been overlooked. In fact, it would appear that a continuous learning model has been increasingly apparent in Canadian community policing initiatives.

Another problem relates to the disengagement of members (both civilian and uniform) within the police department when they are not directly involved with the community policing initiatives. As Silverman (1995) indicates:

> Confusion and ambiguity surrounding role responsibility has been repeatedly noted in reference to the implementation of community policing. This is particularly applicable in regard to the internal clientele not directly involved in the community policing program. (p. 43)

How is it possible to maintain that vital connection with the community when your role responsibility within the police service is directed towards administrative functions? For those who labour "behind-the-scenes," it must be very difficult to see how your daily efforts contribute to the problem-solving, co-productive partnership that is the hallmark of community policing. This is a significant matter that has not been fully addressed in theory or in practice.

In making the transition to community policing, it is clearly a mistake for the senior management of a police organization to simply direct that new modes and orders take shape, without careful planning, education, and implementation. Kratcoski and Blair (1995) emphasize this point, with particular reference to smaller jurisdictions:

> Another dynamic of community policing programs in small cities is that most of the programs are introduced through administrators who hear about it from professional contacts outside the local community. In two of the communities, the supervisors attempted to implement their programs through departmental directives. Officers tend to interpret top-down implementation as heavy handed, even when some effort was made to discuss the concept in an open meeting. (p. 103)

Toch and Grant (1991) have discovered that change, in any direction, but certainly toward the innovative requirements of community policing, is hampered by the strictures of the conventional police budget process:

> One perennial limiting force is the budget process. A police department's existence depends upon negotiating the next year's budget and this process is heavily skewed toward "more of the same." Many administrations have both an "A" and a "B" budget. "A" budgets are based on case and work load projections established from prior years and methods of operation; "B" budgets are

proposals for innovations in operations. Generally, "B" budgets are considered only after "A" budgets have been negotiated, if funds are still available. (p. 67)

In fact, it would appear that moving to community policing cannot immediately be supported by the argument that it will be cheaper. As indicated by Rosenbaum (1998):

> In my judgment, cities can expect to spend considerably more on community policing in the next few years if police administrators continue to follow the current implementation plan. For the many organizations that simply want cosmetic changes . . . cost should not be a major concern. But for those that are determined to introduce fundamental changes in the police organization and function, the present course of action is certain to bring added costs. (p. 20) [emphasis in original]

In order to make community policing viable and palatable in the context of existing and anticipated pressures, police leaders must be clear about the pitfalls and significant issues they will need to address. Wycoff (1985) assists in this regard by identifying some of the major pitfalls of community-oriented policing:

- Illegal policing;
- Inequitable policing;
- Politicization of the police;
- Corruption; and
- Police intrusion into private areas.

Murphy (1985) has highlighted further significant issues that must be seen as problematic in the context of community policing:

- What is a community?
- Who represents the community?
- Community standards versus individual rights.
- Community role in policing.
- Expanded role of the police.

LEADER IN COMMUNITY POLICING

Chris Braiden, Consultant & retired Superintendent, Edmonton Police Service

What is your understanding of "community policing" and its importance to the evolution of policing in Canada?

As I noted in an essay I prepared in 1987, entitled: "Community policing: nothing new under the sun,":

Chris Braiden LEADER IN COMMUNITY POLICING continued

Try to get someone to define Community Policing for you . . . and there is silence. Most seem to see it as a new "thing" in policing. My position is that they are wrong on both counts. It is neither new, nor is it a "thing." It's much deeper than that. I believe it is simply a re-emergence of the founding philosophy on which Peel built his public police in 1829. So let us first get rid of the notion that we have a new product on the block. That is why it will never be realized as an "add-on" to the conventional model. It won't stick. Unfortunately that is what most police departments have tried to do with it; stick a new box on the edge of the organizational chart, put a few people in it and announce the birth of Community Policing.

In another article, "Ownership: who paints a rented house?" I have stated:

Community policing is perhaps the most topical issue in policing today. It is also misunderstood by most. It seems to mean all things to all people. Volumes have been written and library shelves are bending under their weight. One thing seems certain; whatever it is, everyone seems to like it. Many people are hitchhiking on its wave of popularity. I don't like hitchhikers; they remind me of the guys who lean on scrums. They are more interested in advancing their own cause than that of policing. They apply community policing to everything they do outside of fundamental patrol work. Often, all it means is changing the sign on the door of bureaucracy.

Outline your own contribution to the development of "community policing" in Canada.

In 1993, Chris Braiden retired as Superintendent from the Edmonton Police Service. Today he is President of Chris Braiden Consulting Inc. Chris grew up in the town of Mullingar, Ireland and left school at the age of thirteen. He was determined to become a "copper," but before he realized this goal, he was rejected in three countries: in Ireland, because he couldn't speak Gaelic; in England, because he was missing a finger; and, in Canada, initially, because he's colour-blind.

His eventual entry into policing was less than spectacular. The chief of a small town interviewed him on a Friday afternoon, and he started to work on Monday the following week, without a uniform or training. Two years later, in 1964, this town was annexed by Edmonton along with its employees. Edmonton never chose Chris, it inherited him.

Edmonton is a city of approximately 650 000, with about 1200 police officers. Along the way, Chris served in all ranks and aspects of policing. As well, he carried out assignments with provincial and federal governments in Canada as special advisor on policing matters. He has consulted and lectured internationally (including Japan, France, Australia, and China) and extensively across the North American continent. Chris has also advised police commissions and provincial and state governments. He spent six years of service transforming the Edmonton Police Service around the philosophy of community policing. This

Chris Braiden LEADER IN COMMUNITY POLICING continued

work has been evaluated by the Police Executive Research Forum (Washington, D.C.) and the Canadian Research Institute for Law and the Family, and has been described as the best of its kind in North America.

Chris believes nothing great is ever achieved without a passion for a cause. His personal cause is to transform the culture of policing. His personal view on policing is that if it is done right it is a calling, a vocation, an art. If it is done wrong, it is nothing but a job. For Chris, there's a huge difference between peacekeeping and law enforcement. The former is a cause, the latter, merely a function. Others have described Chris as being a voice for change and a true pioneer in transforming the culture of policing.

What is your view of the future of "community policing" in Canada?

When tied to a strategy for Problem-Oriented Policing (especially as defined and explained by Herman Goldstein), a philosophy or mindset of community policing should become the norm for innovative and imaginative policing in Canada. Community policing, rightly understood, can remove the shackles on the minds of officers and civilians in policing who have been programmed to follow rigid procedures for too long. While community policing is often associated with foot patrols, it doesn't really matter what mode of transportation is used, it is the purpose of the patrol that is paramount. My cardinal rule for policing can be summarized very simply:

Police others as you would have others police you.

COMMUNITY GROUPS

The role of the community is clearly pivotal to the success of community policing. Without discovering compelling ways to mobilize and transform community concern for public safety issues into action, the future of community policing is in serious doubt. However, Buerger (1985) raises the issue of settling on an identity for, and the definition of, community:

> Little attention has been given to a definition of the community commensurate with the vast promise embedded in the community policing rhetoric. Even less has been spared for defining the role that can reasonably be expected of "the community," howsoever it should be defined. Though some departments and localities are engaged in honest efforts to make the promises a reality, community policing by and large remains a unilateral action on the part of the police. (p. 270)

And further:

> For all the rhetoric about empowering the community or the community "taking responsibility" for itself, when it comes down to cases the police establishment assigns the community a role that simply enhances the police response to crime and disorder. Rather than "empower" the community to act in its own interest, this model calls upon it to help focus police crime-fighter activities. (p. 271)

Rosenbaum (1998) cites another source of concern that works to undermine the efforts made in favour of community policing:

> . . . the concept of "fear of crime" is another lightning rod for critics who argue that community policing is soft on crime. They claim that switching the focus of policing from fighting crime to maintaining order and reducing fear of crime is simply a "smoke and mirrors" tactic to make citizens feel good about themselves, the police department and their neighborhood. These critics argue that such a "warm fuzzy" approach to policing will divert police attention away from the real job of arresting criminals. (p. 11)

A great deal of community policing effort is being made within the context of municipal police departments. There has been a longstanding point of debate over the unique status of such police agencies, largely due to the overriding significance of police legislation that establishes an accountability regime that is different from other municipal departments. For example, a police service is responsible to its police services board and *not* to municipal council. Many municipalities have worked to align their police departments more closely with the other departments within the municipal corporation. However, the guiding police legislation continues to hold sway. Mastrofski (1998) has noted further impediments regarding the idea of municipal service integration:

> It may be that service integration would be as effective as advocates hope, but that there will be few opportunities to test major initiatives. As compelling as public safety may be as an organizing objective for service integration, it may not be strong enough to overcome bureaucratic inertia and political alliances that are well entrenched in the local political scene. (p. 181)

Murphy and Muir (1985) speak to the potential conflict between public interest and community goals:

> The often competitive and conflictual nature of 'public interest' or 'community goals' requires further analysis and discussion in any future elaborations of the community-based model. Recognition of community variation and plurality of community interests makes it even more important that the police become sensitive to all group interests in the community they serve and devise flexible policies and programs which can meet and mediate community differences. (p. 104)

Friedman laments the shortage of serious community participation (1985):

> Little space has been created for community participation in either national or local dialogue. Little thought has been given to the conditions necessary for effective community participation and almost no resources are committed to the community role. (p. 263)

The Policing in British Columbia Commission of Inquiry (1994) looked at many issues relating to policing and the community. Mr. Justice Wallace T. Oppal, the Commissioner of the Inquiry, included the following community responsibilities for governance of policing in his final report:

1. Providing an effective community voice in policing by
2. Promoting the community's interest in the recruiting and training process,
3. Preserving the identity of the local police force,
4. Providing police services to local residents as a first priority,

5. Setting community-specific enforcement priorities, and
6. Ensuring that local policing garners public trust and support;
7. Managing police services by
8. Operating an efficient police force,
9. Promoting a healthy relationship between the employer and employees,
10. Ensuring that community-specific employment issues are resolved,
11. Achieving economies of scale,
12. Reducing policing costs by sharing resources, equipment and facilities,
13. Developing cost-effective policing methods, and
14. Ensuring that private police do not interfere with the operations of the local police force;
15. Ensuring that community-specific enforcement priorities are implemented,
16. Evaluating the performance of the chief constable and ensuring the assessment of individual police officers, and
17. Ensuring that allegations of misconduct are investigated and adjudicated promptly and fairly; and
18. Superintending police operations by
19. Being kept informed of policing operations, and
20. Ensuring that police officers are held accountable to civilian authorities. (pp. B-49–50)

MUNICIPAL GOVERNMENTS

Municipalities are pressed to make many decisions with regard to service delivery within their jurisdiction that ranges across many functional areas (e.g., roads, sewers, libraries, emergency services, sanitation, transportation, health-care, among others). Policing typically constitutes an expensive municipal service funded largely through the local ratepayers. The tension between a police service and its municipal government often erupts during the annual budget approval process, when significant expenditures are being requested by the police department alongside all of the competing demands of other municipal departments. This tension can be detrimental to experiments of any kind in police service delivery, perhaps especially in the area of community policing. The politics of police budgeting may be seen to interfere with a genuine reappraisal of service delivery options, including the philosophy of community policing. There is a strong emotional content to police budget debates that is intensified by the local media and makes real reform difficult. As Murphy (1993) summarizes:

> This politicization of the budgetary process is exacerbated by police commissioners and city councillors who personally identify with the police and are often actively courted by them. Neighbourhood politicians see that being supportive of police budget requests, especially if it means acquiring new police resources for their neighbourhoods, provides their constituencies with valuable public services. With this kind of emotional and pragmatic politics at work, few city managers, city councillors or mayors find it appetizing to engage the police in public resource battle. The anti-police label so readily applied to even modest questions about resource use and effectiveness, silences even the best intentioned critics. It is easier to give police what they

want and pass on the cost to taxpayers or take the money from some less effectively represented service or constituency. (p. 43)

But for real innovation to take root and develop in ways that can be evaluated and consolidated, it is essential that hard questions be asked about police budgets and their application to alternative approaches. The traditional and the professional models of policing have been found to be wanting in significant ways and, therefore, the community policing approach has been offered as a solution to the problems of crime, safety, and security. Experimentation is necessary to discover if this approach can be adapted in meaningful ways in Canadian police departments. Municipal governments must be prepared to subsidize the pursuit of such experimentation. Of course, there are risks associated with the trial and error of experimentation, and the political environment is not always conducive to such circumstances (Mastrofski, 1998, p. 188–189).

SUMMARY

This chapter has concentrated some attention on the critique of community policing as a theory and its application to the practice of policing. These concerns issue from several different perspectives. However, they combine in a manner that reveals certain weaknesses within this approach. Certainly, those functioning within police departments have much to gain, or lose, by trying to move toward the goals and objectives typically associated with community policing. The police department that embarks on a change process driven by community policing principles must certainly take stock of the risks and liabilities involved in such an undertaking. The existence of criticism should not, however, dissuade police leaders from considering the merits of this approach. The complications associated with its implementation should not discourage those already moving in this direction from continuing on their charted course.

What this chapter reveals is that community policing is not a simple, straightforward system that can be bolted onto the organizational structure and set in motion. It requires careful planning, a degree of organizational readiness, and commitment at all levels of the service. It further requires awareness of the pitfalls, a clearly defined process of implementation that will prepare and support both internal and external players for their various roles. And, finally, it requires an evaluation mechanism that will measure the **quality** of this fundamentally different approach to the policing of communities.

QUESTIONS FOR CONSIDERATION

1. What are the major criticisms raised about community policing by police executives?
2. How could it be argued that community policing is "nothing new"?
3. What are the arguments in favour of using auxiliary, volunteer, or another category of officer to support community policing?
4. Are there serious drawbacks to the use of auxiliary, volunteer or another category of officer in Canadian police organizations?

5. What concerns do scholars raise about the introduction of community policing?
6. Why are police middle managers commonly identified as a source of opposition to the introduction of community policing?
7. What areas of concern have police governing authorities raised around community policing?
8. What do municipal governments find problematic with regard to community policing?
9. What are the "pitfalls" identified by academics with respect to community policing?
10. How could the "pitfalls" identified in question 9, above, be remedied?

REFERENCES

Albritton, James S. (1995). "The technique of community-oriented policing: an alternative explanation." In Kratcoski, Peter C. and Duane Dukes (eds.) *Issues in community policing*. Highland Heights, Ky.; Cincinnati, Ohio: Academy of Criminal Justice Sciences, Northern Kentucky University; Anderson Publishing Co.

Bayley, David (1991). "Strategy." In Chacko, James and Stephen E. Nancoo (eds.) (1993). *Community policing in Canada*. Toronto: Scholars' Press Inc.

Bennett, Trevor (1994). "Community policing on the ground: developments in Britain." In Rosenbaum, Dennis P. (ed.) *The challenge of community policing: testing the promises*. Thousand Oaks, Calif.: Sage Publications.

Blair, I. (1994). "Let the police fund their own expansion." *The Times* (October 1).

Bradburn, Lenna (1995). "Current and future roles of police staff associations." In *Management challenges in 21st century policing: conference proceedings*. Ottawa : Centre for Police Management and Research, Staffordshire University and Canadian Research Institute for Law and the Family, University of Calgary.

Braiden, Chris (1990). "Enriching traditional roles." In Hoover, Larry T. (ed.) *Police management: issues & perspectives*. Washington, D.C.: Police Executive Research Forum.

Brodeur, Jean-Paul (1998). "The assessment of police performance: conclusions." In Brodeur, Jean-Paul (ed.) *How to recognize good policing: problems and issues*. Thousand Oaks, Calif.: Sage Publications and Police Executive Research Forum.

Buerger, Michael E. (1994). "The limits of community." In Rosenbaum, Dennis P. (ed.) *The challenge of community policing: testing the promises*. Thousand Oaks, Calif.: Sage Publications.

Donovan, E. and W. Walsh (1986). "An evaluation of Starret City security services." Pennsylvania State University.

Eng, Susan (1998). Interview between Susan Eng and Paul McKenna, March 4, 1998, Osgoode Hall Law School, York University, Downsview, Ontario.

Friedman, Warren (1994). "The community role in community policing." In Rosenbaum, Dennis P. (ed.) *The challenge of community policing: testing the promises*. Thousand Oaks, Calif.: Sage Publications.

Geller, William A. and Guy Swanger (1995). *Managing innovation in policing: the untapped potential of the middle manager*. Washington, D.C.: Police Executive Research Forum.

Goldstein, Herman (1990). *Problem-oriented policing*. New York: McGraw-Hill, Inc.

Greene, Jack R. and Ralph B. Taylor (1985). "Community-based policing and foot patrol: issues of theory and evaluation." In Greene, Jack R. and Stephen D. Mastrofski (eds.) *Community policing: rhetoric or reality*. New York: Praeger.

Greenwood, P.W. et al. (1975). *The criminal investigation process: volume I: summary and policy implications*. Santa Monica, Calif.: Rand Corporation.

Hann, Robert G. et al. (1985). "Municipal police governance and accountability in Canada: an empirical study." *Canadian Police College Journal*, Vol. 9, no. 1, pp. 1–85.

Inkster, Norman D. (1991). "Policing in the '90s and beyond." In Chacko, James and Stephen E. Nancoo (eds.) (1993). *Community policing in Canada*. Toronto: Scholars' Press Inc.

Johnston, Les (1998). "Late modernity, governance, and policing." In Brodeur, Jean-Paul (ed.) *How to recognize good policing: problems and issues*. Thousand Oaks, Calif.: Sage Publications and Police Executive Research Forum.

Kelling, George L. and Mark H. Moore (1985). "From political to reform to community: the evolving strategy of police." In Greene, Jack R. and Stephen D. Mastrofski (eds.) *Community policing: rhetoric or reality*. New York: Praeger.

Kratcoski, Peter C. and Robert B. Blair (1995). "Dynamics of community policing in small communities." In Kratcoski, Peter C. and Duane Dukes (eds.) *Issues in community policing*. Highland Heights, Ky.; Cincinnati, Ohio: Academy of Criminal Justice Sciences, Northern Kentucky University; Anderson Publishing Co.

Manning, Peter K. (1985). "Community policing as a drama of control." In Greene, Jack R. and Stephen D. Mastrofski (eds.) *Community policing: rhetoric or reality*. New York: Praeger.

Mastrofski, Stephen D. (1985). "Community policing as reform: a cautionary tale." In Greene, Jack R. and Stephen D. Mastrofski (eds.) *Community policing: rhetoric or reality*. New York: Praeger.

Mastrofski, Stephen D. (1998). "Community policing and police organization structure." In Brodeur, Jean-Paul (ed.) *How to recognize good policing: problems and issues*. Thousand Oaks, Calif.: Sage Publications and Police Executive Research Forum.

McConville, M. and D. Shepherd (1992). *Watching police watching committees*. London: Routledge & Kegan Paul.

McLaughlin, Vance and Michael E. Donahue (1995). "Training for community-oriented policing." In Kratcoski, Peter C. and Duane Dukes (eds.) *Issues in community policing*. Highland Heights, Ky.; Cincinnati, Ohio: Academy of Criminal Justice Sciences, Northern Kentucky University; Anderson Publishing Co.

Murphy, Christopher (1985). "The development, impact and implications of community policing in Canada." In Greene, Jack R. and Stephen D. Mastrofski (eds.) *Community policing: rhetoric or reality*. New York: Praeger.

Murphy, Christopher (1993). "Thinking critically about police resources." In Doob, Anthony N. (ed.) *Thinking about police resources*. Toronto: Centre of Criminology, University of Toronto.

Murphy, Chris and Graham Muir (1985). *Community-based policing: a review of the critical issues*. Ottawa : Solicitor General Canada, Programs Branch.

Police Association of Ontario (1996). *Policing in Ontario: building safer communities: submissions to the Ministry of the Solicitor General and Correctional Services*.

Policing in British Columbia Commission of Inquiry (1994). *Closing the gap: policing and the community: the report, volume I*. [S.l.]: The Commission.

Reiner, Robert (1991). *Chief constables: bobbies, bosses, or bureaucrats?* Oxford: Oxford University Press.

Roberg, Roy (1994). "Can today's police organizations effectively implement community policing?" In Rosenbaum, Dennis P. (ed.) *The challenge of community policing: testing the promises*. Thousand Oaks, Calif.: Sage Publications.

Rosenbaum, Dennis P. (ed.) (1994). *The challenge of community policing: testing the promises*. Thousand Oaks, Calif.: Sage Publications.

Rosenbaum, Dennis P. (1998). "The changing role of the police: assessing the current transition to community policing." In Brodeur, Jean-Paul (ed.) *How to recognize good policing: problems and issues*. Thousand Oaks, Calif.: Sage Publications and Police Executive Research Forum.

Sacco, Vincent F. (1998). "Evaluating satisfaction." In Brodeur, Jean-Paul (ed.) *How to recognize good policing: problems and issues*. Thousand Oaks, Calif.: Sage Publications and Police Executive Research Forum.

Sadd, Susan and Randolph Grinc (1994). "Innovative neighborhood oriented policing: an evaluation of community policing programs in eight cities." In Rosenbaum, Dennis P. (ed.). *The challenge of community policing: testing the promises*. Thousand Oaks, Calif.: Sage Publications.

Saldanha, Colin (1995). "Local police boards: invisible, inevitable, invaluable." In *Management challenges in 21st century policing: conference proceedings*. Ottawa: Centre for Police Management and Research, Staffordshire University and Canadian Research Institute for Law and the Family, University of Calgary.

Schwartz, A.T. and S.N. Clarens (1977). *The Cincinnati team policing experiment: a summary report*. Washington, D.C.: Police Foundation.

Shearing, C.D. (1992). "The relation between public and private policing." In Tonry, M. and N. Morris (eds.) *Modern policing: crime and justice: a review of research (volume 15)*. Chicago: University of Chicago Press.

Shearing, C.D. and P.C. Stenning (1981). "Modern private security: its growth and implications." In Tonry, M. and N. Morris (eds.) *Crime and justice: an annual review of research (volume 3)*. Chicago: University of Chicago Press.

Shearing, C.D. and P.C. Stenning (1987). *Private policing*. Newbury Park, Calif.: Sage Publications.

Sherman, L.W., C.H. Milton, and T.V. Kelley (1973). *Team policing: seven case studies*. Washington, D.C.: Police Foundation.

Silverman, Eli B. (1995). "Community policing: the implementation gap." In Kratcoski, Peter C. and Duane Dukes (eds.) *Issues in community policing*. Highland Heights, Ky.; Cincinnati, Ohio: Academy of Criminal Justice Sciences, Northern Kentucky University; Anderson Publishing Co.

Skogan, Wesley G. (1994). "The impact of community policing on neighborhood residents." In Rosenbaum, Dennis P. (ed.) *The challenge of community policing: testing the promises*. Thousand Oaks, Calif.: Sage Publications.

Skogan, Wesley G. and Susan M. Hartnett (1997). *Community policing: Chicago style*. New York: Oxford University Press.

Stenning, Philip C. (1981). *Police commissions and boards in Canada*. Toronto: Centre of Criminology, University of Toronto.

Stenning, Philip C. (1989). "Private police and public police: toward a redefinition of the police role." In Loree, Donald J. (ed.) *Future issues in policing: symposium proceedings*. Ottawa: Canadian Police College

Toch, Hans and J. Douglas Grant (1991). *Police as problem solvers*. New York: Plenum Press.

Trojanowicz, Robert and Bonnie Bucqueroux (1990). *Community policing: a contemporary perspective*. Cincinnati, Ohio: Anderson Publishing Co.

Walinsky, Adam (1995). "The crisis of public order." *The Atlantic Monthly*, Vol. 276, no. 1, pp. 39–54.

Weatheritt, Mollie (1985). "Community policing: rhetoric or reality?" In Greene, Jack R. and Stephen D. Mastrofski (eds.) *Community policing: rhetoric or reality*. New York: Praeger.

Wycoff, Mary Ann (1985). "The benefits of community policing: evidence and conjecture." In Greene, Jack R. and Stephen D. Mastrofski (eds.) *Community policing: rhetoric or reality*. New York: Praeger.

Zhao, Jihong, Quint C. Thurman, and Nicolas P. Lovrich (1995). "Community-oriented police across the U.S.: facilitators and impediments to implementation." *American Journal of Police*, Vol. 4, no. 1, pp. 11–28.

TOWARD MORE LEARNING

POLICE PRACTIONERS

It is appropriate to provide reference to the writings of Chris Braiden, the featured "Leader in Community Policing" in this chapter, for further study. As someone who has a long career as a police practitioner, and, as someone who has taken time to study policing in Canada, his observations and thoughts are extremely valuable, although not uncontroversial. What follows is a listing of Braiden's writings:

Braiden, Chris (1986). "Bank robberies and stolen bicycles: thoughts of a street cop." *Canadian Police College Journal*, Vol. 10, no. 1.

Braiden, Chris (1990). "Community based policing: a process for change: position paper written for the executive officers' team." Unpublished paper available through Chris Braiden Consulting.

Braiden, Chris (undated). "Policing: from the belly of the whale." Unpublished paper available through Chris Braiden Consulting.

Braiden, Chris (undated). "Leadership: it's not what—or where—we think it is." Unpublished paper available through Chris Braiden Consulting.

Braiden, Chris (undated). "Moments of truth: what are they?" Unpublished paper available through Chris Braiden Consulting.

Braiden, Chris (undated). "Ownership: who paints a rented house?" Unpublished paper available through Chris Braiden Consulting.

Braiden, Chris (undated). "Plato's cave." Unpublished paper available through Chris Braiden Consulting.

Braiden, Chris (undated). "Policing: time for a new winejug." Unpublished paper available through Chris Braiden Consulting.

Braiden, Chris (undated). "Thinking about justice." Unpublished paper available through Chris Braiden Consulting.

Braiden, Chris (undated). "Why change?" Unpublished paper available through Chris Braiden Consulting.

Braiden, Chris (1992). "Community policing: nothing new under the sun." In *Community oriented policing & problem solving*. Sacramento, Calif.: California Department of Justice, Attorney General's Office, Crime Prevention Center.

Chris Braiden Consulting Inc.
40 Glen Meadow
St. Albert, Alberta
Canada
T8N 3B6
Fax.: (403) 459-5475

POLICE ASSOCIATIONS

The Police Association of Ontario (PAO) was formed in 1933, and presently represents the interests of nearly 19 000 front-line municipal police personnel. The most recent positions taken by the PAO on issues such as the use of alternative categories of officers, auxiliaries, special constables, and unpaid volunteers can be found on the PAO website, along with a great deal of other valuable information.

Website: http://www.pao.on.ca/

ACADEMIC COMMUNITY

Brodeur, Jean-Paul (ed.) (1998). *How to recognize good policing: problems and issues*. Thousand Oaks, Calif.: Sage Publications and Police Executive Research Forum.

Edited by Jean-Paul Brodeur, a professor at the School of Criminology, University of Montreal, this is a recent collection of essays that examines some of the significant problems and issues relevant to ongoing debates and discussions around community policing in Canada. Of particular interest are those that look in some detail at the following topics:

- changing role of police in modern society;
- the difficulties associated with assessing individual police officer performance;
- community participation in a community policing environment;
- the evaluation of community satisfaction;
- the implementation of community policing as a change strategy;
- restructuring the police organization for community policing; and
- police governance in a modern context.

Rosenbaum, Dennis P. (ed.) (1994). *The challenge of community policing: testing the promises*. Thousand Oaks, Calif. : Sage Publications.

Professor Rosenbaum is the head of the Department of Criminal Justice at the University of Illinois (Chicago), and has edited this excellent work (that contains a valuable "Foreword" by Herman Goldstein) that draws on the insights, research, and evaluations of several leading thinkers in the area of community policing. Topics covered in this volume include:

- Effectiveness, equity, and efficiency in community policing;
- Evaluation of community policing programs in eight (U.S.) cities;
- Organizational change for community policing in six (including Edmonton, Alberta) cities;
- Cultural and structural changes required by community policing;
- Impact of community policing on police personnel;
- Impact of community policing on community residents and their problems;
- Community policing in Canada: an overview (by Barry Leighton);
- Current issues and concerns around community policing; and
- A synthesis of recent research and policy implications of community policing.

CHAPTER TEN

Why are there so few evaluations of Canadian policing? First, there are few academics involved with police research and fewer still pursuing evaluations of policing. Second, by the very nature of community policing, it is difficult to prove that a philosophy works. It is also difficult to achieve closure to the implementation of fundamental organizational and operational change that would allow for an appropriate impact study.

Barry Leighton (1994). "Community policing in Canada: an overview of experience and evaluations."

MEASURING SUCCESS IN COMMUNITY POLICING

LEARNING OBJECTIVES

1. Identify various approaches for evaluating the success of community policing.
2. List several Canadian, and other, police services that have undertaken efforts to evaluate their efforts at community policing.
3. Explain the distinction between performance measurement as it relates to individuals and organizations.
4. Explain the value of public surveys in measuring the effectiveness of police practice in the area of community policing.
5. Differentiate among several approaches to the evaluation of community policing.
6. Identify the purpose of a community safety audit.

INTRODUCTION

This chapter will look at ways of measuring the success of community policing. It is important to quickly distinguish between the measurement of the performance of individuals within a police service and the measurement of overall organizational performance in pursuing specific community policing goals. Individual human performance is measured in specific ways, while organizational effectiveness must be evaluated in a different manner.

Also, this chapter will consider several police organizations that have made serious efforts to assess the effectiveness of their community policing strategy, and will offer a summary of those efforts. There cannot be a true commitment to the philosophy of community policing unless it can be demonstrated that it accomplishes certain clearly defined goals that contribute to public safety and community order. There will be a review of the practice of public opinion surveys that can be effectively used to determine both community safety concerns and community satisfaction with police services.

Finally, some consideration will be given to the use of community safety audits as a means for defining the elements of public concern requiring attention in the context of community policing.

PERFORMANCE MEASUREMENT

It has been established that all organizations owe whatever success they have to those individuals that comprise that organization (Morgan, 1993). It has also been established that the organization of the future will need to constantly encourage those individuals to innovate if it is going to survive (Argyris, 1963). Therefore, it is profoundly important that human performance be seen as a priority for overall organizational effectiveness. Along with the increased focus on public service across Canada has come a renewed interest in establishing better mechanisms for the measurement of performance in those services (Management Board Secretariat of Ontario, 1996; Alberta Treasury, 1996). Taxpayers are interested in receiving value-for-money from all of the services provided at all levels of government (i.e., municipal, provincial, and federal). This interest has resulted in significant effort on the part of those levels of government to increase the effectiveness, efficiency, and economy of public services (see the discussion in Chapter Four, above).

A renewed focus on performance measurement has resulted in some valuable literature that leads to a better understanding of this topic (Sloma, 1980; Kaplan and Norton, 1992; Campbell, 1995). Again, it is always important to distinguish the performance of individuals from organizational performance.

PERFORMANCE CRITERIA

While it is frequently noted that community policing is not simply a program, it frequently makes its appearance in the form of a program that is designed to be an integral part of an overall community policing strategy. In order to determine some useful performance criteria to assess such a program, the following categories may be useful:

- *Program efficiency*—what is the cost per unit of the service?
- *Customer service quality*—is the customer satisfied with the service received?
- *Community impact*—is the expected program outcome achieved? Does the service provided have a positive impact on the target population?

In making a decision to examine the issue of performance measurement, it is valuable to place the matter into a larger organizational context. Such a context could include the elements suggested by Garrett (1996) which are represented in Figure 10-1, below:

- *Strategic planning*—the overall approach to guiding the goals and objectives of an organization through a systematic examination of its strengths, weaknesses, opportunities, and threats across internal and external horizons;
- *Program based management & measurement*—organizations are frequently comprised of several "divisions" which offer a particular range of programs, supported

FIGURE 10–1
MEASUREMENT MODEL

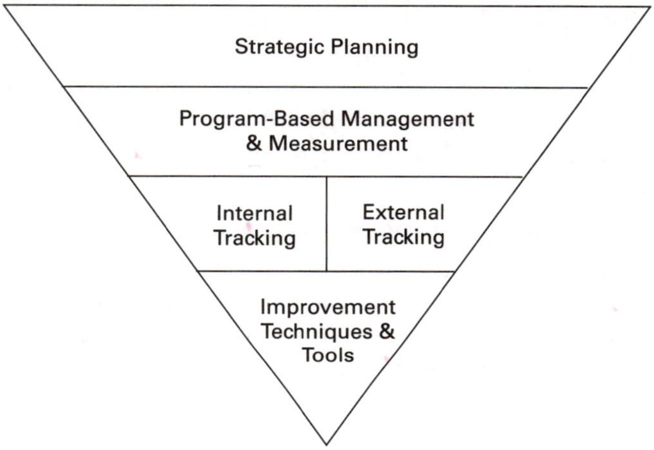

by other organizational units or services (e.g., accounting, audit, human resources, facility management, policy, purchasing, etc.). With such a diversity of programs and supports, it often becomes difficult to establish real program costs and effectiveness measures;

- *Internal tracking*—examining the efficiency of a program on the basis of comparative statistical data that show performance over time and within specific program activities;
- *External benchmarking*—examining the efficiency of a program on the basis of comparisons with other organizations in the same field, or corporations offering a comparable service or program;
- *Improvement tools & techniques*—may include any of the following elements: total quality management, continuous improvement, quality circles, cycle time reduction, business process redesign, employee involvement or empowerment.

Kasul and Motwani (1995) reviewed the literature on performance measurements and arrived at the following organizational requirements for world-class organizations:

- Management commitment;
- Quality;
- Customer service;
- Vendor and material management;
- Advanced technology;
- Facility control;
- Flexibility;
- Price/cost leadership; and
- Global competitiveness.

Clearly the last two categories (i.e., price/cost leadership and global competitiveness) are not directly relevant to public police organizations, unless it is accepted that they should strive to be as cost-effective as possible, given the operational uncertainties attached to the enterprise of policing. It is also feasible to suggest that a modern police organization should demonstrate sufficient "global" leadership that allows it to compete with other departments in terms of the quality and effectiveness of its programs, processes, and services.

Organizations may have a number of reasons for introducing a system of performance measurement. The following three are considered essential (Management Board Secretariat of Ontario, 1996):

- *To provide a mechanism for public accountability*—are the organization's goals and objectives being met?
- *To drive change*—assists in priority setting, redirects efforts of staff to new goals and new ways of doing business. Establishes performance expectations; and
- *To improve management practice*—allows the organization to focus on priorities, assess programs, identify and improve potential problem areas, across both corporate management and operational levels;

Six Steps to Performance Measurement

The following six steps may be seen as a simple approach to addressing the challenge of performance measurement that would be adaptable for a police service introducing community policing:

1. *Documenting goals and objectives*—
   - Establish and understand what is to be accomplished with a listing of goals and objectives;
   - Ensure there is a shared understanding of the goals and objectives, arrive at key, functional definitions;
   - Look to available sources to verify that the organization's goals and objectives are complete (e.g., return to the mission statement, strategic plan, relevant legislation, etc.).

2. *Establishing the desired outcome*—
   - Derive a realistic and achievable outcome from the goals and objectives. This may be the impact quality level or profit to be accomplished;
   - Employ key words to express desired outcomes. Sources of information could include legislation, court decisions, policies and standards, comparisons with similar high-performing organizations, previous levels of satisfactory performance;
   - Recognize that desired outcomes are typically long-term (i.e., more than one year) and may need to be set in addition to medium-term or short-term outcomes;
   - Recognize that a measurement in any new area of performance will necessarily be an estimate. However, it is important to employ the broadest range of resources and experiences available. Comparisons with other areas of performance may be helpful.

3. *Establish indicators leading to performance*—
   - Begin with a limited number of indicators for each objective and desired outcome, add to this list, as necessary. Consider what would be of value to different audiences, both internal and external. Brainstorming can be helpful in this regard. Distinguish between indicators and other factors which you do not have control over or responsibility for and set the latter ones aside.
4. *Decide on an initial list of indicators*—
   - Build an effective performance measurement system upon data and information that is easy to obtain, cost-effective to obtain, readily available, and reliable.
   - Identify groups of indicators that can be combined to demonstrate that goals and objectives are being achieved.
   - Establish the activity level for each indicator. Does it need to be broken out by type of case, region, level of activity, etc.?
5. *Performance measurement using indicators*—
   - Begin performance measurement early in the process.
   - Test initial performance measurements against desired outcomes. Do they fit?
6. *Reporting the results*—
   - Effective reporting of the results of performance measurement can lead to a number of beneficial outcomes—greater accountability, improved morale and encouragement of increased levels of performance, clearer focus on current and future priority goals, increased credibility, reallocation of resources, alternative organizational modes for the delivery of services, business process reengineering, determining other areas of expenditure reduction, future business planning processes.

In order to clarify some of the components that will contribute to a performance measurement system, it is relevant to identify the types of measures needed, that fall under certain general categories: input, process, output, and outcome.

- *Input measures*—refers to the resources allocated to particular programs (e.g., money, personnel, equipment, and supplies). These are important in helping to determine the costs of providing certain services and for assessing the effect on service quality if resource levels are changed (e.g., number of police per population, number of student spaces);
- *Process measures*—includes activity-based measures such as, how "busy" the organization is, which will reflect a level of demand for services (number of hours worked, number of applications received), and efficiency measures which represent the cost of providing service on a per unit basis. Process improvement approaches concentrate on speeding up response time, or reducing per unit costs. Cost-efficiency (i.e., how cheaply a service can be provided) may be considered against cost-effectiveness (i.e., method that provides good results at a reasonable cost). Cost per medical procedure, cost per student.

- *Output measures*—the level of service being provided (kilometres of roads paved, number of grants provided). These measures may be helpful, but they do not indicate if broader goals are being achieved. Number of operations, number of graduates.
- *Outcome measures*—these focus on desired results and can be difficult to develop since they are often found in the future. It may be necessary to measure intermediate steps. Valuable to look at landmarks of achievement (e.g., healthy Albertans, highly-skilled workforce).

In selecting specific measures, it is important to remember the following key criteria that have been identified by the Alberta Treasury (1996):

- *Understandability*—the measure and information are clear and easily understood by the public, and sufficiently explain how performance is being assessed;
- *Relevance*—the measure is an accurate representation of what is being measured. The information presented is timely and directly related to the subject matter;
- *Reliability*—the information is free from error, unbiased, and complete. Also, the results can be duplicated by others using the same information and methodology; and
- *Comparability*—results can be compared to other years or to similar organizations. (p. 15)

Performance measurement is not a simple task and should be approached with due caution, and a recognition that it will evolve and develop over time. As noted by the Management Board Secretariat of Ontario (1996):

> Performance measurement has been described as a journey, not a destination. It is not something that is done once and put away. It is a tool for continually improving work processes, products and services. Implementing a system of performance measurement will not transform services and programs overnight into models of efficiency and customer service. Over time however (2–3 years), a well-designed system will tell us if our assumptions were correct . . . (p. 31)

Oettmeier and Wycoff (1998) have also observed that:

> There is nothing simple about constructing a valid and reliable evaluation process, and few agencies are staffed for the task. Many agencies lack basic planning and research units and those that do have such units seldom have the resources to hire staff with evaluation expertise.

These authors have refined a useful "performance analysis model" that allows for the evaluation of any range of activities that can be applied to policing, especially in a context of community policing. Their model is reproduced as Figure 10-2, below:

The purpose of individual performance evaluation may seem obvious. However, there are three substantive reasons for this effort as identified by Mastrofski and Wadman (1991):

- *Administrative*—to help managers make decisions about promotion, demotion, reward, discipline, training needs, salary, job assessment, retention and termination;
- *Guidance and counseling*—to help supervisors provide feedback to subordinates and assist them in career planning and preparation, and to improve employee motivation;

FIGURE 10-2

A PERFORMANCE ANALYSIS MODEL

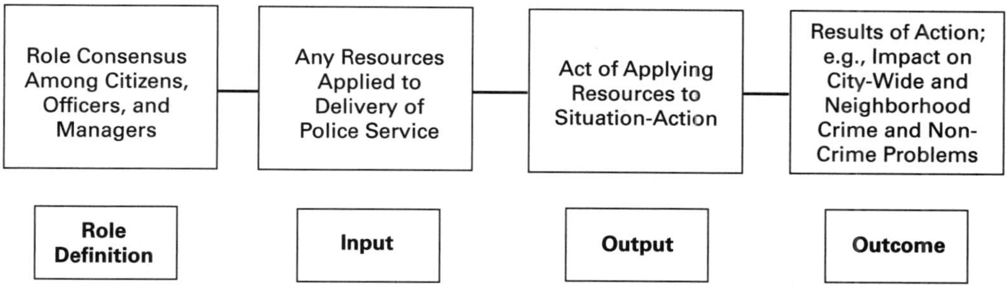

Source: Oettmeier, Timothy N. and Mary Ann Wycoff (1998). "Personnel performance evaluations in the community policing context." In Shelley, Tara O'Connor and Anne C. Grant (eds.). *Problem oriented policing: crime-specific problems, critical issues and making POP work*. Washington, D.C.: Police Executive Research Forum. p. 363. Used with permission.

- *Research*—to validate selection and screening tests and training evaluations, and to assess the effectiveness of interventions designed to improve individual performance. (p. 364)

Oettmeier and Wycoff (1998) have added three further reasons for performance evaluations, based on their own research with the Houston Police Department:

- *Socialization*—to convey expectations to personnel about both the content and style of their performance, and to reinforce other means of organizational communication about the values of the department;
- *Documentation*—to record the types of problems and situations officers are addressing in their neighborhoods and the approaches they take to them. Such documentation provides for data-based analysis of the types of resources and other managerial support needed to address problems and allows officers the opportunity to have their efforts recognized; and
- *System improvement*—to identify organizational conditions that may impede improved performance and to solicit ideas for changing the conditions. (p. 366–367)

Oettmeier and Wycoff (1998) also make a convincing argument for increasing the level of complexity for individual performance evaluation as the employee's length of service increases:

> Early in a career, it is important to determine whether the employee has the requisite knowledge, skills and attitudes (KSAs), the ability, and willingness to do the job. In other words, we want to determine if the employee has the capacity to do police work. Once this is established, performance assessment might more reasonably focus on whether the employee effectively

> uses these KSAs in the field. Evaluation is used to determine whether the officer is consistently doing things correctly. At some advanced stage of the career, assessment could focus on whether the officer does the right thing – in other words, whether the officer is able to select the correct response to fit the service needs of the area for which he or she is responsible. (p. 370)

The value of the kind of "functional and strategic job analyses" done under the auspices of the Ontario Strategic Planning Committee on Police Training and Education (1992) relates directly to this area of discussion. These approaches were aimed at building a model for performance evaluation that takes a long view of the relevant learning requirements for various categories of police personnel, based on an anticipated future for policing in Ontario, and allowing for the listing of general performance standards to be met. Oettmeier and Wycoff go on to note that police organizations are beginning to move beyond an exclusive concentration on individual human performance:

> In addition to recognizing the value of community feedback, community policing has also caused some administrators to question the appropriateness of individual employee evaluations. Some departments are emphasizing focus teams or workgroups rather than individuals. Those retaining individual evaluations may abandon them as a means of differentiating among employees for the purpose of rewards and, instead, use them as a means of helping individual employees identify and meet their own career goals. (p. 376)

The performance criteria designed by the task force in Houston were grounded on the following tasks and activities carried out by Neighborhood-Oriented Policing (NOP) officers, based on the actual work done by a sample of officers during their shifts. Figure 10-3 provides this useful outline:

FIGURE 10-3

HOUSTON POLICE TASKS/ACTIVITIES

1. ***Learn characteristics of area, residents, businesses***
   a. Study beat books
   b. Analyze crime and calls-for-service data
   c. Drive, walk area and make notes
   d. Talk with community representatives
   e. Conduct area surveys
   f. Maintain area/suspect logs
   g. Read area papers (e.g., "shopper" papers)
   h. Discuss area with citizens when answering calls
   i. Talk with private security personnel in area
   j. Talk with area business owners/managers

2. ***Become acquainted with leaders in area***
   a. Attend community meetings, including service club meetings
   b. Ask questions in survey about who formal and informal area leaders are
   c. Ask area leaders for names of other leaders

3. ***Make residents aware of who officer is and what s/he is trying to accomplish in area***
   a. Initiate citizen contacts
   b. Distribute business cards
   c. Discuss purpose at community meeting
   d. Discuss purpose when answering calls
   e. Write article for local paper
   f. Contact home-bound elderly
   g. Encourage citizens to contact officer directly
4. ***Identify area problems***
   a. Attend community meetings
   b. Analyze crime and calls-for-service data
   c. Contact citizens and businesses
   d. Conduct business and residential surveys
   e. Ask about other problems when answering calls
5. ***Communicate with supervisors, other officers and citizens about the nature of the area and its problems***
   a. Maintain beat bulletin board in station
   b. Leave notes in boxes of other officers
   c. Discuss area with supervisor
6. ***Investigate/do research to determine sources of problems***
   a. Talk to people involved
   b. Analyze crime data
   c. Observe situation if possible (stakeout)
7. ***Plan ways of dealing with problem***
   a. Analyze resources
   b. Discuss with supervisor, other officers
   c. Write Patrol Management Plan, review with supervisor
8. ***Provide citizen information about ways they can handle problems (educate/empower)***
   a. Distribute crime prevention information
   b. Provide names and number of other responsible agencies; tell citizens how to approach these agencies
9. ***Help citizens develop appropriate expectations about what police can do and teach them how to interact effectively with police***
   a. Attend community meetings/make presentation
   b. Present school programs
   c. Write article for area paper
   d. Hold discussions with community leaders

10. *Develop resources for responding to problem*
    a. Talk with other officers, detectives, supervisors
    b. Talk with other agencies/individuals who could help
11. *Implement problem solution*
    a. Take whatever actions are called for
12. *Assess effectiveness of solution*
    a. Use data, feedback from persons who experienced the problem, and/or personal observation to determine whether problem has been solved
13. *Keep citizens informed*
    a. Officers tell citizens what steps have been taken to address a problem and with what results
    b. Detectives tell citizens what's happening with their cases

Source: Oettmeier, Timothy N. and Mary Ann Wycoff (1998). "Personnel performance evaluations in the community policing context." In Shelley, Tara O'Connor and Anne C. Grant (eds.). *Problem oriented policing: crime-specific problems, critical issues and making POP work*. Washington, D.C.: Police Executive Research Forum. p. 383–385. Used with permission.

Within the immediate context of problem-solving and the community policing approach, Spelman and Eck (1987) have suggested an added dimension that could be useful for performance evaluations (both at the individual and the organizational levels):

- Totally eliminate the problem;
- Reduce the number of incidents;
- Reduce the seriousness of the incidents;
- Design methods to better handle the incidents; and
- Remove the problem from police consideration.

Police Performance Measures in Great Britain

In Great Britain, there has been a substantial effort to produce realistic performance measures for police departments. At the centre of this effort is the Home Office, which has overall responsibility for the adequacy and effectiveness of police departments across Great Britain (of which there are currently 43). In 1990, the Association of Chief Police Officers (ACPO) made a firm commitment to monitoring police performance and service delivery. This was the theme of their strategic policy document, "Setting the standards for policing: meeting community expectation," which was published in October 1990. From that starting point, the police service worked with the Home Office, Her Majesty's Inspectorate of Constabulary, and Local Authority Associations in the development of clear measures of police performance that could be more easily and consistently collected. This activity dovetailed with the introduction of "Citizen's Charters" in Great Britain during 1991. These charters were government-sponsored documents that published local targets and performance indicators designed to meet the Quality of Service initiative. The Audit Commission was required under the *Local Government Act, 1992*, to determine a set of indicators for local authority services, which include the police, in terms of general efficiency, effectiveness, and economy. A list of forty-five indicators was developed for all police services. The broad categories under which these indicators are grouped is outlined below:

- *Call Management*—because research revealed that nearly six out of ten contacts between the police and the public occur by way of the telephone, it was determined that the manner in which a police service deals with calls is significant for performance evaluation. A series of factors needs to be taken into account in this area, including: overall level of demand on the police, the number of "999" calls made to the police, the availability of officers for dealing with calls, public expectations; geographical and related local factors, and customer satisfaction for service provided. It is noted that speed of response, while important, is only one of several measures of effective service delivery;
- *Crime Management*—this area covers a wide range of police activities, including the application of authorized police powers, crime detection, and the preparation of Crown briefs for the prosecution of suspects. Indicators in this category are intended to ensure that the public is aware of how police are in fact applying their powers. They are intended to demonstrate that police powers are being applied fairly and consistently. There is a recognition that too much attention to clearance rates as a performance indicator could be dangerous, if it resulted in too much emphasis on the detection side of the policing function;
- *Traffic Management*—this area accounts for a large number of contacts between the police and the public, accounting for a large portion of police resources. This area also poses several difficulties in terms of performance measurement because of the many external factors at play, including types of road surface and conditions, weather factors, and the interaction of several other emergency response agencies. Indicators would seek to provide information on: level of road traffic offences and impact on police workload, police effectiveness in addressing drinking and driving offences, and customer satisfaction with the delivery of police services in terms of road and traffic safety;
- *Public Order Management & Police Visibility on Patrol*—this category also reveals some difficulties in terms of accurate measurement of police effectiveness. Public disorder may occur on a scale that is beyond the immediate capacity of a police service to address successfully. There is an expectation that such events will be dealt with quickly, with little disruption or injury, and without repetition. However, these are often isolated events that can come into existence with little or no warning and through factors entirely beyond the control of the police;
- *Community Policing Management*—in light of the fact that the police rely on the support and confidence of the community, it is essential that all police services take

into account the needs and expectations of the public in order to sustain that confidence and support. In this category, the performance indicators consider:

- The extent to which the composition of the police service reflects the community at large as part of its commitment to equality of opportunity;
- The level of complaints against the police.

Consideration is to be given to the number of women and ethnic minority members employed by the police service to indicate the effectiveness of equal opportunity policies. Tracking progress of these members through the ranks of the police service will also indicate the quality of service to the community. Complaints are one measure of a failure to deliver a satisfactory service; and

- *Resources & Costs*—this category includes a consideration of the use of resources for the delivery of police services. Normally, policing costs are calculated as a per capita cost. It is noted that simple comparisons among police services may be misleading, and that they should always take into account several important considerations, including: range and variance of demands on a police service, differing geographical considerations, relevant and distinctive social and economic conditions, locally defined priorities, and others. Availability of officers for duty will also form a part of this category and can be detailed with information about sick leave management, training, court time, workers' compensation leave, etc.

LEADER IN COMMUNITY POLICING

Lenna Bradburn, Chief of Police, Guelph Police Service

What is your understanding of "community policing" and its importance to the evolution of policing in Canada?

Community policing is a philosophy of service delivery that recognizes the joint responsibility of the police and the community to identify and solve crime and social order problems. This concept goes much deeper than crime prevention programs, foot patrols, and other activities that were seen to be adjuncts in the 1980s to "real" policing. It requires a fundamental change in the culture and organizational structure of police services, as this philosophy invokes new values and expectations on the organization and its members. To put this into private sector terminology, it is customer service principles and quality management rolled into one. There are a variety of external and internal factors which have led policing away from the isolation of the traditional model of policing to this joint community integration. This shift is important, as it moves us away from

Lenna Bradburn LEADER IN COMMUNITY POLICING continued

the para-military philosophy which invokes notions of being at war with someone or something, and returns us to the original philosophy of Western policing as described by Sir Robert Peel.

Outline your own contribution to the development of "community policing" in Canada.

I believe my contribution focuses on moving policing from the program delivery component of community policing to the infrastructure issues of the police organization itself. I have spent much of the last eight years on changing the business practices of police services to support this new relationship with the community and the activities we believe front-line officers need to engage in.

What is your view of the future of "community policing" in Canada?

I think the philosophy of community policing will continue to survive, with the operational component becoming more sophisticated. I believe that we will come to the point where our members recognize that community policing does embrace both proactive and reactive policing strategies, and that these activities should be driven by community need.

What other comments would you like to make on this topic?

I cannot underscore enough the importance of needing to change the culture and organizational design of police services. Community policing will not be achieved if the values, traditions, symbols, and practices of a police service are inconsistent with the philosophy. This requires an examination of these issues and the elimination of those values which are inconsistent and the introduction of those which are consistent.

PUBLIC OPINION SURVEYS

Another means of measuring the impact of community policing initiatives is by directly addressing the "consumers" of police services, the public. This approach has proven effective in many jurisdictions and can be an essential component in any strategy for identifying public **needs** with regard to problem-oriented policing, as well as capturing public **satisfaction** with the delivery of service.

In order to undertake a public opinion survey, several elements are important to consider. Survey design requires a degree of skill and precision in order to generate usable results

(Alreck and Settle, 1985; Babbie, 1990; Eck and La Vigne, 1993). Several police departments in Canada, and elsewhere, have used this technique to good effect. Frequently, police organizations seek out the assistance of specialists in this area to lend credibility to the creation of an appropriate survey instrument. In several instances, Canadian police departments have turned to social science faculties and have received both guidance and support for the preparation and analysis of public needs or public attitude surveys. What follows is a brief summary of some key Canadian and American examples in the following jurisdictions:

- Gloucester Police Force;
- Halton Regional Police Service;
- Ontario Provincial Police;
- Reno (Nevada) Police Department; and
- Seattle (Washington) Police Department.

While not all of these surveys speak directly to community policing, it is consistent with an enlightened understanding of the philosophy of community policing that an effort will be made to *systematically* understand the needs of the community and/or its level of satisfaction with services being delivered.

GLOUCESTER POLICE FORCE

Beginning in 1988, preliminary discussions were held between the Gloucester Police and staff from Carleton University. Working groups were established to design and pre-test an appropriate questionnaire. It was determined that a mail-out survey would be more easily administered than a telephone survey. In the spring of 1989, one thousand questionnaires were mailed out to randomly selected citizens in Halton. Responses were received from 390 individuals which represented a 43% response (not including the 100 unopened surveys returned to the department). A copy of this survey instrument, along with an introductory letter from the chief of police, is included in Appendix I. The responses received were processed to capture both quantitative and qualitative data. The survey findings were assessed and categorized as follows:

- Problems in Gloucester;
- Face-to-face contact with the Gloucester Police Force;
- Telephone contact with the Gloucester Police Force;
- Victims of crime;
- Gloucester Police Force in general;
- Comments about, and changes to, the Gloucester Police Force; and
- Demographics of respondents.

The most serious problems raised through this survey pertained to drugs and thefts from homes. Respondents reported feeling safer during the daytime hours, while in the evening hours, they felt a decline in safe feelings, particularly while driving, or in parks. There was a general level of satisfaction with face-to-face contact with police officers in Gloucester.

Halton Regional Police Service

In January 1995, a survey was mailed to 2000 residents of Halton. These randomly selected recipients were asked a series of questions dealing with policing priorities and their expectations for policing services in their community (Halton Regional Police Service, 1995). With 725 completed surveys returned, there was a 36% response rate, which is considered high. Halton Regional Police Service looked upon the results as an "action document" guiding the police and the community towards the formulation of strategies tailored to meet identified needs in Halton. A copy of this survey is included in Appendix I. Among the key findings of this survey were the following observations:

- Five top problems the Halton area were: vandalism, residential break and enters, youth loitering, theft from cars, and auto theft; and
- High percentage (96%) believe that the public and the police share responsibility for providing a safe community.

Ontario Provincial Police

Following the major organizational review of the Ontario Provincial Police (OPP), and in order to establish a benchmark with regard to current level of service, the organization worked with the Environics Research Group to develop a public opinion survey. During a period in April 1995, representatives of the Environics Research Group conducted 1114 telephone interviews with randomly selected Ontarians using the approved survey instrument. The survey questions fell into three main categories:

- Image of the OPP;
- Priorities for the OPP; and
- Benchmark measures.

Image of the OPP

- Top-of-mind associations with the OPP were very positive;
- Slightly less than 6 in 10 Ontarians said they have a great deal of confidence in the OPP;
- Lukewarm ratings were given to the OPP for the quality of services. Respondents who were very familiar with the OPP, as well as Northern Ontarians were more positive;
- Highest ratings were for accident response, with lowest ratings for crime detection and investigation. Again, respondents with experience dealing with the OPP gave more favourable ratings; and
- High rating for aid to victims of traffic incidents and lowest for aid to victims of domestic violence or property crimes.

Priorities for the OPP

- Belief among Ontarians that reacting to, and investigating crimes and patrolling communities and enforcing the law should be high priorities for the OPP;

- When asked about top-of-mind associations with the term "community policing," three in ten Ontarians noted points relating to availability, accessibility, or visibility of the police within the community;
- Six in ten Ontarians stated a preference for a highly visible police service as opposed to one that worked behind the scenes; and
- Six in ten Ontarians agreed that some OPP duties could be performed by civilians.

Benchmark Measures

- Nearly four in ten Ontarians strongly agreed that the OPP is visible, working to address community problems and is delivering policing services that are among North America's best. Furthermore, about three in ten Ontarians strongly agreed that the OPP is sensitive to racial minorities, knowledgeable about communities, prompt, innovative, and open to community involvement.

In response to the question: What comes to mind when you hear the words "community policing"? there was a range of responses that included references to protection, safety, and help (15%), neighbourhood watch or neighbours policing the community (13%), or some form of local/regional policing (12%). Further analysis (Environics Research Group Ltd., 1995) of the responses generated the following background:

> Regional analysis finds that residents of the Greater Toronto Area are more likely than average to say community policing refers to foot patrols. Eastern Ontarians are more likely than average to say it refers to local or regional policing or to police patrolling neighbourhood. Northern Ontarians are slightly more likely than average to say community policing refers to neighbourhood watch or neighbours policing the community.
>
> Demographic analysis finds that Ontarians in the higher education groups are more likely than others to say community policing refers to police involved with or working in community or foot patrols. Those in the higher income groups are also more likely than others to say it refers to police involved with or working in community. Those in lower income groups are more likely than others to say community policing refers to protection, safety or help. (p. 34)

Reno (Nevada) Police Department

When the Reno (Nevada) Police Department reorganized in 1987, it adopted a community-oriented policing model. As part of this change in strategy, the department made a commitment to incorporate public views when determining the direction the Reno Police would take. They used a public opinion survey approach as part of a new awareness of police-citizen interaction.

The results of these surveys are combined with other management information sources to make decisions relating to training, human resource deployment, and program implementation. As is noted in the department's December 1993 survey report (Bigham, 1993):

> We believe that it is crucial to the effectiveness of our managerial efforts to clearly understand and react to the concerns and expectations of the citizens that we serve. Our overall effectiveness over the past years is largely based on implementing the information gained from the surveys. This

survey continues to provide substantial legitimate reinforcement that we are being very effective in listening to and providing quality service to our citizens.

The Reno Police Department obtained assistance from the University of Michigan and the University of Nevada, Reno to ensure the validity and reliability of their survey instrument. Their survey methodology has been recognized as a model by the Police Executive Research Forum.

SEATTLE (WASHINGTON) POLICE DEPARTMENT

In Seattle, Washington, the Seattle Police Department researched the topic of community policing and assembled a series of "core ideas" that could readily be identified as a common part of community policing (Fleissner, 1997):

- Link with the community;
- Expanded or broadened role and function of the police;
- Systematic problem-oriented approach;
- Focus on outcome measures and results; and
- Management and organizational changes.

Fleissner (1997) provides a matrix that incorporates elements of the "stage assessment" approach used within the Seattle Police Department for implementing planning and organizational measurement. Figure 10-4 represents their community policing stage assessment matrix:

FIGURE 10-4

SEATTLE COMMUNITY POLICING STAGE ASSESSMENT MATRIX

Development Stage	Program Components/Elements	
Stage 1—Awareness/ Discovery	*Problem Solving Approach*	*Program Evaluation*
Initial interest in the new style of policing from magazine articles/conferences/exchange of information with other police personnel; take action to learn more about new community policing approach; discuss use of new programs in meetings and formulate tentative new program ideas.	General discussions about how projects are selected are held between police and citizens; concerns about involvement of citizens and security issues; concerns about police workload & limited resources; little or no interaction with the public about problems.	No specific evaluation questions are considered relevant at this stage; basic process evaluations of planning and information gathering timeliness (if formal project designation) are considered.

Stage 2—Experimental/ Exploratory	*Problem Solving Approach*	*Program Evaluation*
Police agency &/or city identify potential applications of community policing model; acquire more information about specific programs & plan/implement pilot efforts; begin limited discussions with other city personnel to familiarize them with the new problem-solvving & community oriented approach; evaluate pilot projects.	Initial problem-solving meetings involve how to set agendas/select problems and procedures; solution approaches reflect little creativity or involvement of other (non-police) resources; feedback on outcomes requested by citizens, but viewed as time consuming by police.	Process evaluations & tracking of activities; report actions/successes of problem solving projects; initial use of "customer" surveys to identify needs/priorities.
Stage 3—Commitment/ Understanding	*Problem Solving Approach*	*Program Evaluation*
More in-depth analysis of information about community policing & requirements; review results of pilot programs; establish open communication links with other city departments & provide educational materials; develop & conduct extensive training programs; expand involvement of citizens/business groups and build trust; police agency, city officials, and other departments analyze required organizational and managerial changes; resources are committed and coordinated implementation begins.	Working relationships among police/citizens/ other city departments built on trust & respect; problem targets selected by community; long-term problems are addressed; level of creative & innovative solutions increases; amount & quality of analysis increases.	Develop outcome/impact measures for performance evaluations tied to goals; use customer surveys to assess priorities & satisfaction with results of city/ police efforts.

Stage 4—Proficiency/ Renewal	*Problem Solving Approach*	*Program Evaluation*
The new approach to police service delivery & community/business partnership becomes ingrained in police agency operations & management style; most city departments are focused on coordinated service delivery; ongoing & automatic processes for service improvement & innovation are operational; innovative managerial approaches and structural changes to the organization are completed based on new mission & vision.	Joint city-wide problem solving program involves regular meetings & open communications process; private sector & regional resources are part of process; coordinated service delivery is city-wide standard.	Continuous process, impact & monitoring part of normal operations; constant "fine tuning" of performance measures with focus on service improvement.

Source: Fleissner, Dan W. (1997). *Community policing stage assessment model for implementation planning and organizational measurement*, p. 62. Seattle, Wash.: Seattle Police Department. Project supported by Grant No. 95-IJ-CX-0044 awarded by the National Institute of Justice. Used by permission.

FUTURE CANADIAN PUBLIC SURVEYS ON POLICING

Melchers and Roberts (1994) have noted that there are several survey research priorities that should be addressed in the future. Each of these suggested topics could be construed in a manner that would add to our existing stock of knowledge about the possibilities and limitations of community policing:

- *Private security & private policing*—determinations about the growth and impact of these forms of security in Canada;
- *Patrol techniques*—explorations of the various types of patrols that have been experimented with in several jurisdictions including: bicycle patrols, foot patrols, and other variations. Surveys could include public attitudes to these approaches;
- *Community-based policing*—there is a shortage of systematic research on what people want to see done in this area;
- *Public knowledge of police activities*—explore the depth of public understanding with regard to police behaviour;

- *Determinants or correlates of public attitudes toward the police*—consideration of the role of various influences on public attitudes toward the police, including the media;
- *Family violence*—as a topic that is often identified for greater police training, it is valuable to survey public attitudes toward police activity in this area;
- *Special interest groups*—efforts to survey the attitudes of specific, vulnerable groups in Canadian society, e.g, women and racial minorities, with respect to their attitudes toward the police, or their experiences in dealing with police organizations;
- *Regional variations*—surveys to gain knowledge about differences in public attitudes toward the police in various parts of Canada; and
- *Variations over time*—survey research that will provide a basis for historical comparisons regarding attitudes toward the police. (pp. 13–14)

COMMUNITY SAFETY AUDITS

In many jurisdictions, the community safety audit is an invaluable tool for establishing targets against which to measure police performance in key areas. Derived from the area of crime prevention, these audits are a true form of partnership between several community groups and a range of government agencies, including the police department. What is shared is a common interest in achieving greater levels of public safety. It is recognized that the police, alone, cannot produce all of the conditions necessary for people to feel safe in their communities. A community safety audit supports a truly valuable, proactive approach to fear reduction. It represents an inventory of items, features, and characteristics that can be modified, corrected, or removed in order to promote greater community safety and security.

The community safety audit focuses on both public and semi-public areas, including:

- Parks;
- Bus stops;
- Streets;
- Workplaces;
- Colleges and universities;
- Underground parking areas;
- School yards;
- Washrooms in shopping malls;
- Urban transit system;
- Pedestrian walkways;
- Laundry rooms;
- Parking lots;
- Recreation centres.

A detailed *Neighbourhood Crime Prevention Safety Audit Guide* from the City of Edmonton is included as Appendix II.

SUMMARY

This chapter has attempted to cover various ways in which a police organization's commitment to community policing can be measured. It considers the relevance of personnel performance evaluation in an organization that has made the transition to community policing, and looks at the various approaches that can be taken in assessing the organization's overall effectiveness in the implementation of a strategy of community policing. The literature in support of these forms of performance measurement has been cited by way of assistance and guidance.

Also, the value of conducting public opinion and/or satisfaction surveys has been discussed at some length, with several examples in which surveys have been instrumental in guiding, grading, or redirecting a department's community policing strategy.

Finally, the importance of community safety audits has been cited as one means of establishing an inventory of considerations against which community policing efforts may be assessed. The community safety audit is an important tool for encouraging public input and understanding around significant community safety issues.

QUESTIONS FOR CONSIDERATION

1. What are the levels of assessment for evaluating community policing?
2. What are the benefits of conducting public opinion surveys?
3. What are the areas examined in a community safety audit?
4. How does a police organization develop approaches for the evaluation of team performance?
5. What are the tasks or activities that could form the basis for personnel performance evaluation, particularly in the context of a problem-oriented approach to policing?
6. What are the components of an effective public survey?

REFERENCES

Abbey-Livingston, Diane and David S. Abbey (1982). *Enjoying research: a 'how-to' manual on needs assessment*. Toronto: Queen's Printer for Ontario.

Alberta Treasury (1996). *Measuring performance: a reference guide*. Edmonton: Alberta Treasury.

Alreck, Pamela L. and Robert B. Settle (1985). *The survey research handbook*. Homewood, Ill.: Richard D. Irwin.

Argyris, Chris (1965). *Organization and innovation*. Homewood, Ill.: Richard D. Irwin & the Dorsey Press.

Babbie, Earl (1990). *Survey research methods. 2nd ed.* Belmont, Calif.: Wadsworth.

Bigham, Steve (1993). *Attitude and public opinion survey, Reno Police Department*. Reno, Nev.: Reno Police Department.

Brodeur, Jean-Paul (ed.) (1998). *How to recognize good policing: problems and issues*. Thousand Oaks, Calif.: Sage Publications and the Police Executive Research Forum.

Brown, Gregory P. and Donna McAlpine (1996). "Judging performance: a brief introduction." Paper presented to the Ministry of the Solicitor General and Correctional Services business planning process performance measures workout, Tuesday, August 13, 1996.

Cahn, M. and J.M. Tien (1980). *An evaluation report of an alternative approach in police response: the Wilmington management of demand program*. Cambridge, Mass.: Public Systems Evaluation Inc.

California Department of Justice. Attorney General's Office. Crime Prevention Center (1992). *COPPS: community oriented policing & problem solving*. Sacramento, Calif.: California Department of Justice.

Campbell, M.D. (1995). *Outcome and performance measurement systems: an overview*. National Academy of Public Administration, Alliance for Redesigning Government.

Carter, Neil, Rudolf Klein, and Patricia Day (1992). *How organizations measure success: the use of performance indicators in government*. London: Routledge.

Centre of Criminology. University of Toronto (1996). *A selection of readings for a workshop on performance measures in policing and corrections organized for the Ministry of the Solicitor General and Correctional Services by the Centre of Criminology, October 1996.* Toronto: The Centre.

Chicago Community Policing Evaluation Consortium (1997). *Community policing in Chicago, year four: an interim report*. Chicago: Illinois Criminal Justice Information Authority.

Clairmont, Donald (1990). *To the forefront: community-based zone policing in Halifax*. Ottawa: Canadian Police College.

Cordner, Gary W. (1985). *The Baltimore county citizen oriented police enforcement (COPE) project: final evaluation*. Baltimore: Criminal Justice Department, University of Baltimore.

Cordner, Gary W. (1986). "Fear of crime and the police: an evaluation of a fear reduction strategy." *Journal of Police Science and Administration*, Vol. 14, pp. 223–233.

Decker, Scott H. (1981). "Citizen attitudes toward the police: a review of past findings and suggestions for future policy." *Journal of Police Science and Administration*, Vol. 9, pp. 80–87.

DiIulio, John J. Jr. et al. (1993). *Performance measures for the criminal justice system: discussion papers from the BJS-Princeton project*. Washington, D.C.: U.S. Dept. of Justice, Bureau of Justice Statistics.

Eck, John E. and Nancy G. La Vigne (1993). *A police guide to surveying citizens and their environment*. Washington, D.C.: Bureau of Justice Assistance.

Eck, John E. and Nancy G. La Vigne (1994). *Using research: a primer for law enforcement managers. 2nd ed.* Washington, D.C.: Police Executive Research Forum.

Environics Research Group Ltd. (1995). *Ontarians' attitudes toward the OPP*. Toronto: Environics Research Group.

Fleissner, Dan W. (1997). *Community policing stage assessment model for implementation planning and organizational measurement*. Seattle, Wash.: Seattle Police Department.

Fowler, Floyd J. Jr. and Thomas W. Mangione (1983). *Neighborhood crime, fear and social control: a second look at the Hartford program*. Washington, D.C.: U.S. Government Printing Office

Friedman, Robert R. (1987). "Citizen attitudes toward the police: results from an experiment in community policing in Israel." *American Journal of Police*, Vol. 6, pp. 67–94.

Furstenberg, Frank and Charles Wellford (1973). "Calling the police: the evaluation of police service." *Law and Society Review*, Vol. 7, pp. 393–406.

Garrett, Michael (1996). "Proxies for profit" or performance measurement in municipal government." Unpublished paper presented at the roundtable on business planning and organization design: performance measurement, Friday, September 27, 1996, Whitney Block, Toronto.

Goldstein, Herman and Charles E. Susmilch (1981). *The problem-oriented approach to improving police service: a description of the project and an evaluation of the concept*. Madison, Wis.: Law School, University of Wisconsin.

Goldstein, Herman and Charles E. Susmilch (1982). *Experimenting with the problem-oriented approach to improving police service: a report and some reflections on two case studies*. Madison, Wis.: Law School, University of Wisconsin.

Great Britain. Home Office (1993). *Home Office circular no. 17/1993: performance indicators for the police*. London: The Home Office.

Great Britain. Home Office (1995). *Home Office circular 8/1995: performance indicators for the police and core statistics for chief officers' annual reports*. London: The Home Office.

Greenberg, Stephanie W., William M. Rohe, and Jay R. Williams (1982). *Safe and secure neighborhoods: physical characteristics and informal territorial control in high and low crime neighborhoods*. Washington, D.C.: U.S. Government Printing Office.

Greenberg, Stephanie W., William M. Rohe, and Jay R. Williams (1986). *Informal citizen action and crime prevention at the neighborhood level: synthesis and assessment of the research*. Washington, D.C.: U.S. Government Printing Office.

Halton Regional Police Service (1995). *Public survey 1995*. Oakville, Ont.: Halton Regional Police Service.

Hayeslip, Davis W. and Gary W. Cordner (1987). "Effects of community-oriented patrol on police officer attitudes." *American Journal of Police*, Vol. 6, pp. 95–119.

Hoover, Larry T. (ed.) (1996). *Quantifying quality in policing*. Washington, D.C.: Police Executive Research Forum.

Hoover, Larry T. (ed.) (1998). *Police program evaluation*. Washington, D.C.: Police Executive Research Forum.

Horne, David G. and Lester N. Thompson (1989). *Gloucester Police Force public needs survey: preliminary findings, report number one*. Ottawa: Carleton University.

Hornick, Joseph P. et al. (1989). *An evaluation of the neighborhood foot patrol program of the Edmonton Police Service*. Edmonton: Canadian Research Institute for Law and Family.

Kaplan, Robert S. and David P. Norton (1992). "The balanced scorecard: measures that drive performance." *Harvard Business Review*, Vol. 70, no. 1, (January/February) pp. 45–55.

Kaplan, Robert S. and David P. Norton (1993). "Putting the balanced scorecard to work." *Harvard Business Review*, Vol. 71, no. 5, (May/June) pp. 134–142

Kasul, Ruth A. and Jaideep G. Motwani (1995). "Performance measurements in world-class operations: a strategic model." *Benchmarking for Quality Management & Technology*, Vol. 2, no. 2, pp. 20–36.

Kelling, George L. (1992). "Measuring what matters: a new way of thinking about crime and public order." *The City Journal*, Vol. 2, (Spring) pp. 21–34.

Klein, John F., James R. Webb, and Joseph E. Disanto (1978). "Experience with the police and attitudes toward the police." *Canadian Journal of Sociology*, Vol. 3, pp. 441–446.

Lavrakas, Paul J., S.F. Bennett and B.S. Fisher (1987). "The neighborhood anti-crime self-help program evaluation: some preliminary findings on community organization and police interaction," paper presented at the annual meeting of the American Society of Criminology.

Lavrakas, Paul J. and James Kushmuk (1986). "Evaluating crime prevention through environmental design: the Portland commercial demonstration project." In Rosenbaum, Dennis P. (ed.) *Community crime prevention: does it work?* Beverly Hills, Calif.: Sage Publications.

Lewis, Lois (1988). *Citizen survey*. Calgary: Calgary Police Service.

Management Board Secretariat of Ontario (1996). *A framework for performance management in government programs and services*. Toronto: Management Board Secretariat.

Mastrofski, Stephen D. (1989). "Police knowledge of the patrol beat: a performance measure." In Bennett, Richard R. (ed.) *Police at work: policy issues and analysis*. Beverly Hills, Calif.: Sage Publications.

Mastrofski, Stephen D. (1996). "Measuring police performance in public encounters." In Hoover, Larry T. (ed.) *Quantifying quality in policing*. Washington, D.C.: Police Executive Research Forum.

Mastrofski, Stephen D. and Robert C. Wadman (1991). "Personnel and agency performance management." In Geller, William A. (ed.) *Local government police management*. Washington, D.C.: International City Management Association.

Mathie, Robert C. (1989). "Examining performance: some suggestions on measuring performances for evaluating police work." *Policing*, Vol. 5, no. 4, (Winter), pp. 287–304.

McElroy, Jerome E., Colleen A. Cosgrove and Susan Sadd (1989). "An examination of the community patrol officer program (CPOP) in New York City," unpublished report. New York: Vera Institute of Justice.

Melchers, Ron and Julian Roberts (1994). *Public opinion and policing in Canada: a review of recent findings and selected social trends and demands on policing*. Ottawa: University of Ottawa, Department of Criminology.

Meyer, Christopher (1994). "How the right measures help teams excel." *Harvard Business Review*, Vol. 72, no. 3 (May/June), pp. 95–103.

Morgan, Gareth (1993). *Imaginization: the art of creative management*. Newbury Park, Calif.: Sage Publications.

Murphy, C. and S. Lithopoulas (1988). *Social determinants of attitudes toward the police: findings from the Toronto community policing survey*. Ottawa: Solicitor General Canada.

Oettmeier, Timothy N. and Mary Ann Wycoff (1994). "Police performance in the nineties: practitioner perspective." *American Journal of Police*, Vol. 13, no. 2, pp. 21–49.

Oettmeier, Timothy N. and Mary Ann Wycoff (1998). "Personnel performance evaluations in the community policing context." In Shelley, Tara O'Connor and Anne C. Grant (eds.) *Problem oriented policing: crime-specific problems, critical issues and making POP work*. Washington, D.C.: Police Executive Research Forum.

Ontario. Ministry of the Solicitor General. Strategic Planning Committee on Police Training and Education (1992). *A police learning system for Ontario: final report and recommendations*. Toronto: Ministry of the Solicitor General.

Payne, Dennis M. and Robert C. Trojanowicz (1985). *Performance profiles of foot versus motor officers*. East Lansing, Mich.: National Neighborhood Foot Patrol Center, Michigan State University. (Community policing series, no. 6)

Peak, Kenneth J. and Ronald W. Glensor (1996). *Community policing and problem solving: strategies and practices*. Upper Saddle River, N.J.: Prentice Hall.

Petersilia, Joan, Allan Abrahamse, and James Q. Wilson (1990). "A summary of RAND's research on police performance, community characteristics, and case attrition." *Journal of Police Science and Administration*, Vol. 17, no. 3, pp. 219–226.

Portland State University. School of Urban and Public Affairs (1995). *Assessment of community policing performance within the Portland Police Bureau Domestic Violence Reduction Unit*. Portland: Portland State University.

Rea, Louis M. and Richard A. Parker (1992). *Designing and conducting survey research: a comprehensive guide*. San Francisco: Jossey-Bass.

Robbins, Stephen P. and Nancy Langton (1999). *Organizational behaviour: concepts, controversies, applications. Canadian edition*. Scarborough, Ont.: Prentice Hall Canada.

Rogerson, Philip (1995). "Performance measurement and policing: police service or law enforcement agency?" *Public money & management*, (October-December), pp. 25–30.

Rosenbaum, Dennis P. (ed.) (1994) *The challenge of community policing: testing the promises*. Thousand Oaks, Calif.: Sage Publications.

Rosenbaum, Dennis P. and Terry L. Baumer (1981). *Measuring fear of crime: a set of recommended scales*. Evanston, Ill. : Westinghouse Evaluation Institute.

Rosenbaum, Dennis P., D.L. Lewis, and J.A. Grant (1985). *The impact of community crime prevention programs in Chicago: can neighborhood organizations make a difference?* Final report of the Ford Foundation. Chicago: Department of Criminal Justice, University of Illinois at Chicago.

Sadd, Susan and Randolph M. Grinc (1993a). *Issues in community policing: an evaluation of eight innovative neighborhood-oriented policing projects: volume I of a report submitted to the National Institute of Justice*. New York: Vera Institute of Justice.

Sadd, Susan and Randolph M. Grinc (1993b). *Issues in community policing: descriptions of programs in eight cities: volume II of a report submitted to the National Institute of Justice.* New York: Vera Institute of Justice.

Schwartz, Alfred I. and Sumner N. Clarren (1977). *The Cincinnati team policing experiment: a summary report*. Washington, D.C.: Police Foundation.

Skogan, Wesley G. and Susan M. Hartnett (1997). *Community policing, Chicago style*. New York: Oxford University Press.

Sloma, R.S. (1980). *How to measure managerial performance*. New York : Macmillan Publishing.

Stephens, Darrel W. (1996). "Community problem-oriented policing: measuring impacts. In Hoover, Larry T. (ed.) *Quantifying quality in policing*. Washington, D.C.: Police Executive Research Forum.

Themes and variations in community policing: case studies of community policing. (1996). Washington, D.C.: Police Executive Research Forum.

Trojanowicz, Robert C. (1983). "An evaluation of a neighborhood foot patrol program." *Journal of Police Science and Administration*, Vol. 11, pp. 410–419.

Trojanowicz, Robert C. and Bonnie Bucqueroux (1992). *Toward development of meaningful and effective performance evaluations*. East Lansing, Mich.: National Center for Community Policing, Michigan State University.

U.S. National Institute of Justice (1997). *Measuring what matters, part two: developing measures of what the police do: a report from the Policing Research Institute*. Washington, D.C.: National Institute of Justice.

Walker, Sandra Gail, Christopher R. Walker, and James C. McDavid (1992). *The Victoria community police stations: a three-year evaluation*. Ottawa: Canadian Police College.

Weatheritt, Mollie (1993). "Measuring police performance: accounting or accountability?" In Reiner, Robert and Sarah Spenser (eds.) *Accountable policing: effectiveness empowerment, and equity*. S.l.: Institute for Public Policy Research.

Weisburd, David and Jerome E. McElroy (1988). "Enacting the CPO role: findings from the New York City pilot program in community policing." In Greene, Jack R. and Stephen D. Mastrofski (eds.) *Community policing: rhetoric or reality*. New York: Praeger.

West Yorkshire Police Authority (1996). *Policing plan for West Yorkshire 1995-1996*. Wakefield: West Yorkshire Police Authority.

Whitaker, Gordon P. et al. (1982). *Basic issues in police performance*. Washington, D.C.: U.S. Dept. of Justice.

Wycoff, Mary Ann and Timothy N. Oettmeier (1993). *Evaluating patrol officer performance under community policing: the Houston experience. Technical report*. Washington, D.C.: Police Foundation.

Yin, Robert K. (1986). "Community crime prevention: a synthesis of eleven evaluations." In Rosenbaum, Dennis P. (ed.) *Community crime prevention: does it work?* Beverly Hills, Calif.: Sage Publications.

TOWARD MORE LEARNING

The best means of arriving at a full understanding of the process and products of performance measurement is to consider its application in depth. There are several references that provide this detailed level of insight. Because community policing as an integral part of policing practice is still relatively new, it is extremely helpful to turn to works that have considered the evaluation (i.e., performance measurement) end of the spectrum in specific applications.

PERSONNEL PERFORMANCE

Oettmeier, Timothy N. and Mary Ann Wycoff (1998). "Personnel performance evaluations in the community policing context." In Shelley, Tara O'Connor and Anne C. Grant (eds.) *Problem oriented policing: crime-specific problems, critical issues and making POP work*. Washington, D.C.: Police Executive Research Forum.

This is an excellent summary of the work done by the authors in researching the topic of personnel performance evaluations, as well as their engagement with the application of the principles of evaluation in a context of community policing in the Houston (Texas) Police Department.

Reiner, Robert (1998). "Process or product?: problems of assessing individual police performance." In Brodeur, Jean-Paul (ed.). *How to recognize good policing: problems and issues*. Thousand Oaks, Calif.: Sage Publications and the Police Executive Research Forum.

A recent essay that examines the longstanding deficiencies in research dealing with individual police performance assessment. Beginning with a consideration of policing in Great Britain, the author moves to consider available research in other jurisdictions, including Canada, concluding that individual police performance, by its nature, cannot be "precisely calibrated" and therefore efforts to make it so would be counterproductive.

ORGANIZATIONAL ASSESSMENTS

Greene, Jack R. (1998). "Evaluating planned change strategies in modern law enforcement: implementing community-based policing." In Brodeur, Jean-Paul (ed.). *How to recognize good policing: problems and issues*. Thousand Oaks, Calif.: Sage Publications and the Police Executive Research Forum.

Chicago Community Policing Evaluation Consortium (1997). *Community policing in Chicago, year four: an interim report*. Chicago: Illinois Criminal Justice Information Authority.

Fleissner, Dan W. (1997). Community policing stage assessment model for implementation planning and organizational measurement. Seattle, Wash.: Seattle Police Department.

Great Britain. Home Office (1993). Home Office circular no. 17/1993: performance indicators for the police. London: The Home Office.

Great Britain. Home Office (1995). Home Office circular 8/1995: performance indicators for the police and core statistics for chief officers' annual reports. London: The Home Office.

The two preceding publications provide a detailed outline of the performance indicators that have been adopted in the forty-three police services in England and Wales. These indicators have resulted from considerable effort and collaboration among a number of organizations and agencies interested in policing and local government standards. They have been instructive in the development of similar quality standards in Canada, especially Ontario.

Hoover, Larry T. (ed.) (1998). *Police program evaluation*. Washington, D.C.: Police Executive Research Forum.

Sadd, Susan and Randolph M. Grinc (1993a). *Issues in community policing: an evaluation of eight innovative neighborhood-oriented policing projects: volume I of a report submitted to the National Institute of Justice*. New York: Vera Institute of Justice.

Sadd, Susan and Randolph M. Grinc (1993b). Issues in community policing: descriptions of programs in eight cities: volume II of a report submitted to the National Institute of Justice. New York: Vera Institute of Justice.

The two publications cited above, when considered together, provide a thorough synopsis of the evaluation of Neighborhood-Oriented Policing (NOP) projects in the following cities:

- Tempe (Arizona);
- Norfolk (Virginia);

- New York (New York);
- Hayward (California);
- Portland (Oregon);
- Prince George's County (Maryland);
- Houston (Texas); and
- Louisville (Kentucky)

Skogan, Wesley G. and Susan M. Hartnett (1997). *Community policing, Chicago style*. New York : Oxford University Press.

PUBLIC SURVEYS

Halton Regional Police Service (1995). *Public survey 1995*. Oakville, Ont.: Halton Regional Police Service.

Horne, David G. and Lester N. Thompson (1989). *Gloucester Police Force public needs survey: preliminary findings, report number one*. Ottawa: Carleton University.

Hornick, Joseph P. et al. (1989). *An evaluation of the neighborhood foot patrol program of the Edmonton Police Service*. Edmonton: Canadian Research Institute for Law and Family.

Melchers, Ron and Julian Roberts (1994). *Public opinion and policing in Canada: a review of recent findings and selected social trends and demands on policing*. Ottawa: University of Ottawa, Department of Criminology.

Murphy, C. and S. Lithopoulas (1988). *Social determinants of attitudes toward the police: findings from the Toronto community policing survey*. Ottawa: Solicitor General Canada.

CHAPTER ELEVEN

Different societies make different arrangements in regard to weights, measures, and money; these arrangements do not contradict one another. But if different societies hold different views regarding the principles of justice, their views contradict one another. Differences regarding things which are unquestionably conventional do not arouse serious perplexities, whereas differences regarding the principles of right and wrong necessarily do.

Leo Strauss (1953), *Natural Right and History*, p. 100.

MADE IN CANADA—EXPERIMENTS AND LESSONS LEARNED

LEARNING OBJECTIVES

1. Identify police services across Canada applying community policing theories and practices.
2. State the key elements that have been identified in implementing a community policing model.
3. Explain some of the lessons learned in the process of implementing community policing in Canada.
4. Develop a detailed understanding of various approaches to community policing in Canada.

INTRODUCTION

With all of the discussion about the philosophy of community policing, it is helpful to turn attention to those police organizations that have attempted to apply it within their operations. It has been stated on many occasions that, correctly understood, community policing is not a program. However, for the substance of this approach to have any real meaning as a viable means of police service delivery, it must take shape in some concrete manner. This chapter looks in considerable detail at those police organizations across Canada that have applied the theory of community policing in their overall mission. Clearly, this selection is by no means comprehensive, and there are continuous refinements occurring even within the departments highlighted in the chapter. However, this chapter should serve to substantiate the observation that many Canadian police organizations are making significant efforts at implementing some form of community policing within their practice. This chapter attempts to explore a limited but representative range of Canadian experience in this mode of modern policing.

HALIFAX, NOVA SCOTIA

With its long history, the Halifax Police Department gained a reputation as an honest, efficient, but fairly stolid and conservative police force (Clairmont, 1990). However, change came to this department, as it does to so many other police services (Clairmont):

> In 1985, subsequent to the appointment of a new chief of police, a five year plan of change was introduced into the Halifax Police Department. The change focused around the introduction of community-based zone policing but also included civilianization and upgrading in training and technical (especially computerization) areas. On the surface at least the changes could profoundly affect the organizational structure, operational dynamics and the subculture of the department. (p. 11)

The new chief had the support of the chairman of the board of the police commission, as well as the city manager. This level of support would prove to be instrumental in making the substantial changes to this department. As Clairmont (1990) notes:

> The quasi-military character of the police organization appears evidenced in the generation of the innovative plans and in the implementation process. Beyond the chief, the innovating group at the Halifax Police Department essentially involved the youngest inspector, two sergeants and a constable. Its appears that most top Halifax Police Department management personnel were not significantly involved in these initial processes nor <u>in their view</u> were they well-informed about the innovation or encouraged to provide criticisms or alternatives. (p. 20) [emphasis in original]

Before the introduction of community-based policing, the Halifax Police Department demonstrated the following features (Clairmont, 1990):

- From 1969 amalgamation the department had undergone substantial growth, but without structural or organizational change.
- Movement to a decidedly "crime fighting" role a degree of "distancing" from the public.
- Growth in the number of civilian members and more engagement of civilian governing authorities buffer provided by the local police commission.
- Solid reputation of the Halifax Police Department, responded rapidly and competently to calls for service.

Planning for the change to community-based zone policing in Halifax began in early 1985. As part of the planning process, consultants were engaged to make recommendations with respect to change within the Halifax Police Department. As part of their final report, the consultants (William M. Mercer, 1990) noted that:

> Since the early 1970s, police forces throughout North America have initiated programs of change based on different versions of community-based policing. The team policing programs in the United States provided the first models. In the mid-1970s, Calgary, Vancouver, and the RCMP in the lower mainland of British Columbia, began to build their versions of zone or community-based policing. Since the 1970s, other major police forces such as Metropolitan Toronto, Edmonton, Halton Region, and the RCMP have proceeded to implement versions of community-based policing. (p. 8)

The Mercer consultants, after researching the circumstances of the Halifax Police Department, recommended a five-phase evolution to a new model of policing:

- *Phase I*—command based policing with a focus on response and specialists;
- *Phase II*—community-based police I with a focus on zones and programs;
- *Phase III*—community-based police II, refocus and problem-solving in the zones;
- *Phase IV*—"corporate" contemporary policing with a focus on institutionalization and accountability; and
- *Phase V*—renewal for breakthrough policing for the mid to late 1990s.

The consultants also recommended that the model of policing function as an integrated "system" with substantial coherence among the following key components within the police service:

- Vision;
- Leadership;
- Roles Focus;
- Structure;
- Management Systems;
- Operating Systems & Technology;
- Human Resources Programs;
- Rewards & Incentives; and
- Resources & Costs.

The first zone (Alpha) was in operation by June 1986, with Bravo and Charlie zones coming into place in the winter and spring of 1987. The new approach was in effect throughout the entire city between June 1987 and January 1988. The introduction of this innovative approach was the result of three key people (Clairmont, 1990):

> . . . there were three immediate catalysts and persons who gave shape to the rather amorphous demand for organizational change; the new chief, the chairman of the board and the city manager. Their broad strokes were further elaborated and pieced together with the Halifax Police Department by the innovating group. (p. 86)

The Halifax Police Department experiment is a useful lesson in collaboration and the critical, if not indispensable, need to engage appropriate allies in the pursuit of large-scale change.

Training was seen as being essential to the introduction of community-based zone policing in Halifax. Indeed, many other jurisdictions had met with failure because they did not properly train their members in the new philosophy. The Halifax Police Department spent several months putting together a detailed training curriculum for their "Alpha" group. Below is an outline of that group's training:

OUTLINE OF TRAINING FOR THE ALPHA GROUP (THEMES COVERED IN SEQUENCE)

Week One: (NCOs only)—

- Chief's introduction, Policing Today (2), Organizational Expectations and Definitions (2), Leadership, Communication and Delegation, Decision-Making,

Participatory Management, Zone Deployment (2), The Rand Study, Crime Analysis (2), Case Management, Investigative Procedures (3), Review.

Week Two: Full Complement—

- * Policing Today (4), Community-Based Policing (4), Note Taking, Drugs, Crime Analysis, Technology, * Court and Prosecution Procedures (4), Crime Scenes (4).

Week Three: Full Complement—

- Traffic Road Blocks, Surveillance (2), Foot Patrol, Motorized Patrol, Accident Investigation (2), * Public Speaking (8), Team Policing and Crime Prevention (4).

Week Four: Full Complement—

- Sexual Assault and Child Abuse, Duty Rosters, Homicide and Suicide, Fraud, * Victim Consultation (4), Interrogation Procedures, Municipal Ordinances, Criminal Code, * Agency Referral (2), Case Management (4).

Week Five: Full Complement—

- Preliminary Investigative Procedures (6), Problem Solving (6), Job Analysis, Informant Recruiting, Public Speaking (2), Departmental Policies (2), Review, Graduation.

* Asterisks indicate that non-Halifax Police Department personnel were involved in the training session. Numbers following a session theme indicate how many one and one half hour periods were spent on the topic; if only one then no number is given.

The training program developed gives some idea of the scope of the transformation anticipated in the Halifax Police Department. As Clairmont (1990) elaborates:

> Halifax Police Department's organizational change under Community Based Policing was indeed extensive. New roles were created and other roles significantly redefined. Divisions were eliminated or radically altered in mandate and manpower and new ones developed. Zones replaced the platoon/districts system with attendant implications for supervision, constables' work patterns and departmental decision-making. Promotion routines and criteria were changed and civilianization increased. Perhaps most importantly organizational change appeared to be directed at the way the police organization actually operated. Considerations relating to "corporate style" and quality of working life (e.g., feedback, planning, job enrichment, participative policy-development and decision-making) were both radical and contingent. (p.117)

Following the efforts within the Halifax Police Department to move toward a community-based zone policing model, a new challenge took shape in April 1996. This was the "shotgun marriage" of three existing police departments: the cities of Halifax and Dartmouth, and the Town of Bedford, to create the new Halifax Regional Police Service (McKinnon, 1997). This new organization was a blend of three different histories and three separate approaches to policing. This presented the senior managers with a problem requiring a great deal of sensitivity in order to arrive at a successful amalgamation. It was recognized that a quick and dirty approach would not be suitable, or acceptable, especially in light of distinctive collective agreement issues across the three amalgamating services.

In November 1997, two faculty members from Dalhousie University, Chris Murphy and Don Clairmont, facilitated a three-day program dealing with the formulation of a strategy for the Halifax Regional Police Service. There was a conscious effort to avoid the term "community policing" during this exercise. The effort was to work toward getting feedback from members on progress to date, and to identify outstanding issues. By way of commentary, coming out of these sessions, there is some sense that the Halifax Regional Police Service has not been successful at "institutionalizing" the goals of the community-based policing project. The process was perhaps accelerated too quickly and, therefore, it is now valuable to concentrate some effort at embedding the community-based policing approach within the organization.

There is a sense that police systems are typically weak in the area of accountability, and internal and external accountability processes are in potential conflict. It has also been observed that there is an absence of accountability measures to fully support community policing.

In the area of performance evaluations, it is not unusual for supervisor (i.e., sergeant) evaluations to be constructed in such a manner as to undermine the philosophy of community policing being advanced. As in any major change process, there were people in the organization who were in non-compliance with the new philosophy of policing. These people were not taken through a behavioural change process. In retrospect, more time might have been invested in supporting the NCOs within the organization. This gap needs to be remedied within the Halifax Regional Police Service because it has created a degree of resentment. There is recognition that the first attempts at "institutionalizing" the new philosophy have not been successful. It is acknowledged that the culture within the department needs to be strengthened in ways that advance the community policing model over the military model. Presently, the Halifax Regional Police Service is working to establish a "matrix approach to management practices."

The Halifax Regional Police Service does not present its approach as a program or project, but as a philosophy of policing, a fully integrated approach to service delivery. However, it remains important to recognize that command and control requirements still exist. It is essential to avoid the problems that are created when members are responsible for everything, but in charge of nothing. Community policing is best seen as a generic form that operates across a continuum. Accordingly, the Halifax Regional Police Service philosophy is stated as follows:

The primary objectives of Halifax Regional Police—

- Respond to calls for service;
- Enforcement of all statutes federal, provincial, and municipal;
- Crime solving through investigations of offenses, collection and analysis of evidence, gathering information from informants, and other sources such as Crime Stoppers, and surveillance;
- Crime prevention through public service talks, messages and promotion of crime prevention programs such as Operation Identification, Neighbourhood Watch, etc.;

- Referral to other social or government agencies;
- Public education through programs to inform and educate the public about problems in their neighbourhoods, or talks on timely topics, encouraging community involvement with police.

MONTREAL, QUEBEC

The Montreal Urban Community (MUC) was created on January 1, 1970 and police services for the municipalities that made up the MUC were combined in 1972. Employing approximately 4170 officers and 675 civilians, it represents the largest municipal police department in Quebec and the second largest in Canada.

During the early 1980s the Montreal Urban Community Police began to develop crime prevention strategies that could be seen as the start of a shift to a model of community policing. In 1991, the Montreal Urban Community Police adopted a five-year strategic plan highlighting some of the following elements:

- Prevention programs touching upon violence and narcotics;
- Actions aimed at reinforcing the feeling of security of citizens;
- Development of an equal opportunity program to better reflect the community's social and cultural reality;
- Community crime prevention program;
- Community relations programs;
- Development of mechanisms for consultation and collaboration in order to rally citizens and partners to the fight against crime; and
- Improvement of communications with the community.

When a new director (i.e., the equivalent of a chief of police) was appointed in January 1994, an initiative began to bring about major changes in the MUCP's organization. The objective of this initiative was:

> To turn the [Montreal Urban Community Police] into a better police service, more effective and efficient, respected and valued by the community, of which employees are proud.

As a result of this project, the MUCP devised a new model of policing for its community (Montreal Urban Community Police, 1996):

> The Neighbourhood Policing model can improve the quality of life of citizens by favouring the emergence of a more dynamic neighbourhood life. Citizen involvement in finding solutions to problems and in projects to counter crime stimulate social interaction and can reinforce social controls. Geographical responsibility and a service approach will also encourage mobilization of citizens. Finally, the establishment of partnerships with the community and local organizations as well as coordination with other municipal services will facilitate joint actions at all levels. By reuniting the various forces present, the Neighbourhood Policing model will play a primary role in creating a quality neighbourhood life. (p. 32)

Figure 11–1 represents a chart of a typical "neighbourhood station" operating within the MUCP's new model:

FIGURE 11-1

CHART OF A "NEIGHBOURHOOD STATION" OPERATION

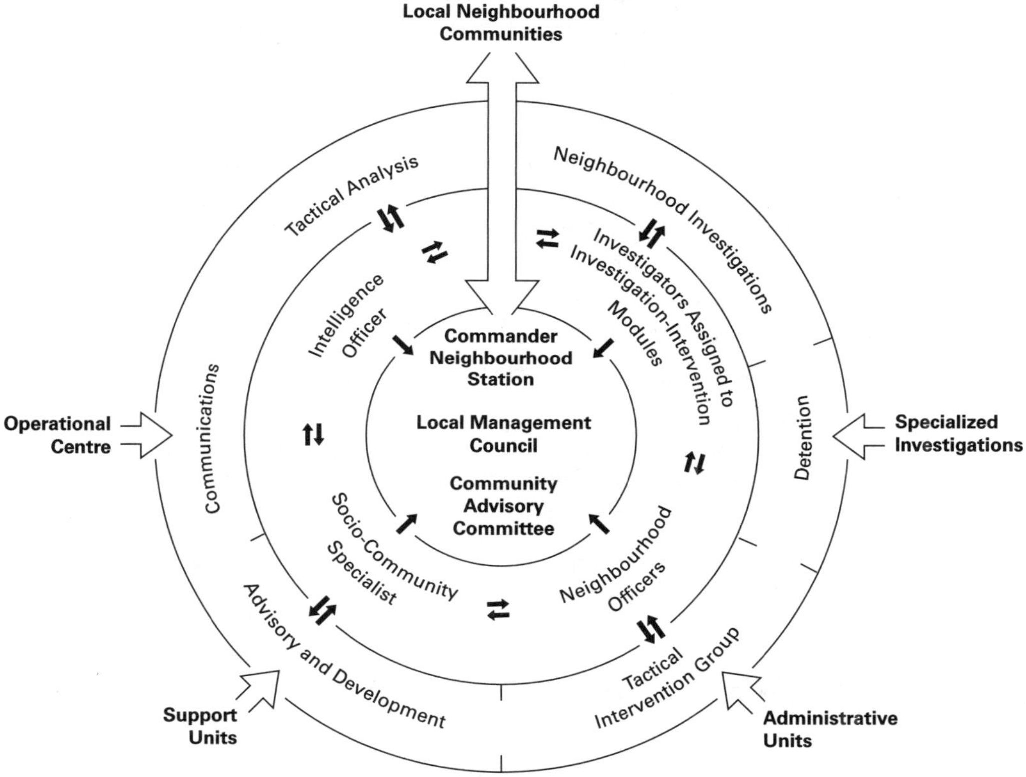

Source: Montreal Urban Community Police (1996). *Neighbourhood policing: together for better service*. Montreal: Montreal Urban Community Police, p. 52. Used with permission.

The MUCP realized that there would be much work required in achieving success under this model and identified the following conditions that have to be met in order to advance community policing:

- *Communications*—particularly during the transition period, including an internal and external strategy that would work to inform, educate, and reassure people regarding the substance of the organizational change;
- *Training*—recognizing the need for new skills among "rank" officers (i.e., commanders and superintendents) as well as patrol officers. Everyone requires a clear understanding of the underlying principles of problem solving, partnering with the community, and the service approach to police operations. Also, a training program designed for civilian staff;
- *Human resource management*—essential to put in place a mechanism for the evaluation of individual and collective knowledge and skills. Review staff assignment with

the new organization. Human resource management systems to be realigned to focus on new skills needed to meet the objectives of change;

- *Resistance to change*—anticipate inevitable resistance in various forms and ensure strong senior management commitment to assist in overall project success;
- *Call response*—adopt and apply the recommendations of the reengineering team that considered calls for service, prior to the opening of the twenty-three neighbourhood stations;
- *Technology*—introduce needed technology to support the decentralization of police activities. Integrated communications and information systems;
- *Partnerships with unions*—implementation will not be successful without a thorough examination of the collective agreement; and
- *Design of neighbourhood stations*—these twenty-three stations and two operational centres must be fully operational within the established timeframe.

As noted in the report on the new model (Montreal Urban Community Police, 1996):

> The success of Neighbourhood Policing also rest [sic] on the capacity of each and everyone to ensure passage from a culture characterized by its removal from the community to a culture oriented toward service to citizens. The Neighbourhood Policing model already contains elements favouring the emergence of a new organizational culture. Changes to the structure, the development of criteria for accountability, the focus on partnership with external groups and on a service approach as well as the recruitment of personnel motivated by the will to serve will favour the transition. (p. 94)

KINGSTON, ONTARIO

Since 1849 the city of Kingston has had its own police service. It is a unique jurisdiction in that it includes several federal penitentiaries, and is situated in close proximity to both Montreal and Toronto. Furthermore, it has a high proportion of people who are on some form of social assistance within its boundaries. In spite of these elements that might contribute to problems, Kingston is a historically rich community with a strong tourist base, as well as a highly regarded university, and a strong presence of Canadian military institutions (e.g., the Royal Military College, and the Canadian Forces Base, Kingston).

With the appointment of a new chief of police in August 1995, considerable effort has been made in the area of community policing. This focus was complicated by issues within the external environment that had a direct bearing on the department's ability to pursue its community policing initiatives. These issues, discussed below, serve to demonstrate that police organizations operate within a deeply political context that impacts significantly on service delivery. In 1994, efforts began to effect an amalgamation in this area. The Ontario government, through its Ministry of Municipal Affairs, had undertaken a program of municipal restructuring, which, for example, resulted in the creation of the Toronto "megacity," and other amalgamations throughout the province. In the Kingston area the "old" city of Kingston and the townships of Kingston and Pittsburgh were to be combined into a new, single jurisdiction (Downey and Williams, 1998). While there was opposition to this initiative, on January 1, 1998 the "new" city of Kingston came into existence.

The complication presented by this amalgamation from a policing standpoint was substantial. Each of the three amalgamating areas had a different approach to the delivery of policing services:

- the old city of Kingston was policed by its own municipal police service, accountable to the Kingston Police Services Board;
- the former Pittsburgh township was receiving "free" OPP policing (as were some 575 other municipalities throughout Ontario). However, all these jurisdictions were going to be required to pay for their policing, as a result of "equitable financing" amendments to the *Police Services Act*; and
- the former Kingston township was being policed by a local detachment of the OPP under the terms of a contract between the township and the provincial Solicitor General. This contract was due for renewal in September 1999.

Therefore, the first key decision point with respect to policing services for this new municipal entity turned on who would be responsible for policing the new community. The range of options available to a municipality is set out in section 4 of the *Police Services Act*. Complicating the process was a determination on the part of the Ontario Provincial Police Association, which represented the labour interests of OPP members, that a costing process be pursued that would place the Kingston Police in direct competition with the local detachment of the OPP. This problematic, and divisive, scenario was avoided when the new Kingston city council voted unanimously to support the expansion of the existing Kingston Police into the new jurisdiction (Kingston Policing Services Steering Committee, 1998). However, in order to ensure that policing in the new city would address the needs of the municipality, a joint session of city council and the Kingston Police Services Board established the Kingston Policing Services Steering Committee, with the following objective:

> The establishment of a framework for police services in the City of Kingston that will enhance public safety and proactive service delivery with broader involvement of the community, leading to the integration of emergency services, while reducing overall policing costs.

In order to achieve this objective, the Kingston Policing Services Steering Committee undertook extensive internal and external consultations aimed at generating insight and feedback. One important component of this process of consultation was the preparation and release of a discussion paper followed by a "town hall" meeting at which interested groups or individuals were invited to make submissions to the steering committee.

The Kingston Policing Services Steering Committee undertook extensive research to ensure that the new "paradigm" of policing would reflect modern, innovative practices. To that end the following elements were taken into consideration:

- Requirements of the *Police Services Act* (as amended by Bill 105);
- The "Quality Assurance Plan" of the Ontario Association of Chiefs of Police;

- Benchmarks and "best" practices in other jurisdictions;
- Literature review; and
- General audit principles.

By December 1998, the Kingston Policing Services Steering Committee had completed its round of consultations and the research necessary to put together its final report. This report included findings in the following topic areas:

- Efficiency & effectiveness;
- Proactive & reactive services;
- Broader community involvement;
- Integration of emergency services; and
- Other areas of study (i.e., administrative & operational issues).

Based on all of the background data gathered by the Kingston Policing Services Steering Committee, a new "paradigm" for policing in the city of Kingston was proposed. The Proactive Service Delivery Model is summarized below:

FIGURE 11–2

PROACTIVE SERVICE DELIVERY MODEL

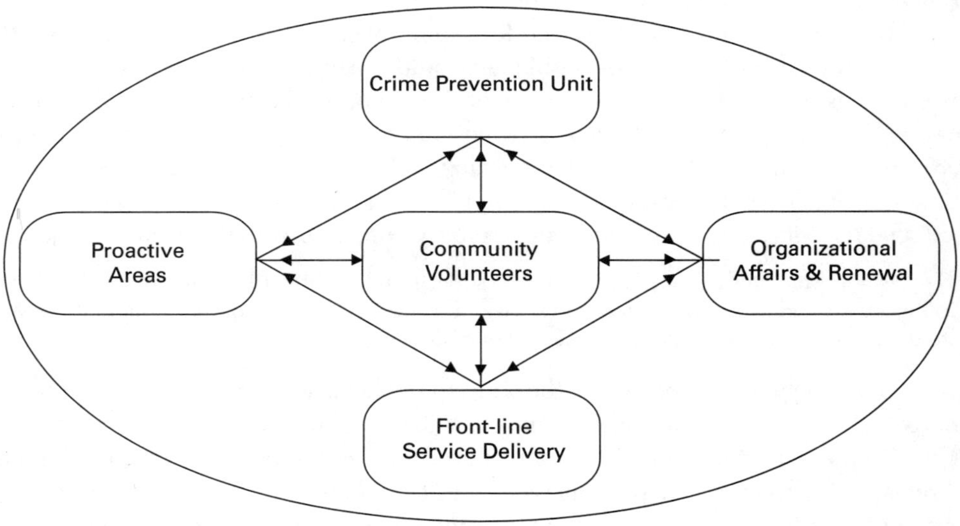

Reproduced with permission of the Kingston Police Services Board.

Also, a new organizational chart was developed around the substance of this model. Figure 11–3 shows the old organizational chart for the Kingston Police, and Figure 11–4 contains the approved new structure.

FIGURE 11-3

OLD ORGANIZATIONAL STRUCTURE FOR THE KINGSTON POLICE

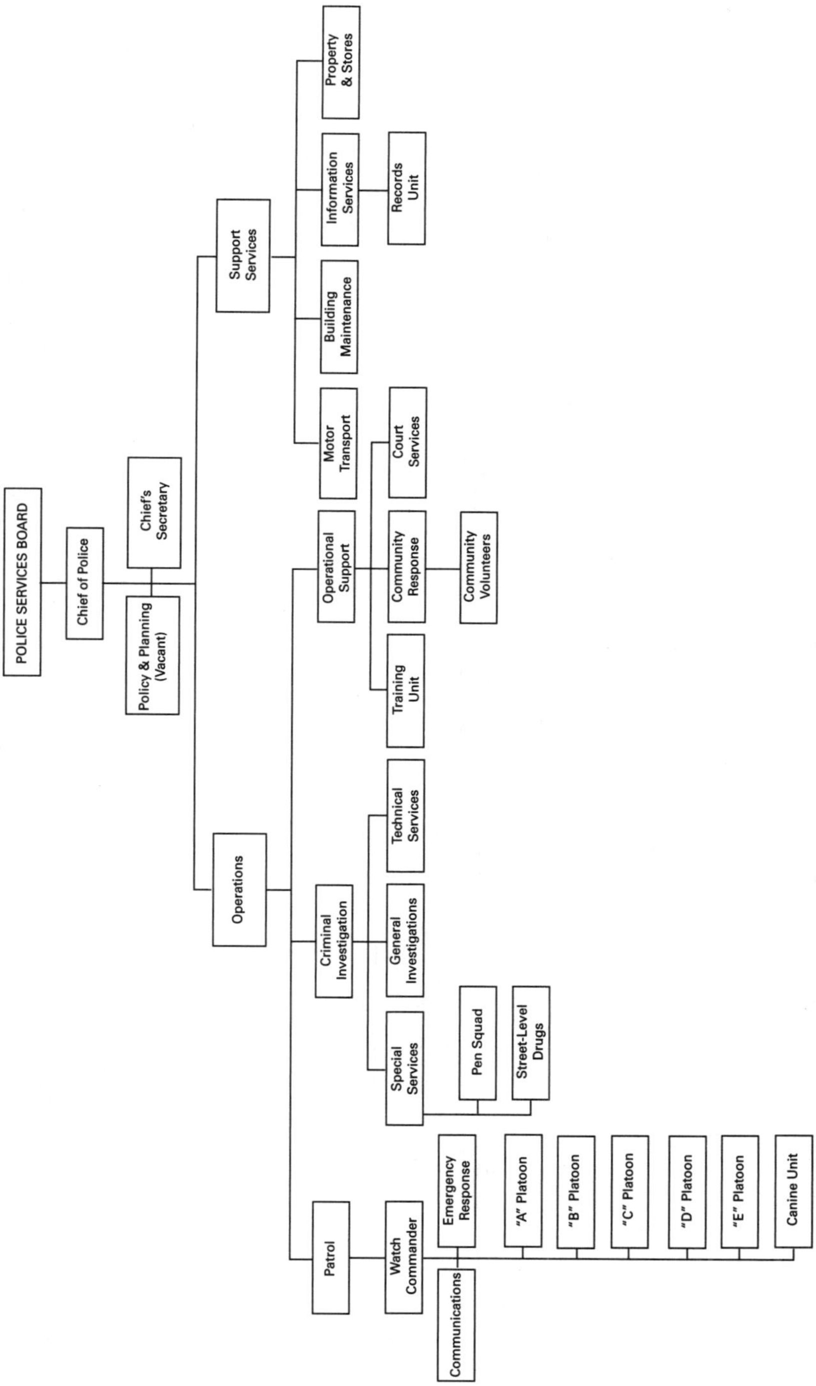

Reproduced with the permission of the Kingston Police Services Board.

FIGURE 11–4

PROACTIVE SERVICE DELIVERY MODEL FOR THE KINGSTON POLICE

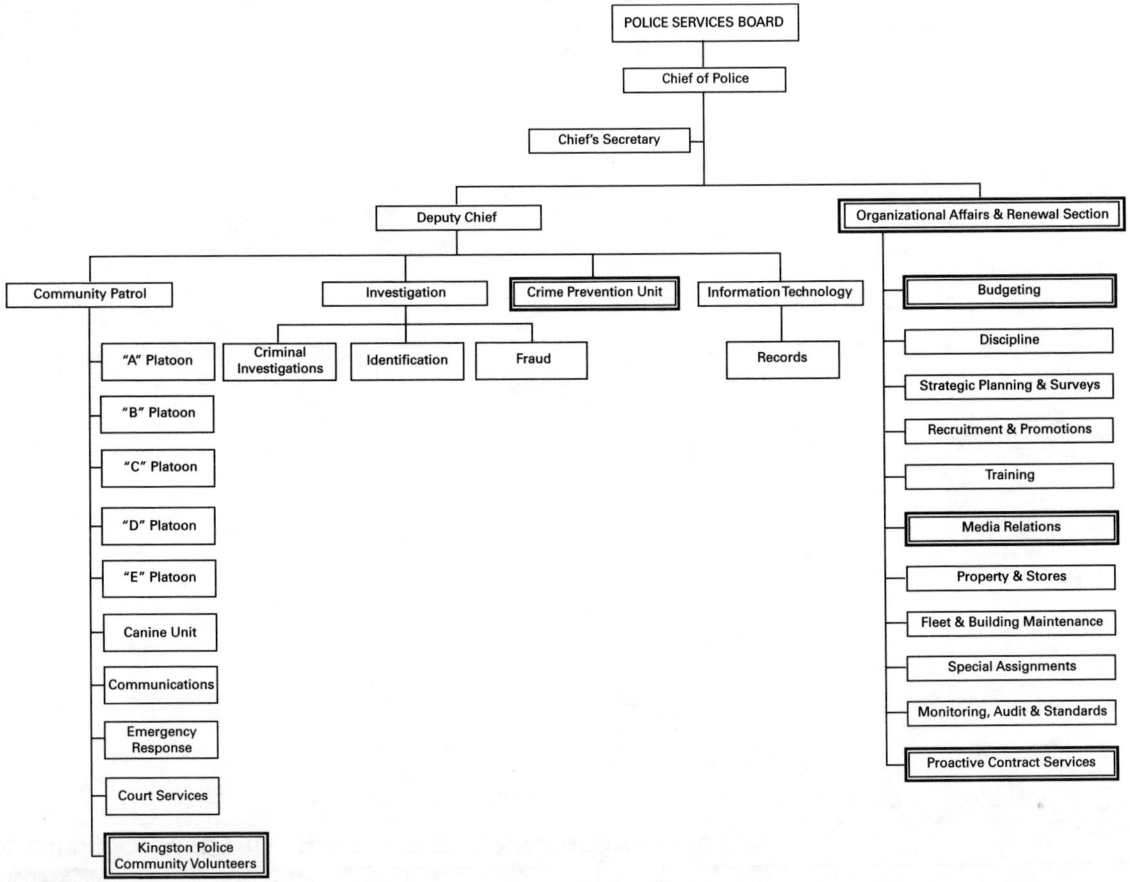

Reproduced with the permission of the Kingston Police Services Board.

While the Kingston Police have been involved with community volunteers and community development over the past few years, it is particularly important following the approval of the proposed paradigm of policing by the Kingston city council on February 23, 1999, that substantial work now begin in terms of moving toward the chief of police's vision for the department:

> A Kingston Police organization in which all employees enhance the community's quality of life through the prevention of victimization by practising compassion, understanding, conflict resolution, and problem solving.

Figure 11–5 provides an overview of the components of that vision:

FIGURE 11–5

THE NEW VISION FOR THE KINGSTON POLICE

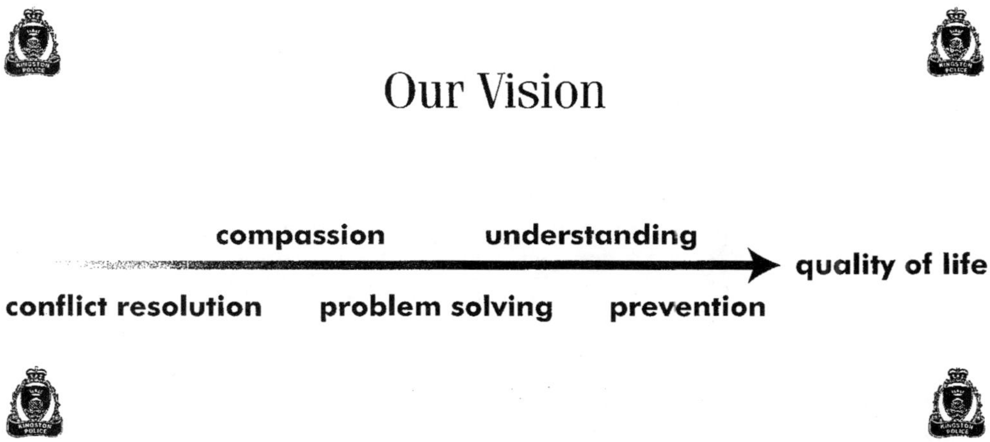

Reproduced with the permission of the Kingston Police Services Board.

BELLEVILLE, ONTARIO

Community policing was started in September 1993 with the appointment of a new chief of police. This resulted in the establishment of a storefront operation in the spring of 1994. The positive community response to this initiative led to the opening of four additional offices during that year.

Belleville is divided into five zones. This division resulted from the research undertaken by the Community Police Implementation Committee, made up of civilian and uniform members of the police service. Initially, the Belleville Police Service focused attention on the downtown core and assigned a Community Policing Officer to this central zone. Community Police Committees were formed to work with the Community Policing Officers in each zone. Also, the Belleville Police Service began to work at developing an extensive volunteer base that continues to grow. These volunteers are seen as the "driving force" in the community and participate in zone decision-making and the implementation of local initiatives, including the following:

- *Volunteer manual*—a guide to assist volunteers with the procedures for minor reports and other relevant details;
- *Crime watch*—involves the preparation of a form containing basic information on serious occurrences (e.g., hit-and-run, abductions, missing, or lost children, robberies) that is sent by facsimile to local radio stations for immediate broadcast within the community;
- *Hallowe'en safe streets*—use of community volunteers working in pairs to patrol small areas to promote child safety. Volunteers are equipped with portable radios and fluorescent vests;
- *Business watch*—trained volunteers visit businesses in their zone and provide tips on security improvement and crime prevention. Community Policing Officers contact businesses and invite them to seminars on frauds, theft (internal and external), shoplifting, and counterfeit currency;
- *Residential watch*—similar to the business watch program, above, designed for apartment/condominium/home owners;
- *Crime awareness program/Neighbourhood watch*—beginnings of program have been established with Community Policing Officer working closely with Block Captains in order to pass along pertinent safety information;
- *Youth program*—involves opening up local high school gymnasia on Saturday nights, providing a place for youth to play basketball, floor hockey, or volleyball. Program is run by volunteers and students. This program has helped develop rapport between the Community Policing Officer and the local student body;
- *Child print book*—seeks to establish and maintain a current inventory of children in the Belleville area. The costs are sponsored locally. The program coincides with school picture day. An extra photograph is taken of children for inclusion in the book, along with the child's fingerprints;
- *Refrigerator emergency cards*—provide medical and emergency information for the benefit of emergency room personnel;
- *Skateboard park*—exploring the establishment of a park and skateboard ramp for safe use by skateboarders;
- *Seniors*—with 22.5% of Belleville's population over 60 years of age, there are resulting demands on police services. Officers attend meetings to share statistical information and provide safety tips;
- *Community volunteer watch*—zone volunteers patrol in their own vehicles during peak hours and have cellular phone contact with the Belleville Police Service; and
- *Operation "red nose"*—volunteers cooperating with the police to help keep impaired drivers off the roads. These volunteers drive impaired drivers home once contacted by the police.

GUELPH, ONTARIO

During 1996, the Guelph Police Services Board, in collaboration with the Guelph Police Service, began steps toward implementing a strategic planning process for the department. There were efforts at external and internal consultations aimed at surfacing key issues impacting on the Guelph Police Service, now and into the future (Guelph Police Service, 1997). As a result of these efforts, in November 1997, the police service presented a long-term vision that includes a commitment to neighbourhood policing.

The Guelph Police Service also established a new organizational framework to facilitate the implementation of neighbourhood policing. The concept of neighbourhood policing in Guelph rests upon five components:

- *Problem-solving*—with officers becoming familiar with specific neighbourhood needs through continuous local interaction;
- *Geographical responsibility*—front-line officers to be assigned to specific neighbourhoods for an extended period. Teams will be expected to arrive at deployment decisions in order to meet community needs. Each team will be assisted by an Area Manager, who will have a degree of autonomy with regard to local deployment decisions;
- *Improve quality of life*—establish specific quality of life measures in order to measure success in terms of neighbourhood policing. Areas that could be included are: personal safety (both real and perceived), traffic safety, protection of property, opportunities for youth, among others;
- *Partnership*—the neighbourhood teams are expected to establish and maintain a range of partnerships with residents, business owners, other service providers, and community groups; and
- *Member recognition/career enrichment*—increasing opportunities for members to learn and apply new skills. Neighbourhood teams will play a key role in identifying learning requirements in support of their cross-functional role. Recognition to be closely tied to achieving neighbourhood objectives through specific job performance.

The city of Guelph was examined in order to redefine existing police boundaries. As a result, three "Primary Neighbourhoods" were established:

- *North-West Neighbourhood*—bounded by the Speed River to the south and east;
- *North-East Neighbourhood*—bounded by the Speed River to the west and the Eramosa River to the south; and
- *South Neighbourhood*—bounded by the Eramosa and Speed Rivers to the north.

Within these "Primary Neighbourhoods," smaller, local neighbourhoods will be defined that mirror local resident perceptions and existing city service boundaries. These boundaries will be flexible over time.

THUNDER BAY, ONTARIO

At the end of 1994, the Thunder Bay Police Services Board requested the Thunder Bay Police to develop a report that would explore community-based policing and identify obstacles to its implementation. The committee tasked with this assignment began with the goal of arriving at a working definition of "community-based policing." This group also tackled the challenge of examining the current status of policing in Thunder Bay and reviewing practices in other jurisdictions for possible insight.

It was recognized that the Thunder Bay Police had followed a professional model of policing, with its emphasis on reactive, incident driven services. There had been some experimentation with zone policing. However, due to cost considerations and a lack of full commitment, this project was suspended. In order to begin the process of thinking about the application of community-based policing in Thunder Bay, the committee arrived at the following definition (Thunder Bay Police, 1995):

> Community policing is a new philosophy of policing based on the concept that police officers and private citizens working together in creative ways can help solve contemporary community problems related to crime, fear of crime, social and physical disorder and neighbourhood decay. The philosophy is predicated on the belief that achieving these goals requires that police develop a new relationship with law-abiding people in the community, allowing them a greater voice in setting local police priorities and involving them in efforts to improve the overall quality of life in their neighbourhoods. It shifts the focus of police work from handling random calls to solving community problems. (p. 6)

The committee looked at successful community policing initiatives in other jurisdictions, including:

- New York City, New York;
- Houston, Texas;
- Newark, New Jersey; and
- Edmonton, Alberta.

In light of the research available, members of the Thunder Bay Police began to consider the potential obstacles to the implementation of community policing. Those obstacles were divided into the following general categories:

- *Senior management*—including the chief of police, deputy chief, senior officers, and the police services board. A lack of continuity within these levels could undermine any community-based policing project. Absence of a shared understanding of community-based policing was seen as highly detrimental, as was any adherence to an exclusively reactive mode of policing;
- *Police services board*—the political and economic motivation of this body removes it from actual, operational policing issues. The governing body must be able to understand the need for community-based policing and be willing to commit the necessary resources for successful implementation;
- *Police association*—tendency of this group to resist change, especially in the context of their collective agreement. Adherence to the professional model and general

view that policing is "crime fighting" and response to calls for service. Requirement to change the overall definition of policing;
- *Officer selection*—requirement for a broader range of knowledge, skills, and abilities in police officers. Recruitment practices must be modified accordingly;
- *Community*—willingness of neighbourhoods to participate in community resolution of problems. May be a requirement for changed expectations on the part of the public with regard to crime and who holds responsibility for its control;
- *Budget*—increased staffing and equipment will be required to introduce neighbourhood policing. Fiscal restraint poses a large problem;
- *Training*—absence of a formalized training program to introduce the new philosophy and outline the anticipated police officer role. Combine theory and practice to provide a well-rounded understanding of community-based policing and show evidence of its effectiveness;
- *Professional model based reward system*—existing performance review system places an emphasis on reactive performance and does not anticipate the value of preventative police performance. A suitable method of personnel appraisal needs to be developed;
- *Promotions/Leadership*—existing promotional process does not acknowledge the value of community-oriented qualities. Process needs to advance individuals with neighbourhood policing initiative into senior positions; and
- *Internal and external communications*—strategic communication to effect necessary changes within the organization. Participative approach to the development and implementation of neighbourhood policing.

Combining the results of their research in other jurisdictions, and anticipating some of the significant obstacles to successful implementation of community-based policing, the committee made the following recommendations (Thunder Bay Police, 1995):

1. That the Police Services Board pursue a community based policing program that includes the fundamental elements as described in the Neighbourhood Policing Proposal.
2. That the community based policing committee attend Edmonton Police Service to see Neighbourhood Policing in operation to clarify some of the mechanics relating to the overall program and the role of other sections of the Police Service.
3. That a pilot test be conducted at two sites. One being largely commercial and the other residential.
4. That officers selected to carry out the pilot test be knowledgeable, person oriented, problem solving people committed to the philosophy of community based policing.
5. That a management style that encourages initiative and creativity be applied with regard to the constables assigned to the test sites.
6. That arrangements be made with the Edmonton Police Service to expose members to a two week practical training experience involving Neighbourhood Policing.

7. That performance measurements be determined and that the test sites be compared on a pre and post pilot basis and that the results be used to determine whether the program should be expanded and/or changed.
8. That the measurements used to conduct the evaluation be consistent with preventative policing objectives and agreed to by all stake holders before the evaluation is conducted. (pp. 33–34)

PEEL REGION, ONTARIO

Peel Region was the result of municipal amalgamation undertaken in 1973. The existing Peel Regional Police came into being in January 1974 with the combination of the five local police departments, namely: Mississauga, Port Credit, Streetsville, Brampton, and Chinguacousy. Planning was seen as central to the overall effectiveness of the new Peel Regional Police, and this organization has been thoroughly committed to an extensive and careful planning process for the preparation of its administrative and operational strategies.

With their serious devotion to strategic planning, the Peel Regional Police has been able to anticipate both the major growth patterns within the region, as well as the changing operational needs within the various communities. For example, some of the early subdivisions within the region (for example, Meadowvale and Erin Mills) have experienced significant change. During the 1970s, the families in these areas were mostly young, with growing children. Their parents were largely new homeowners struggling with mortgages and other kinds of payments. As a result, the calls for service were frequently reactive in nature, such as domestic disputes. With time, these calls have been transformed to ones dealing with young offenders and delinquency (Bain, 1998). High concentrations of teenagers have created a new type of challenge for the police.

In 1997, a new chief of police was appointed. The chief promptly committed the Peel Regional Police to the following key challenges and priorities:

- enhancing the role and performance of front-line officers;
- forging stronger communications with the Police Services Board and Regional Council;
- improving internal communication, implementing the Statement of Organizational Character;
- continuing and strengthening investment in projects that enrich Peel Region and Peel Regional Police as well as taking the lead on developing key community partnerships; and
- enhancing the training and support of management.

The Peel Regional Police, again as part of a conscientious program of environmental assessment tied to its cycle of strategic planning, has undertaken a series of public surveys to ensure that it is meeting community expectations. Of additional significance is this department's "Statement of Organizational Character." This statement clearly expresses the service orientation of the Peel Regional Police, and is presented below as Figure 11–6 :

FIGURE 11-6

Statement of Organizational Character

We, the members of the Peel Regional Police, dedicate ourselves to developing and maintaining an environment which promotes a satisfying and effective service.

We recognize the importance and value of our members and are committed to treating everyone in our organization with mutual respect, support and trust through the following: ***Leadership, Teamwork, Development, Communication, and Professionalism.***

Through our efforts we will build an organization in which there is a shared sense of community, belonging and ownership.

Leadership

- Leadership is not only a function of rank. All members have opportunities to assume leadership roles within the organization.
- Encourage innovation and recognize that if mistakes are made, constructive feedback will promote learning.
- When planning, consult with members.
- Decentralize the decision-making process by supporting and giving members the authority to make decisions within their area of competence.
- Provide coaching by sharing knowledge and experience.

Teamwork

- Work together to achieve common goals.
- Developing co-operative partnerships within the organization.
- Treat everyone in a fair and consistent manner.
- Respect the opinions and contributions of all members.

Development

- Recognize performance and acquired skills.
- Encourage and assist members to reach their full potential.
- Offer career development, learning opportunities, guidance and support for skills enhancement for all members.
- Respect the decisions individuals make to achieve professional and personal balance in their lives.

Communication

- Be responsible, open, honest and less formal.
- Listen and respond to each other.
- Encourage direct lines of communication.
- Make information accessible to all levels of the organization, ensuring that it is current and timely.

Professionalism

- Demonstrate high ethical standards, accept responsibility and be accountable.
- Be proud, dedicated and self-disciplined.
- Do the right things - be effective.
- Do things right - be efficient.
- Commit to continuous learning and improvement.

For further information contact: THE POLICY CENTRE - Mr. John MacNeil, Manager - Organization Development Bureau, Peel Regional Police, 7750 Hurontario Street, Brampton, Ontario, L6V 3W6 Phone (905) 453-3311 Ext. 4050 or FAX (905) 453-4048.

Source: Peel Regional Police. Used with permission.

Priorities for the Peel Regional Police are established annually by means of a "Policing Priorities Opinion Survey." In 1996, the surveys identified the following priority issues (Peel Regional Police, 1997):

- Crime prevention . . . includes public education and awareness, crime prevention through environmental design and crime prevention through physical design, youth and ethnic liaison, addressing hate crime, target hardening, neighbourhood/school/business liaison, partnerships with community-based crime prevention associations and participation in public policy formulation.
- Crimes of violence . . . includes sexual and other assaults, robberies, threatening, domestic disputes and violence in schools.
- Crimes against property . . . includes break and enters, all thefts, vandalism, mischief and frauds.
- Drug control includes public education and awareness, prevention, prosecution and participation in public police formulation.
- Lawless public behaviour . . . includes disturbance, incidents dealing with disruptive and emotionally disturbed persons, liquor offenses and public drunkenness.
- Traffic management . . . includes impaired driving, general and directed enforcement, traffic safety, public education and traffic engineering. (pp.37–38)

HALTON REGION, ONTARIO

The story of community policing within the Halton Regional Police Service has been recently summarized in some considerable detail (Cooke-Scott, 1998). Dating back as far as 1975, the Halton Regional Police Department experimented with "team policing" in the city of Burlington. The aim was to bring front-line officers into closer contact with the public. The chief of police at that time was interested in approaches that would contribute to crime prevention. This philosophy was consolidated with the department's decision to hire a social worker to function full-time within the organization, a first in Canada.

In 1982, a so-called "pro-active policing squad" was assigned to an area experiencing the highest crime rate. In 1983 this proactive approach was expanded throughout Burlington. The Halton Regional Police also introduced "community conference committees" designed to provide a forum for the public to gather together and raise concerns with the police regarding community problems. Patrol officers were instructed to devote half of their shift to addressing proactive (i.e., community) policing. When this divided approach proved to be problematic, the platoons were split into categories of proactive and reactive officers. Further complications with this alternative resulted in the introduction of a "Community Directed Patrol" division, which included the assignment of "village constables" to specific areas.

In order to consolidate these new approaches, the senior management of the Halton Regional Police moved to establish a clear vision that would express the department's philosophy. This philosophy stated an adherence to a "combined strategy of preventative, proactive and reactive policing programs, using the concept of the Constable Generalist" (Cooke-Scott, 1998). A Halton Regional Police document goes on to define some of these important terms (Halton Regional Police, 1992):

- Preventative policing: the recognition, anticipation and appraisal of a crime risk and initiating the necessary action to remove or reduce it.
- Proactive policing: the specially planned response to a particular pattern of crime or situation of concern that may develop in specific areas at certain times.
- Reactive policing: the capability of responding to calls for service relating to crime, traffic and social order complaints.

During 1991, the Halton Regional Police Services Board engaged an outside consultant to facilitate strategic planning for the department. This process resulted in the identification of specific goals relating to community policing:

- Increasing the understanding of the needs of the community and service delivery issues;
- Increasing public understanding of policing; and
- Ensuring that the service is responsive to the needs and concerns of the community.

In 1994, the Halton Regional Police Service undertook a major Organizational Review Project. This initiative is summarized in a detailed report prepared by the project team (Halton Regional Police Service. Organizational Review Project, 1995). This project came into effect because of major reductions in the department's budget in 1993, along with the prospect of further provincial funding cuts. The overall theme of the Organizational Review Project was:

> . . . [T]o "eliminate, combine or automate" those activities which take away from our ability to effectively deliver services or which mean more paperwork and bureaucracy. (p. 1)

The project's final report contained more than 170 recommendations for change including the following suggestions under the category "Community Policing":

1. Develop and inter-platoon "team" approach for policing services utilizing larger geographic areas (four recommended per district, but to be determined given the varied characteristics and needs).
2. Continue the concept of Resource Officers and review present and future relationships between the concept and the role in community policing. (p. 2)

Also, under the category of "Community Support Services" the following recommendations were made:

- Establish Education Services Unit consisting of a Sergeant (new), D.A.R.E. officers, School Resource Officers, Safety Village staff.
- Create a Community Policing Coordinator Sergeant (new) who is responsible for coordinating, developing and maintaining central depository for community programs.
- Expand to full time Victim Services Coordinator (Civilian), to coordinate VSU, help with program support and development of complete guidelines for volunteer recruitment, standards, programming and training.

- Move media Sergeant to Community Support to enhance support of public relations activities within Support Services. Rename the function as Public Affairs Unit.
- Maintain centralized traffic, marine unit and auxiliary to be known as Public Safety Unit. Staff the Marine Unit over summer months with officers returning to District or other service duties for the winter.
- All of these units be grouped as the Community Support Unit to be managed by a Staff Sergeant reporting to a senior officer responsible for direction of Community Policing programming and development.
- Maintain Emergency Services at current strength with Canine and eventually E.D.U. reporting to the Community Policing Support senior officer. (pp. 3–4)

Among the ten "Task Force" project teams established under the auspices of the Halton Regional Police Services' Organizational Review Project one was devoted to Community Policing. This group examined the following key topics and issues in order to arrive at the final recommendations:

1. Principles of policing (which included comment on the mission statement).
2. A survey of the citizens of Halton, designed to obtain feedback on expectations, experiences and priorities.
3. An internal survey of the members of the organization.
4. An analysis and review of the current system and method of delivering service.
5. The idea of an appropriate, effective and efficient format providing the front line delivery of service.
6. Support required for the adopted concept. (p. 7)

This study group came to the strong conclusion that a "team" approach was best suited to the service delivery philosophy adopted by the Halton Regional Police Service. While these more broadly based teams were recommended, they were not intended to alter existing communities, or displace the community consultation groups already in existence. The new teams were to be seen as an additional resource with the following potential benefits:

- Provides ownership;
- Uses natural and real geographic barriers;
- Ensures, or is conducive to, consistency;
- Includes and possibly enhances the "Village Constable" application;
- Increases resources for a problem solving environment;
- More flexibility;
- Empowerment;
- Increases staff development;
- Thorough reporting;
- Integrates response;
- Removes isolation;
- Enhances consultation; and
- Reduction in need for specialization. (pp. 8–9)

The return to team policing in the Halton Regional Police Service, while not a new concept, is seen to be a comfortable fit for this police department. As noted by retired chief of police, Peter Campbell (Halton Regional Police Service, 1997):

> The model of team policing described herein is built on those previous initiatives, on new knowledge and on the foundation that growing community policing practices give us. The idea came from "the troops" who participated in an organization-wide renewal project. The concept, the details and the training have all come from the bottom up.

In January 1999, a new chief of police was sworn in to assume leadership of the Halton Regional Police Service. With a change at the top of the organization, there may be further developments or refinements in the department's approach to team policing. This story continues to evolve.

TORONTO, ONTARIO

Toronto has the largest municipal police department in Canada, with 4906 uniformed officers and 1986 civilian personnel. In the early 1980s, Toronto was experimenting with the mini-station approach to community policing (Lambert, 1988). There was an effort during this period to gauge citizen satisfaction with the police, and knowledge of police services, especially in the context of the mini-stations in operation (Murphy, 1988; Murphy and de Verteuil, 1986).

Another feature that has been instrumental in helping the Toronto Police move toward a community policing model relates to their detailed planning process. This has normally included public discussions aimed at clarifying police priorities with community input. A significant advance was made in this context in 1991, with the release of a key document; *Beyond 2000: the strategic plan of the Metropolitan Toronto Police*. This document evolved from the work of an internal strategic planning group that was engaged in the conduct of a detailed environmental assessment (Metropolitan Toronto Police. Strategic Planning, 1990). The planning group undertook extensive research in order to establish clear goals and objectives for the department. Their main areas of focus were as follows:

- New trends or patterns of crime or public order;
- Changes in population density or character;
- Technological changes or requirements;
- New services organized or required;
- Legislative changes anticipated or required;
- Developing community priorities;
- Other areas of concern or anticipated changes; and
- Other means of deployment or more effective use of personnel.

As noted in the introductory remarks of Susan Eng, the chair of the Metropolitan Toronto Police Services Board at the time of the *Beyond 2000* plan, there was going to be a greater emphasis on neighbourhoods and communities in the "new" plan (Metropolitan Toronto Police, 1991):

> Paradoxically, modernizing also means getting back to basics. The sense of safety in the streets depends very much on how well we address local, neighbourhood concerns. Ultimately, local communities and their neighbourhood police officers will set the policing priorities for their neighbourhoods. In many ways, the officer of the future will be the "Police Chief of your neighbourhood."

This document was intended to provide a framework for guiding the Metropolitan Toronto Police toward a future that would exemplify the principles of community-based policing in a context of Neighbourhood Policing. The expectation was that Neighbourhood Policing would enhance the front-line officer's role and focus on problem-solving, crime prevention and order maintenance, in addition to the requirements for law enforcement and the investigation of crimes. There was a recognition that such a strategy would place new demands on uniform members (Metropolitan Toronto Police, 1991):

> Police officers will develop enhanced skills in areas such as **problem solving, crisis intervention, investigative methods, crime prevention and interpersonal communication**, and use these new skills in their efforts to realize the vision. Front line supervisors will be responsible for facilitating the efforts of the neighbourhood officers, for providing resources, and for ensuring effectiveness in problem solving and crime prevention efforts. (p. ix) [emphasis in original]

The Metropolitan Toronto Police had a strategy that would allow the combination of their zone policing experience from the early 1980s with the community based policing initiatives introduced near the end of that decade. In an effort to clarify its understanding, the Metropolitan Toronto Police included a statement of principles regarding community based policing (Metropolitan Toronto Police, 1991):

> Principles of Community Based Policing
>
> 1. The main purpose of a community-based philosophy is local problem-solving to improve public safety. The specific strategies must be determined in accordance with the problems and resources in the community.
> 2. The community includes the full range of residents, business people, agencies, etc. in a geographic area or neighbourhood and the safety needs of all segments of the neighbourhood must be addressed.
> 3. The essential factors in the implementation of Community Based Policing are a problem-solving orientation and a consultative approach to obtain input from the public.
> 4. Within the Community Based Policing framework, a wide range of strategies (i.e., programs, activities) are appropriate. These may include, for example, attendance at public meetings to provide information or to obtain feedback; special programs in schools to build closer ties between police and youth; mini stations (mobile or fixed); community foot patrols; special outreach to victims of domestic violence, and community liaison committees. None of these programs standing alone guarantees the success of Community Based Policing. It is the ongoing consultative process to identify and prioritize local problems that is essential.
> 5. The police force should not and cannot address all public safety needs on its own. Community residents, agencies and organizations offer a resource to help set priorities

and to help solve problems, particularly when police resources are scarce. All problems cannot be given equal attention.

6. A large number of policing problems overlap with other concerns, such as social, economic, health and environmental problems. Public and private agencies with mandates in these areas must be drawn upon.
7. Community foot patrols are valuable in reaching the community but are not the only way to implement this policing orientation. Divisions without resources to dedicate to area foot patrols can nevertheless develop a wide range of other strategies which will enhance problem-solving and local input without fully dedicated resources.
8. Community Based Policing is not an add-on, nor is it "soft policing". It should enhance the focus on serious local problems and facilitate effective solutions. These may well be high-profile, targeted enforcement strategies.
9. Successful Community Based Policing will result in more effective use of all resources because resources will be allocated in accordance with high priority concerns. (p. 23)

On March 1, 1995, a new chief of police was named for the Metropolitan Toronto Police. The chief's appointment came during a time of considerable turmoil and, also, coincided with the presentation of the final report of the Beyond 2000 Restructuring Task Force. This report (Metropolitan Toronto Police, 1995) was designed to provide the Metropolitan Toronto Police with ". . . a blueprint for the organizational changes necessary to realize its vision of community policing." The focus on community policing is expressed throughout this report. Particularly, there is the following specific response to the question: What is Community Policing?:

> It is important to emphasize that community policing means a change in police practices, but not in the objectives of policing. It is a change in the means rather than the ends.
>
> Community policing is <u>not</u> public relations. Community policing is <u>not</u> simply a program, nor do community programs in and of themselves mean that a police force has adopted community policing. Community policing is a philosophy, one that must be translated by the executive and senior management into a way of doing business. The programs that are developed must be operational means to philosophic ends.
>
> Community policing offers a strategic alternative to traditional crime control policing practices, providing a framework for focusing, organizing, delivering, and assessing police services. (pp. 2–3) [emphasis in original]

Again, because policing in Canada does not exist in isolation from the larger political context, this topic became part of the deliberations of the Greater Toronto Area (GTA) Task Force appointed by the Ontario government. While the commissioners assigned to the Task Force were primarily concerned with making recommendations for a new model of local government in the form of a city-region, they did make certain determinations about policing in the Greater Toronto region. Essentially, the Task Force dwelt on four main issues in their final report (Greater Toronto Area Task Force (Ont.), 1996):

- *Improving community policing*—recognizing that the *Police Services Act* makes reference to the need for cooperation between police service providers and their communities in its "Declaration of Principles," the Task Force encouraged police departments to increase their capacity to respond to a diverse population and offer greater public accountability. Public partnerships should be built that give communities input into ". . . police methods, priorities, resources, and new initiatives."

- *Maintaining civilian control of police forces*—it was noted that there is a trend to greater civilian control of policing through governing authorities interested in accountability and cost-effectiveness;
- *Enhancing budget accountability*—the Task Force questioned the effectiveness of requiring budget appeals to the Ontario Civilian Commission on Police Services and noted that full accountability and full approval for police budgets could properly rest with municipal councils;
- *Making police more effective*—there were concerns raised about the exchange of information among police departments, their failure to maximize savings through pooled resources, and deficiencies in the training and education standards. It was noted that some departments may be too small to properly afford adequate technology.

The prospect that the five regional police services (i.e., Metropolitan Toronto Police, Peel Regional Police, Halton Regional Police, York Regional Police, and Durham Regional Police) in the GTA might be fashioned into one "mega" police force never materialized. However, it prompted a quick and decisive reaction from those police departments in question.

ONTARIO PROVINCIAL POLICE

As early as 1989 the Ontario Provincial Police (OPP) began to formulate its overall mission, philosophy, and plans in the context of community policing (OPP, 1993). The organization presented its principles in the following terms:

> Community policing is the fundamental principle under which all our policing services will be delivered. To this end we shall:
>
> - Involve the people of the community in the identification of crime, traffic and social order problems;
> - Provide policing services that are consistent with the identified concerns, expectations and needs of the community;
> - Participate with other concerned agencies and interest groups to effectively address police and community concerns;
> - Promote public commitment and active involvement in the resolution of crime, traffic and social order problems;
> - Create an open and responsive working environment which encourages commitment and creativity in the design and application of community policing services;
> - Increase our effectiveness through the application of innovative reactive and proactive policing services.

This understanding of community policing was further refined by the OPP and was published in its *Framework '93* document, as follows:

The Philosophy—Community Policing
The System—Generalist policing supported by excellence in traffic management, crime prevention, detection, investigation and administrative services.
The Mechanism—The right people, with the right equipment, in the right organizational structure.

This clear strategic commitment to community policing required specific application within the OPP. As a result of a review by the Provincial Auditor in 1994, the following recommendations were offered in this regard:

> To enhance effectiveness measurement, better use of statistical information is required for problem identification and analysis in order to develop more specific strategies to address problems. A thorough evaluation of the strategies developed is needed to assess their effect in resolving identified problems.
>
> To improve effectiveness reporting, the OPP should require detachments to conduct;
>
> - More comprehensive analysis of community problems, and
> - More thorough evaluation of actions carried out to address them.

Beginning in 1994, the OPP embarked on a major Organizational Review Project that would apply process reengineering to the ten "businesses" identified (OPP, 1995). The overall objectives of this project reveal a serious intention to advance the principles of community policing:

- Ensure that our organizational design and processes facilitate the implementation of our strategic plan and directions;
- Institutionalize community policing by designing an organization that will provide the appropriate police service delivery mechanisms;
- Improve efficiency and effectiveness and thereby the ability to deliver services at reduced levels of spending;
- Ensure that the structure is consistent with that of a modern, innovative organization which would:
- Serve to attract and keep the best people;
- Facilitate their development;
- Derive the maximum benefit from their efforts; and
- Provide a productive and fulfilling working environment.

One of the sub-projects within the Organizational Review Project dealt with the Prevention of Victimization. As a result of the work of the sub-project team assigned to this area, four key recommendations were put forward:

- Adoption of a problem-solving model;
- Increase victim assistance by supporting the Ministry of the Solicitor General's Victims Crisis Referral Service (VCARS);

- Implement the revised Uniform Crime reporting survey; and
- Establish a Community Policing Development Centre.

Through careful discussion and analysis of the concept of community policing, including the role of the public police (Hastings, 1995), the OPP arrived at a better understanding of the theoretical framework within which to place their organizational strategy regarding community policing. The result presented a three-stage evolution that would, ultimately, result in a true community policing approach. Figure 11–7 outlines the OPP's framework for this evolution of community policing:

FIGURE 11–7

OPP'S FRAMEWORK FOR THE EVOLUTION OF COMMUNITY POLICING

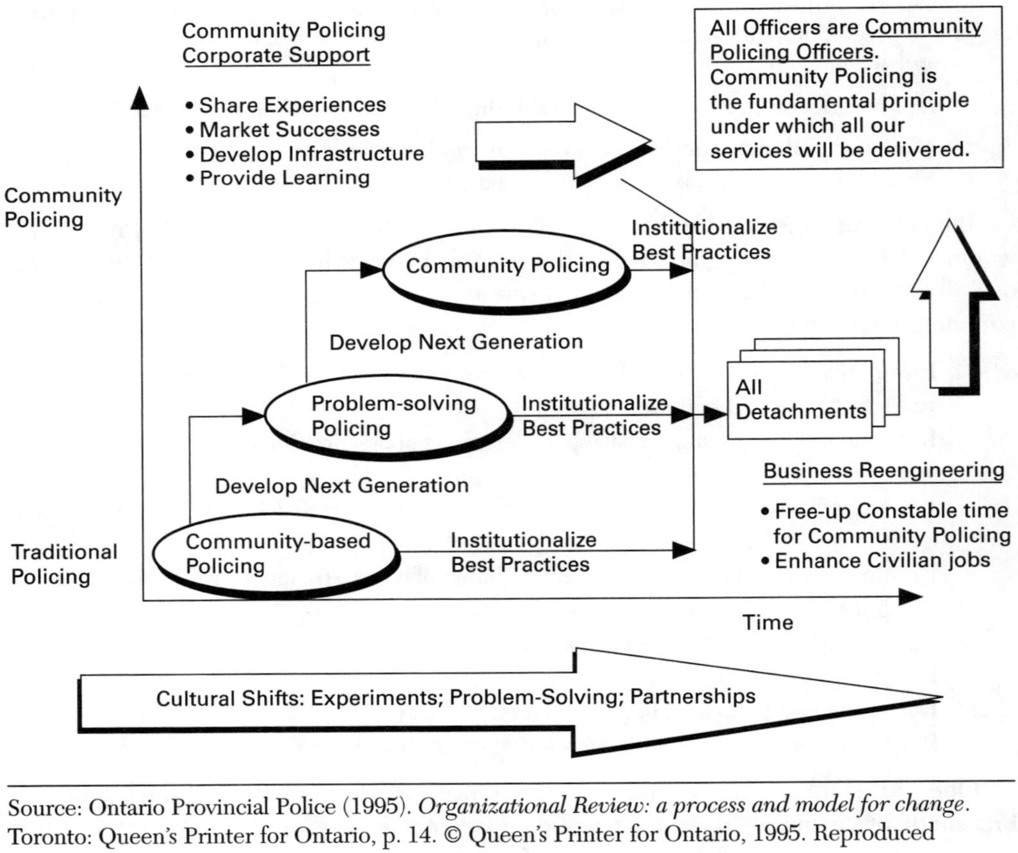

Source: Ontario Provincial Police (1995). *Organizational Review: a process and model for change*. Toronto: Queen's Printer for Ontario, p. 14. © Queen's Printer for Ontario, 1995. Reproduced with permission.

The specific levels of this evolutionary model move from a traditional approach towards a fully functional approach to community policing. The details of those levels are outlined below:

- *Community-based policing*—central to this level is the effort to recruit the public as volunteer participants in the delivery of traditional types of reactive police services. Accordingly, this level entails no real change in the mandate of the public police and requires little organizational reform or enhanced resources;
- *Problem-oriented policing*—this approach involves a shift to a proactive response to recurring community problems. There is a recognition that all incidents may not fall into neat problem clusters, and places a priority on the preventative aspects of policing and focuses efforts on problem resolution. This level entails a change in the police role as well as an expansion of their mandate; and
- *Community policing*—at the fullest extent, this level involves a full partnership between the police and the community. Through the "co-production" of order, these partners work together to define public safety problems and collaborate on the design, delivery, and assessment of policing services.

The OPP adopted a problem-solving approach (for more detail on this approach, see the earlier discussion in Chapter Six) that adapts the SARA model popularized in several police departments across North America. A further refinement resulting from the OPP's Organizational Review Project was the creation of its Community Policing Development Centre. This new entity was placed at a significantly high level within the reworked organizational structure, reporting to a Deputy Commissioner (specifically, the Provincial Commander for Corporate Support). The original mission of the Community Policing Development Centre (CPDC) is outlined below:

> The mission of the CPDC is to ensure that the OPP has the organizational capacity to operationalize Community Policing. It will contribute to this goal by:
>
> a) Enhancing our ability to do problem-solving in partnership with the community;
> b) Serving as a resource for improved service delivery; and
> c) Promoting the development of a learning organization.

Furthermore, the specific role of the CPDC was determined to be to:

- Enhance the capacity of the OPP to assess its role in policing, and to select and implement the most effective Community Policing strategies and programs;
- Monitor, assess, and evaluate current Community Policing and problem-solving efforts within the OPP in order to identify best practices;
- Identify best practices from other police services and communities or agencies, and to promote their relevance for the work of the OPP; and
- Serve as a catalyst for the design, implementation, and delivery of innovative Community Policing practices within the OPP.

In extending their understanding of the three levels of community policing, the OPP structured some of the various roles and responsibilities across the entire organization. Figure 11–8 includes that breakdown:

FIGURE 11–8

OPP ROLES & RESPONSIBILITIES UNDER COMMUNITY POLICING

Position	Community-based Policing	Problem-oriented Policing	Community Policing
Constable	Communicate & mobilize participation	Analyse & solve problems	Mobilize or catalyze partnerships
Supervisor	Communicate & mobilize participation	Facilitate & coach team problem-solving	Facilitate & coach partnership process
Middle Management		Manage change	Structure the process
Senior Management		Lead change	Advocate public security

Source: Ontario Provincial Police. © Queen's Printer for Ontario, 1995. Used with permission.

In order to ensure success for the community policing approach within the OPP the following essential management supports were also identified:

- Encouragement of innovation and creativity;
- Support for empowerment;
- Freedom to fail;
- Recognition of problem-solving initiatives (i.e., rewards);
- Facilitation of learning and development; and
- Foster interagency and community partnerships.

With the massive organizational transformation of the OPP came the challenges of implementation. However, the OPP remained committed to the philosophy of community policing and worked toward the staffing and operation of the Community Policing Development Centre. As a result of extensive consultations, this unit published a key document in 1997, entitled: *"How do we do it?" manual of community policing*. This manual was designed to be a guide for OPP detachments and local community policing committees. It sets out the current operating definition of community policing within the OPP:

> The delivery of police services, resulting from a community and police partnership, that identifies and resolves issues to maintain social order.

This manual contains the following parts:

- Part I—Introduction;
- Part II—Community Partnerships;
- Part III—Managing and Recruiting Volunteers;
- Part IV—Committee Management;
- Part V—Committee Procedures;
- Part VI—The Service Delivery Process;
- Part VII—Problem Solving and Planning

The OPP continues to operate under the philosophy of community policing and, as an organization, has invested considerable effort in arriving at an understanding of the substance of this model of police service delivery.

In 1998, the Provincial Auditor again examined the OPP on behalf of the Ontario taxpayer and, as a result of the examination, made several findings with respect to the introduction of community policing. The Provincial Auditor offered the following observations in this area:

> The OPP has established a process for identifying and prioritizing policing services to solve community problems and meet service expectations. It has developed measures for monitoring, evaluating and reporting on community policing including: the implementation of the new service delivery model; community satisfaction with services; levels of reported and unreported victimization rates; and community policing successes with community partners.
>
> We noted several examples of effective community policing initiatives for problem-solving and reporting that had been carried out by some detachments. However, we found that the processes and measures described above had yet to be implemented in most of the detachments we visited.
>
> While the OPP has indicated that it is implementing a more integrated community policing approach to service delivery, we noted that approximately eight years have elapsed since community policing was first required by the Police Services Act. It has taken the OPP a considerable length of time to develop community policing service delivery processes and related officer training.

Accordingly, the Provincial Auditor made the following recommendations pertinent to continuous improvement in the area of community policing within the OPP:

> In order to be more effective in serving the community through community policing activities, the Ontario Provincial Police should:
>
> - Fully implement the process developed for identifying and prioritizing policing services to meet community service expectations;
> - Identify and disseminate best practices in community policing among detachments; and
> - Measure the effectiveness of community policing activities against established criteria.
>
> The Ontario Provincial Police should also periodically evaluate the progress of community policing implementation, taking corrective action where necessary to ensure the implemented initiatives are effective in helping communities achieve the objectives of reducing crime and victimization. (pp. 235–236)

CALGARY, ALBERTA

The philosophy of community-based policing has been guiding the operation of the Calgary Police Service since the early 1970s (Perry, 1997). Under this philosophy, the city was grouped into districts representing large geographical areas, and further subdivided into zones that were made up of several smaller communities. Officers assigned to these areas were expected to develop a sense of ownership and responsibility for their zone. They would also be in a position to better understand community concerns. This local approach has evolved considerably over the years in response to changing needs and the substantial growth in the size and diversity of the city.

Zone and team deployment structures were proposed by consultants to the Calgary police in early 1972. This was intended as a community-based approach that could replace the traditional platoon system. In addition, the consultants made recommendations in the following areas in order to support a decentralized approach to police service delivery in the city:

- Supervision;
- Communications;
- Job satisfaction;
- Training;
- Promotions; and
- Other organizational factors.

Evaluations of the first zone policing initiative were completed in 1977. These evaluations revealed that the public was satisfied with policing services and felt that officers were more involved with the community. However, there remained an expressed need for greater police involvement in programs geared to youth and better engagement with other social agencies (Perry, 1997). There were other deficiencies identified including a degree of organizational fragmentation and communication difficulties between zones. Many constables assumed that community relations would be done entirely by Community Policing Officers. There was no consistent sense that this function was everyone's responsibility. Also, there were certain weaknesses seen in the areas of leadership and supervision, and these were the cause of some further difficulties. Deficiencies in the training of supervisory skills were seen as the primary cause of this problem. Finally, the challenge of dealing with "non-performers" was also identified.

In order to deal with these issues, the Calgary police provided some basic leadership training. Also, district Community Policing Officers were relocated to a centralized Community Policing Unit, and zones were aligned to properly balance the workload within the city's boundaries.

In the early 1980s Calgary was the victim of economic recession. Layoffs resulted from these financial difficulties, even while service demands were continuing to grow. The Calgary Police also experienced a technological change that had the predictable effect of removing patrol officers from regular contact with the public; namely, the installation of a computer-assisted dispatch (CAD) system.

In 1985, the zone policing system was subjected to yet another round of evaluations. The front-line officers continued to believe that zone policing was superior to the old platoon system. There was a sense that the community was getting a better service and that patrol officers had higher levels of job satisfaction which clearly contributed to good officer morale. However, there was not a direct and immediate connection between zone policing and crime prevention. The preventative dimension remained something that was outside of "real police work" (Perry, 1997). Supervision continued to be an issue, essentially because zone sergeants were not engaged on the street level with their constables, and, therefore, were not available to deal with performance issues directly.

With the appointment of a new chief of police in 1989, a renewed mandate was established to "revitalize Calgary's community based preventive policing system." The chief, using a task force approach to this challenge, undertook to:

- Revise the Calgary Police Service mission statement to include an identifiable commitment to community based policing;
- Increase problem solving strategies at the street level (SARA plans); and
- Research and implement a system of differential response to calls-for-service.

This undertaking led to a new definition of community based policing that was approved by the Calgary Police Commission (Beattie, 1990):

> An interactive process between the police and the community to mutually identify, resolve, or prevent community problems/concerns with the goal of improving the quality of life.

Performance measures conducted during this period revealed that zone sergeants were engaged in administrative duties nearly fifty percent of the time, with only seventeen percent being devoted to community crime prevention activities. Alternatively, front-line personnel were devoting approximately forty-five percent of their time to preventive policing. A service-wide training program was introduced to address problem-solving strategies within the department. The model applied is based on the SARA format discussed in Chapter Six. In 1992, a differential response strategy was implemented which was designed to screen calls-for-service and provide more efficient and effective police response.

The Calgary Police Service underwent a major organizational review process in 1994. Superintendents within the department were invited to develop initiatives that would improve effectiveness, efficiency, human resource development, and quality of service in their areas of responsibility. This resulted in more than one hundred submissions within the following categories:

- Leadership;
- Supervision;
- Accountability;
- Officer development;
- Quality of investigations; and
- Community involvement.

The Calgary Police Service determined that it would be valuable to experiment with a team deployment strategy for a fixed period in order to address the identified concerns. A pilot project was established in District 4 to improve front-line core services. This pilot project ran from April 1995 until September 1996. In order to monitor and evaluate this team policing model, it was divided into the following components:

- Deployment (human resource issues and service delivery);
- Investigative functions;
- Community involvement;
- Financial costs; and
- Police facilities (utilization and future planning).

With the careful assessment of this team deployment model, the Calgary Police Service determined that it could broadly apply this approach, with refinements that allow for a cohesive and viable policing option. As reported by Perry (1997):

> The Calgary Police Service is committed to community policing as an effective means to "optimize safety in the City of Calgary." One of the primary principles of community policing is to develop a strong partnership with the community and to jointly solve issues which underlie criminal activity and social disorder. As an organization, we have nurtured this relationship and there is confidence that utilizing a team deployment structure is a long-term strategy that will positively influence the environment in which we live. The Calgary Police Service cannot alleviate crime simply by responding to calls for service but must ensure that there is a balance between proactive and re-active policing strategies. (p. 35)

EDMONTON, ALBERTA

The Edmonton Police Service has an established reputation as an early adopter of community policing principles. This is largely due to the efforts of retired Superintendent Chris Braiden (see "Leaders in Community Policing" profile in Chapter Nine) and the support he was given by several enlightened chiefs of police. There has been considerable attention given to the Edmonton experiment in community policing, even in American publications (Hawkins Jr., Carl W., 1996).

In 1990 a Community Based Policing Team was formed to consider a change in policing in the city of Edmonton. On January 6, 1992, the Edmonton Police Service moved to a new model of service delivery based on decentralized full service community stations. There was an enormous effort made to define an approach to community based policing that would fit the Edmonton Police Service. As noted by Braiden (1995):

> We have kicked around the idea of Community Based Policing (CBP) for four years, yet a common understanding of what it means in real terms has never been struck between us. There is nothing for us to get our teeth into and agree, or disagree, upon. It needs to be translated into identifiable components so that we have at least a 'slab of marble' from which to begin our sculpture. Until this is done, by someone, we will continue to discuss in the abstract and spin our wheels. (p. 78)

The Edmonton Police Service was aware that its officers were becoming detached from citizens. This was a direct result of their pursuit of efficient, "professional" police methods featuring rapid response to calls for service. A study of police calls in 1988 pinpointed twenty-one "hot spots" requiring increased levels of service. The Neighborhood Foot Patrol Program was introduced in which members began to walk the beat in these troubled areas. This encouraged more contact and trust between the officers and community members.

Braiden developed the vision for the change to a community-based policing model (Braiden, 1995) that was approved by the chief of police and executive officers. The Edmonton Police Service embarked on an organizational review that included analysis of workload and units. As a result, $2.5 million was reallocated from the Edmonton Police Service budget to the operational budget. Sixty-eight positions within the organization were reassigned to front-line policing.

The organization adopted a simple declaration to guide its thinking: "Committed to Community Needs." It also adopted a modified differential response model that placed an emphasis on decentralized reporting outlets, problem-solving, and broader public involvement. Police stations were identified within the community for retooling. All members of the Edmonton Police Service were included in a two-day session dealing with organizational changes, problem-solving, and other issues relevant to community policing. By 1995, the Edmonton Police Service had twelve full-service community police stations designated in the high traffic areas of the city.

As noted in the department's own literature (Edmonton Police Service, 1995):

> Community policing in Edmonton began with the development and implementation of the Neighborhood Foot Patrol Program in the late 1980s. However, elements of the community policing model had been part of police practice many years earlier. For example, from the time of the creation of the Edmonton Police Department until the 1960s, walking the beat was a key component of policing and a primary developmental experiment for new officers.

There was a focus on problem-solving and the work of Herman Goldstein was referred to in arriving at this approach to recurring incidents. The approach in Edmonton also resulted in a new collaborative leadership style that emphasized coordination and organization of police services (Dent, 1993). Not wanting to operate in a vacuum, the Edmonton Police Service, and Chris Braiden, look far and wide for inspiration:

> Translating this vision of community based policing into action resulted in the development and implementation of the Neighborhood Foot Patrol Program (NFPP) in 1988. The program was adapted specifically for Edmonton by Superintendent Chris Braiden from similar programs in Flint, Michigan, Detroit, Michigan, and Houston, Texas.

From the introduction of the new service delivery model in January 1992, the Edmonton Police Service continued to consolidate its adherence to community based policing with the opening of several community stations throughout the city. The change in the organization's mentality called for new terminology (Edmonton Police Service, 1995):

From	**To**
Edmonton Police Department	Edmonton Police Service
Task Force or SWAT Team	Tactical Section
Strike Force (anti-crime unit)	Coordinated Crime Unit
Complaint Line (communications)	Dispatch Center
Patrol Divison	Response Division
Beats	Neighborhood Foot Patrol

In January 1994 a statistical review of the 1991-1993 period indicated a twenty-five percent (25%) reduction in crime in Edmonton. The following month the provincial government reduced its grants by fifty percent (50%), and indicated that all grants would be phased out by 1997. This placed the Edmonton Police Service in a position where they would have to restructure.

One of the important achievements of the Edmonton Police Service came in October 1994 when the department received international recognition for their "Call Path Chart." The Webber Seavey Award, sponsored by the International Association of Chiefs of Police (IACP) and Motorola, is given for innovation and excellence in policing. The "Call Path Chart" developed by the Edmonton Police Service offers a detailed graphical representation of the various decisions and processes associated with handling calls for police service.

In February 1995, John Lindsay was appointed as chief of police in Edmonton. The new chief immediately entered into a planning process to ensure that there was a continued commitment to community policing. A statistical review covering the 1991-1994 period indicated that crime had decreased thirty-five percent (35%) in Edmonton. With growing interest in the whole topic of community policing, the Edmonton Police Service sponsored a Community Policing Conference in Banff, Alberta in April 1995 and a second conference in November 1995. As a result of their substantial efforts, the following highlights are worth noting (Edmonton Police Service, 1995):

- Over 100 site visits from police agencies worldwide;
- Over 220 information packages on community policing sent to police agencies and academic centers worldwide;
- District Area Surveys and City Area Surveys are conducted to compare 1991 and 1994 public views of police service in Edmonton;
- Community network survey of Edmonton Police Service members shows that 150 members surveyed are involved with over 400 committees and agencies;
- An employee survey is being developed to measure attitudes, knowledge, and practice of problem solving.

In more detail, Figure 11–9 represents the Edmonton Police Service's new service delivery model in 1995. It is based on the main components—receiving, responding, recording, and resolving:

FIGURE 11-9

SERVICE DELIVERY MODEL OF THE EDMONTON POLICE SERVICE (1995)

New Service Delivery Model

Receiving Principles	*Responding Principles*	*Reporting Principles*
Public comes to us for service Increased contact with police Decentralized reporting Emphasis on local problems Face-to-face contact with public Customer service	Personal safety and security Quality of life Duty Bound Maintain viable response Every member responds to calls for service	Alternatives to 'full' reporting Reduce redundant information gathering Expedite the reporting function Record only crime specific data Automate basic reports
	Applying the Service Model Delivery	
Police and public education EPS listing in phone directory Talking information phone services Decentralized reporting Utilize facsimile reporting Utilize: pre-recorded messages phone call management/reporting appointment schedule answering machines/pages mail in/drop off reports Access police services through: Communications Neighbourhood Foot Patrol Community Stations Division Stations	Ownership members Turf members Neighbourhood Foot Patrol School Resource Officers Community stations Division stations On street managers Communications Division Dispatch 'appropriate' unit Reassign people to the street Seasonal workload adjustments Create an 'ideal' staffing model	Police respond and write a report Other unit creates report Other agency creates report Police referrals to other agencies Public comes to us for service Volunteer/clerk assistance Unlimited future: computer networking remote links computer facsimiles

	Resolving Principles and Techniques	
Problem solving Develop problem identification	Reduce repeat call for service Capture problems solved	Internal and external recognition Ownership
	Functions of Service Model Delivery	
Priority based on community need Develop partnerships with: community affected agencies Reward/recognize hard work Support: risk taking initiatives and creativity	Implement ownership plan to involve: police citizens Analyze problems and trends Share information Emphasize proactive problem solving approaches Community building	Create performance measures to focus on problem solving attempts and effectiveness Provide quality customer service both internally and externally Specialize where necessary Generalize when possible

Source: Edmonton Police Service (1995). *Community policing in Edmonton 1995: the vision continues. 3rd ed.* Edmonton: Edmonton Police Service, p. 33. Used with permission.

The Edmonton Police Service has identified the following significant issues that it will need to address as it prepares for the future:

- training;
- victim offender mediation project;
- technology;
- performance and workload measures;
- organizational renewal; and
- problem solving.

In March 1997, Chief John Lindsay made the following comments indicating the continuing adherence of the Edmonton Police Service to community policing (Lindsay, 1997):

> Our vision of the future must include a strong commitment to the basic proposition that police organizations are really accountable to the communities that we serve, instead of just getting the reported crime down. Our business is not really about problems, it's about solutions arrived at through meeting common goals . . .
>
> Yes—you must lose sight of traditional shores and be ready to anticipate the fierce gales and stormy waters in the path of policing. But, you now have a sense of direction, your navigational gear is readied, and your sails are trimmed for the changing winds of community needs and organizational pressures.

In June of 1998, Chief John Lindsay made a proposal with regard to a review of the cost-effectiveness of the services provided by the Edmonton Police Service. After receiving the support of the Edmonton Police Commission for this undertaking, the consulting firm KPMG was engaged to assist in the identification of "best-value gains" that could be introduced to enhance community policing initiatives within the city. The final report for this project revealed that there is no consensus among members of the Edmonton Police Service regarding the meaning of community policing, or what it should involve. Accordingly, it is seen as being valuable to develop a clear and "strong consensus of support among police personnel for community policing principles." The organizational review of the Edmonton Police Service arrived at seven strategic directions for the police department to follow that would allow resources to be reallocated in order to enhance community policing (KPMG, 1998).

VICTORIA, BRITISH COLUMBIA

In 1987, under the leadership of Chief Bill Snowdon, the first of five Victoria Community Police Stations (Co.P.S.) was opened. With the assistance of a grant from the Donner Foundation, an evaluation project was undertaken to assess the program's effectiveness (Walker, Walker, and McDavid, 1992). While the city of Victoria had normally been considered a safe place, an examination of three nearby municipalities (i.e., Esquimalt, Saanich, and Oak Bay) revealed that crime rates were from eighty-five percent (85%) to one hundred and eighty percent (180%) higher in Victoria. Indeed, Victoria was exceeding cities such as Vancouver and Toronto with respect to certain types of crime (Walker, 1987; Walker and Walker, 1989). To address these problems, and the serious concerns they raised, the Victoria Police developed a strategy to introduce five Community Police Stations. To direct and inform this strategy, a Mayor's Task Force on Policing was established to operate from April to June 1986, with the following three priorities:

1. Identifying community concerns that involve police services;
2. Establishing community-policing programs that will enhance crime prevention; and
3. Reviewing the internal operation of the police department to maximize morale, communication, and effectiveness of deployment of personnel.

As a result of public meetings, the task force came to a fairly clear understanding (Walker, Walker, and McDavid, 1992):

> It became apparent that what was needed was a static police presence in each neighbourhood as well as a means for the community to become involved in their own protection and crime prevention. As a result, it was decided that the community police station concept could readily serve as a suitable strategy for the development of a closer police-community relationship. Furthermore, the developed program had to ensure a major input from the public in three ways: the identification of crime related concerns; the development of programs to deal with those concerns; and, the delivery of those programs. Anything less would be simply a re-invention of a police controlled model for crime prevention. (p. 4)

Between September 1987 and February 1989 five Co.P.S. centres were opened. A three-year formative evaluation began in September 1988 and concluded in August 1991.

In order to arrive at a direct statement of purpose and objectives for the Co.P.S. program, Victoria Police senior personnel and the evaluators agreed upon the following statements (Walker, Walker, and McDavid, 1992):

Purpose—

To bring the police and the community more closely together to work in making Victoria a safer and better place to live.

Outcome Objectives—

1. To reduce the occurrence of crime.
2. To reduce the fear of crime.

Implementation Objectives—

1. To identify and develop creative methods for reducing crime problems.
2. To increase community accessibility to police services.
3. To increase community involvement in the development and delivery of Develop a coherent internal and external communications strategy; crime prevention and identified community programs and services.
4. To increase program delivery and services to the elderly population.
5. To increase program delivery and services to school aged children.
6. To increase the reporting of intelligence information.

In general, the Co.P.S. program had an impact on the delivery of policing services in Victoria. However, the evaluators determined that it failed to meet the majority of the stated objectives and, because it was not applied throughout the entire department, it led many to increase their adherence to traditional, reactive modes of policing. As the evaluators have observed (Walker, Walker, and McDavid, 1992):

> If the new administration is committed to the idea of connecting the reactive with the proactive approach then every effort should be expended to educate all personnel about the Community Policing philosophy, including its practical applications; and involving all staff in re-working the current model of service delivery to include this dual approach. Before the public can be encouraged to join in with the police in a true co-partnership, the Department must ensure that it has forged a partnership within its own ranks which promotes an acceptance of both approaches working towards a common goal. (p. 80)

SUMMARY

This chapter has considered several jurisdictions across Canada where experiments have been conducted in the application of community policing. Unlike a pure science experiment, in which a hypothesis can be tested under ideal circumstances, and where variables are monitored and closely controlled, these police organizations have had to venture into unknown territory while continuing to provide a vital and visible public service. The difficulties involved in moving to a model of community policing, which typically requires large-scale organizational change, are enormous. It is, indeed, remarkable that the challenge of change has been accepted by so many Canadian police departments, given the risks and virtual absence of assured success. It seems clear that traditional, professional, reactive approaches are no longer viable within the current context of policing.

The window of opportunity offered by the introduction of a new chief of police appears to have been capitalized upon in many Canadian jurisdictions. The leadership required in this context is significant, and the task is made considerably less onerous when there are leaders at all levels within the organization. Several examples in this chapter demonstrate the profound value of having change agents within the police structure, who can act as catalysts and drivers for ongoing innovation.

There are several key ingredients for successful change in police organizations as they move toward the implementation of community policing. While nothing is certain in this regard, many of the organizations featured in this chapter exemplify the commitment and creativity required for such an undertaking. Attention to the following elements is recommended:

- Executive support;
- Shared vision or understanding of community policing;
- Establishment of achievable goals;
- Identify agents of change within the organization;
- Create learning opportunities throughout the organization around the change to community policing;
- Facilitate acceptance by front-line officers, supervisors, and the police association with regard to change;
- Establish supports for the implementation of community policing;
- Develop capacity to fully understand the communities and their needs & expectations;
- Ensure that the organization has assessment and evaluation capabilities to determine effectiveness of community policing initiatives; and
- Develop performance measurement systems that are specifically tailored to community policing practices.

QUESTIONS FOR CONSIDERATION

1. What are the common features of the Canadian police services that have introduced some form of community policing?
2. What other Canadian police departments have introduced community policing?
3. Why does the introduction of community policing frequently coincide with the appointment of a new chief of police?
4. What are some other considerations that might move a Canadian police department to consider the introduction of community policing?
5. What are the major obstacles to the introduction of community policing in Canada?
6. How does an innovative Canadian police service maintain the proper balance between the proactive and reactive services it offers to the community?

REFERENCES

Beattie, J. (1990). *Community-based policing task force report*. Calgary: Calgary Police Service.

Braiden, Chris (1990). "A process for change: a position paper written for the executive officers team, July 1990." Edmonton: Edmonton Police Service.

Calgary Police Service (1991). *Enhanced community policing: course training standard*. Calgary: Calgary Police Service.

Clairmont, Donald (1990). *To the forefront: community-based policing in Halifax*. Ottawa: Canadian Police College.

Clairmont, Donald (1993). *Halifax police department and community based policing*. Halifax: The Atlantic Institute of Criminology.

Cooke-Scott, L.A. (1998). "Community-based policing in Ontario: lessons from the Halton Regional Police Service." *Canadian Public Administration*, Vol. 41, no. 1 (Spring), pp. 120–146.

Dent, Peter (1993). *The Westmount community police station: a case study of leadership and community based policing*. Edmonton: Center for Criminological Research, Department of Sociology, University of Alberta.

Don, Kim and Dave Veitch (1996). *Community-based policing in Edmonton: prepared for the 1995–96 Fraser Institute/Financial Post Economy in government competition*. Edmonton: Edmonton Police Service.

Downey, T.J. and R.J. Williams (1998). "Provincial agendas, local responses: the "common sense" restructuring of Ontario's municipal governments." *Canadian Public Administration*, Vol. 41, no. 2 (Summer), pp. 210–238

Edmonton Police Service (1995). *Community policing in Edmonton 1995: the vision continues*. Edmonton: Edmonton Police Service.

Edmonton Police Service (1997). *The police plan: policing for results*. Edmonton: Edmonton Police Service.

Fitch, Leanne J. (1995). Reviving the Peeler?: a case study of organizational change and community-based policing. M.A. thesis, Department of Sociology, University of Alberta.

Geldart, Sheldon L. (1995). "Community based policing: using self-efficacy to combat the fear of change." *RCMP Gazette*, Vol. 57, no. 4, pp. 21–27.

Greater Toronto Area Task Force (Ont.) (1996). *Greater Toronto: report of the GTA Task Force*. Toronto : Queen's Printer for Ontario.

Guelph Police Service (1997a). Moving forward . . . the year 2007. Guelph: Guelph Police Service.

Guelph Police Service (199b). *Moving forward . . . the year 2007: working with our communities: strategic directions of the Guelph Police Service*. Brampton: Peel Regional Police Service.

Halton Regional Police Service (1982). *Halton Regional Police Service policing philosophy*. Oakville, Ont.: Halton Regional Police Service.

Halton Regional Police Service (1995). *Halton Regional Police Service public survey*. Oakville, Ont.: Halton Regional Police Service.

Halton Regional Police Service (1997). *The team concept: policing in Halton*. Oakville, Ont.: Halton Regional Police Service.

Halton Regional Police Service. Organizational Review Project (1995). *The challenge of change*. Oakville, Ont.: Halton Regional Police Service.

Harder, Tony (1991). *The big picture*. Edmonton: Edmonton Police Service.

Hastings, Ross (1994). "Community policing: a basis for organizational change." Paper presented to the Ontario Provincial Police, Organizational Review Project on January 13, 1994.

Hawkins Jr., Carl W. (1996). "Ready, fire, aim: a look at community policing in Edmonton, Alberta, Canada." In *Themes and variations in community policing: case studies in community policing*. Washington, D.C.: Police Executive Research Forum.

Hornick, Joseph, P.J. Duggan, and Denise LeClaire (1993). *Community policing in Edmonton*. Edmonton: Edmonton Police Service.

Juschka, B. and N. Dempson (1994). *District 4 training needs analysis for team policing*. Calgary: Calgary Police Service.

KPMG (1998). *1998 organizational review of the Edmonton Police Service*. Edmonton: Edmonton Police Commission.

Kingston Policing Service Steering Committee (1998). *Policing the city of Kingston: building on strength, securing our future*. Kingston: the Steering Committee.

Koller, Katherine (1990). *Working the beat: the Edmonton neighborhood foot patrol*. Edmonton: Edmonton Police Service.

Lambert L.R. (1988). "Police mini-stations in Toronto: an experience in compromise." *RCMP Gazette*, Vol. 50, no. 6, pp. 1–5.

LeClaire, D. and J. Hornick (1993). *Community policing in Calgary*. Calgary: Canadian Research Institute for Law and the Family.

Lindsay, John (1997). "Policing for the future: a voyage through uncharted seas." Speech presented by Chief John Lindsay at the Banff Creating Change course, March 1997.

Metropolitan Toronto Police (1991). *Beyond 2000: the strategic plan of the Metropolitan Toronto Police*. Toronto: Metropolitan Toronto Police.

Metropolitan Toronto Police. Restructuring Task Force (1994). *Beyond 2000: final report*. Toronto: Metropolitan Toronto Police.

Metropolitan Toronto Police. Strategic Planning (1990). *1990 environmental assessment and force goals & objectives for 1991*. Toronto : Metropolitan Toronto Police.

Montreal Urban Community Police (1996). *Neighborhood policing: together for better service*. Montreal : Montreal Urban Community Police.

Murphy, C.J. (1988). "Community problems, problem communities, and community policing in Toronto." *Journal of Research in Crime and Delinquency*, Vol. 25, no. 4, pp. 392–410.

Murphy, Chris and Graham Muir (1985). *Community-based policing: a review of the critical issues*. Ottawa: Solicitor General of Canada.

Murphy, C. and J. de Verteuil (1986). *Metropolitan Toronto community policing survey*. Ottawa: Solicitor General of Canada.

Ontario Provincial Police (1993). *Framework '93: mission statement, policing philosophy, and management plans of the Ontario Provincial Police*. Toronto: OPP.

Ontario Provincial Police (1994). *Commissioner's directional statement 1994*. Toronto: OPP.

Ontario Provincial Police (1995). *Directions '95: "facing the challenges; finding the opportunities"* Orillia, Ont.: OPP.

Peel Regional Police (1990). *Strategic plan for the 1990s*. Brampton: Peel Regional Police.

Peel Regional Police (1997). *Strategic plan 1997: leadership and action*. Brampton: Peel Regional Police.

Police Regional Police. Organizational Development Burearu (1997). *1996 strategic plan feedback report*. Brampton : Peel Regional Police.

Perry, D. (1997). *Structure and functions of community policing in Calgary: an evolutionary process* [draft]. Calgary: Calgary Police Service.

Provincial Auditor for Ontario (1998). *1998 annual report of the Provincial Auditor of Ontario to the Legislative Assembly*. Toronto: Queen's Printer for Ontario.

Stevens, Nigel (1993). *Future issues in policing*. Edmonton: Edmonton Police Service.

Strauss, Leo (1953). *Natural right and history*. Chicago: University of Chicago Press.

Thunder Bay Police (1995). *Proposal to implement neighbourhood policing: community approach*. Thunder Bay, Ont.: Thunder Bay Police.

Walker, C. (1987). "The community police station: developing a model." *Canadian Police College Journal*, Vol. 11, no. 4, pp. 273–318.

Walker, C. and S.G. Walker (1989). *The Victoria community police stations: an exercise in innovation*. Ottawa: Canadian Police College.

Walker, Sandra Gail, Christopher Walker, and James C. McDavid (1992). *The Victoria community police stations: a three-year evaluation.* Ottawa: Canadian Police College.

Wasson, David (1977). *Community-based policing: a review*. Ottawa : Solicitor General of Canada.

William M. Mercer Ltd. (1990). *Final report: Halifax Police Department community-based policing evaluation project*. Toronto: William M. Mercer Ltd.

TOWARD MORE LEARNING

An excellent source of additional information dealing with the various police services discussed in this chapter can be found in the individual websites maintained by those departments. There are several search engines that will lead the user to relevant law enforcement sites. Many police organizations invest a considerable amount of time and effort in keeping their websites up-to-date.

Halifax, Nova Scotia: www.police.halifax.ns.ca/
Montreal, Quebec: www.spcum.qc.ca/
Kingston, Ontario: www.police.kingston.on.ca/
Belleville, Ontario: www.police.belleville.on.ca/
Toronto, Ontario: www.mtps.on.ca/
Halton Region, Ontario: www.worldchat.com/hrp/
Peel Region, Ontario: www.peelpolice.on.ca
Guelph, Ontario: www.police.guelph.on.ca/
Thunder Bay, Ontario: www.tbaytel.net/tbps/
Ontario Provincial Police: www.gov.on.ca/opp/
Calgary, Alberta: www.gov.calgary.ab.ca/police/
Edmonton, Alberta: www.police.edmonton.ab.ca/

CHAPTER TWELVE

. . . and he should act like those prudent archers, who, when the target they are aiming at seems too far off, aware of the capacity of their bow, set their sight a good deal higher than the desired target, not to reach such a height with their arrow but rather to be able, with the help of aiming high, to reach their target.

Niccolo Machiavelli, The Prince, VI (Translated and edited by Mark Musa)

FUTURE PROSPECTS FOR COMMUNITY POLICING—WILL PROBLEM-SOLVING DISSOLVE CRIME?

LEARNING OBJECTIVES

1. Identify aspects of community policing that have proven effective in addressing crime and disorder.
2. Locate sources of continuing information dealing with community policing, problem-solving, and crime prevention.
3. List police, and other, agencies that are supporting community policing, problem-solving, and crime prevention efforts.
4. Apply the transformative model, outlined in this textbook, to the future challenges facing policing in Canada.
5. Identify critical issues that remain to be addressed relevant to community policing, problem-solving, and crime prevention.
6. Distinguish between the adversarial and the restorative approaches to justice.
7. Differentiate public policing from private policing.

INTRODUCTION

This chapter concludes the textbook with a consideration of the future of community policing in Canada, and looks at the prospects of problem-solving for addressing crime and disorder. In this context, it will be possible to reiterate what has been termed the "transformative" approach to policing (outlined in Chapter One), and to suggest that it offers some promise for the purpose of public safety and security.

While far from exhaustive, this chapter will also look at recent commentary relating to community policing, problem-solving, and crime prevention in order to assist in understanding the breadth of appeal of these approaches. With so much activity geared toward this new paradigm of policing, it is important to take seriously the reports of practitioners, the critical analyses of academics, as well as the fundamental reactions of the multitude of "communities" that are the presumed beneficiaries of that new paradigm.

This chapter will look at the challenge of change in the philosophy and practice of policing in the real world frame of reference in which police services must be delivered, twenty-four hours a day, seven days a week. One American police officer has observed that changing a police organization is much like trying to rebuild a car while you're driving down the freeway. Unfortunately, police leaders have no choice but to balance such change with the rigorous demands of ongoing service delivery. The pressures and predicaments that are brought to bear on police organizations are frequently unpredictable in their specific impact and ramifications. What is predicable is that policing in Canada lies at the very heart of civil society, and that the capacity of community policing (in all its manifestations) to address the problems of crime and disorder has a profound correlation to the quality of life enjoyed in our neighbourhoods.

As it becomes imbedded in the learning, the literature, and the liturgy of modern policing, community policing should be seen as being here to stay. While debates will certainly continue to arise over specific definitions, and the merits of one application over another, the central tenets of this philosophy, as reviewed throughout this textbook, will continue to hold the interest and attention of police leaders, policymakers, academics, and communities. Its success as a "new" way of looking at the policing enterprise has entered the larger public consciousness, where it will remain as a "theory in use." The idea that police organizations cannot, and should not, hold a monopoly on providing safety, security, and order has become part of the accepted wisdom of community policing. The real (and implied) exclusivity of earlier models of policing (i.e., the political, reform, or professional models), in which a distance between the police and their public was seen as essential for those models to function properly, has given way to an acceptance (at least in theory) that policing in Canada must become increasingly *inclusive* if it is to be effective in meeting real social needs. This transformation is, of course, easier to speak about than to implement. The years ahead will reveal the extent of true commitment that exists to the philosophical underpinnings of community policing, and will draw upon the creativity of police leaders, public figures, private sector agencies, scholars, and communities.

As one of the more fully articulated forms of community policing, problem-oriented policing (Goldstein, 1990) holds much promise for the future and is staunchly defended by many of those who belong to what may be called a new breed of police practitioner/scholars. These are police officers who have often performed the fullest range of operational, investigative, and administrative duties in their departments, and have risen to significant levels within their respective organizations. Furthermore, they have acquired substantial formal education (including graduate and post-graduate degrees) which has provided them with a strong, rigorous theoretical basis for evaluating and articulating the future of policing, consistent with the principles covered throughout this textbook. As noted in the discussion on continuous learning addressed in Chapter Seven, these police practitioner/scholars have contributed substantially to the acceptance, adaptation, and evolution of community policing in Canada. As Kenney (1999) observes:

> This movement known widely as problem-oriented policing (POP), is now the dominant form of police innovation, perhaps the most popular topic today among practitioners and students of policing alike. (p. 1)

The genesis of problem-oriented policing, through the formative efforts of Herman Goldstein, has allowed the more generic term "community policing" to be set aside as required in order to avoid confusion, or guilt-by-association, when applied to superficial programs that have adopted this label. Problem-oriented policing is being pursued with an aggressive vigour, particularly in the U.S., and is now being applied to the issues of youth gangs, crimes in the schools, and the management of calls for police service (Shelley and Grant, 1998). As these applications are refined, reviewed, and evaluated, we will begin to see how successful they are in striking at the root causes of crime and disorder. Problem-oriented policing is a method that appears to have substantial intellectual, financial, professional, political, and public support.

However, it must be recognized that nothing significant happens in policing in Canada without repercussions. An assumption was frequently made that by adopting notions of empowerment, problem-solving, and partnership, police departments would be able to release the hidden potential of their front-line officers, allowing them to go out and "do" community policing. Even in those conscientious departments in which training, education, support, and processes were in place to guide this transformation, success was not always guaranteed. Based on years of careful observation and considerable experience, Lunney (1999) notes that the front-line officer plays a role in the problem-solving approach, but more is required for the success of this approach:

> Frontline officers' participation in the problem-solving paradigm is an absolute necessity. They are the department's eyes and ears, uniquely positioned to help in the scanning process and to contribute to the analyses, response and assessment. But only the team leader is positioned to provide the judgment, consistency and drive that ensures consistent success . . .
>
> Team leaders have all the elements lacking at the first level:
>
> A developed sense of policing philosophy;
> An understanding of the departmental mission and goals;
> Control over resources, including time and their team's collective efforts;
> Clear, legitimate authority to commit the department to a course of action within the considerable limits of their resources;
> Power to take reasonable risks;
> Mature judgement and superior experience in managing people; and
> Knowledge of community resources. (p. 6)

For substantive results, police departments will have to work diligently at developing and deploying effective team leaders imbued with the capacities identified by Lunney, above. Several police organizations in Canada have already begun to tackle the difficult task of refining team leadership and team development (notably, Calgary Police Service, the Ontario Provincial Police, Halton Regional Police Service, among others). If problem-solving, as a unique approach to policing, is going to enjoy a secure future, it is essential that it become the *modus operandi* for the **entire** organization. McPherson (1999) points out that many police organizations have limited its application to front-line operations, thereby undermining its overall credibility as a meaningful way of doing business:

> Problem solving is fine as long as it's relegated to the patrol officer or a specialized team. Just don't ask the entire organization to change the way it does business. We may have to tweak a few systems here and there to support the officers' efforts, but the rest of us investigators, support staff, managers, administrators can concentrate on getting the real work done. (p. 8)

This theme of avoidance is relevant when considering the future of community policing in Canada. When bolted onto the larger organization, community policing may be seen as a high profile and pleasing element for police executives to point to and publicize. However, it has been established that for community policing to work it must permeate the entire design of the organization. The principles of community policing cannot simply be assigned to a program, or to a division within the department. If community policing is to have any longevity it must be adopted as an integral part of the culture of the organization.

If police executives and managers have frequently escaped the rigors of problem-solving in the reordering of their departments, police detectives have often been conspicuously absent when the community policing transformations were taking place. Wycoff (1999) has reported briefly on her study of U.S. detectives, conducted through the Police Executive Research Forum on behalf of the National Institute of Justice, in which the following areas were considered:

> How community policing agencies are structuring the investigation function;
> How are they integrating the investigative function with other police services? and
> How they have managed/are managing the change process with this function.

Wycoff (1999) has noted the special skills brought to bear by detectives in avoiding the requirements of community policing, not to mention other organizational innovations:

> Detectives usually are a highly organized work group, and they are commonly perceived as conservative, insular and elitist. Administrators who try to change investigators' roles may encounter substantial resistance, no matter how minor or practical the change. (p. 9)

A word of caution is warranted when community policing is used as a kind of Trojan horse to introduce programs that revert to a more aggressive enforcement style. This is not to suggest that such high enforcement tactics are never warranted (Kelling and Coles, 1996), only that they are not part of a community policing approach to policing. As noted by Welter (1999):

> Some agencies are applying traditional reactive policing methods by using so-called community policing—bicycle teams, tactical teams, walking teams and other specialized units. The teams engage in community policing by asking residents for information about enforcement priorities,

> but do little to involve them in prevention or problem-solving tactics. Many problem locations need heavy enforcement on the front end in order for any intervention to have a chance for sustained success. But to call these units community policing teams without practicing a scientific problem-solving method is a disservice to the COPPS [community-oriented policing problem-solving] effort. (pp. 10–11)

This leads to a more general observation about the need for profound change, *transformative* change, in the way policing is conducted. There are a range of opinions regarding the depth and degree of the change needed in Canadian policing. However, it is submitted that police personnel, and the organizations they occupy, must be prepared for substantive modifications, particularly if they have previously pursued a reactive model of policing. There may be some solace in the fact that these types of changes are impacting on every kind of organization. Drucker (1988) has sage advice for those facing radical corporate change:

> They will have to change old habits and acquire new ones. And the more successful a company has been, the more difficult and painful this process is apt to be. It will threaten the jobs, status, and opportunities of a good many people in the organization, especially the long-serving, middle-aged people in middle management who tend to be the least mobile and to feel most secure in their work, their positions, their relationship, and their behavior. (p. 50)

What further complicates the process of change for police leaders who seek to move their organizations toward the philosophy and principles of community policing, or even to simply restructure and reorganize for greater efficiency and effectiveness, is the long-standing stability of police organizations as public institutions. This is true in Canada and the U.S. and until very recently, with the growing number of municipal and regional amalgamations, has always obtained. In the private sector, it is possible to create a sense of urgency around a major change project. Change is required to stay competitive. Change is required to ensure the company's continued existence. Change is required in order to bring a new product to market in record time. These arguments have some persuasive power in the private sector. The public police typically hold a monopoly on the services they provide in their jurisdiction. As Fraser et al. (1998) suggest in their summary of a major process mapping project being funded by Motorola in the U.S. and Great Britain:

> How can organizational improvements be characterized as urgent in organizations whose personnel routinely deal with life-or-death situations? In such circumstances, an administratively defined crisis is not likely to resonate. (p. 40)

Change is profoundly difficult for individuals. Therefore, organizations must move carefully through the change process if they are going to achieve positive results. Community policing, in spite of all the accolades it has received to date, continues to represent massive change for many police officers and civilian employees, and should be managed accordingly. Again, although speaking specifically about process reengineering (which, incidentally, was the "methodology" applied by the OPP during its organizational review project to assist with the "institutionalization" of community policing), Fraser et al. (1998) are instructive:

> Perhaps the most profound causes of failure in process reengineering relate to the mood and morale of the organization's employees. The prospect of substantial change itself is threatening

to many managers and employees because they cannot predict how these changes might ultimately effect their roles, authority or status in the organization. In organizations with long histories of distrust between line employees and management, process reengineering is often received skeptically, because employees fear that the ultimate goal of management is merely to lay off workers. (p.52)

BROADENING THE BAND—ERASING THE THIN BLUE LINE

Community policing will not survive as a meaningful approach to policing unless and until there is a genuine engagement of the "community." Although it is not possible in this space to offer a precise description of how a police organization should go about defining its communities, there has been sufficient research and reporting on community policing initiatives to support the assertion that the public police have not moved far enough along the continuum of community partnership to really consolidate the expectations of the philosophy of community policing that seeks the "co-production" of order.

Critics of the public police hasten to offer the view that private policing has been more successful at linking with their various communities (Stenning, 1989). Critics of the public police in "liberal democracies" (like Canada) have spoken forcefully about the weakened underpinnings of state-sponsored security generally (Shearing, 1995). Clearly, the promise of community policing will not be kept if Canadian police services fail to transform themselves in ways that lead to some form of co-production of the social order. The evidence is overwhelming in this regard. The approaches of the early models of police are by no means utterly bankrupt, and it would be misguided to dispense with the operational, investigative, tactical, and administrative wisdom that has accrued over time. However, the research has revealed, with a fairly high degree of certainty, that contemporary police organizations do not possess the capacity to provide adequate policing for all the safety and security concerns of our modern society. The problems are systemic, and the "system" being referred to is society at large. In discussing the concept of restorative justice, which parallels the philosophy of community policing, the Nova Scotia Department of Justice (1998) observes that:

> Restoring public confidence, reducing recidivism, and increasing public and victim satisfaction will take enormous effort over many years. It will require a rethinking and perhaps a retooling of not only our justice system, but of our educational, health and social services systems. It will require a comprehensive, multi-disciplinary, multi-departmental, and even multi-governmental strategy to prevent crime. It will require smarter, more effective ways of making our communities safer. (p. 2)

Increasingly, there are articulate voices counting down on the public police. The presence of growing problems with crime and disorder make these voices relevant. Police departments continue to struggle when confronted with specific events that highlight their failure to fully and fundamentally adhere to community policing principles. The following examples represent but a few of the major flashpoints that have seriously threatened public trust in the public police:

- The beating of Rodney King by patrol officers with the Los Angeles Police Department;
- The pepper spraying of demonstrators at the OPEC conference in Vancouver, British Columbia, by RCMP officers;
- The findings of former Chief Justice Lawrence Poitras in his extensive review of the Sûreté du Québec, which revealed an insular fraternity of officers that protect their own at all costs, including breaking the law;
- The inquest into the shooting death of schizophrenic Edmond Yu, by officers with the Toronto Police Service, with several recommendations, including one that all 5000 Toronto officers be given substantial crisis-resolution training;
- The findings and recommendations of Sir William Macpherson's commission examining the failures of the police investigation into the death of Black teenager, Stephen Lawrence, who was beaten by a gang of White youths outside of London, England;
- The findings and recommendations of the review of the Bernardo investigation by Mr. Justice Archie Campbell;
- The findings and recommendations of the review of the investigation of Guy Paul Morin, by Mr. Justice Fred Kaufman.

These failures by the public police (in Canada, and elsewhere) which are often precipitated by the egregious behaviour of only a few officers, threaten the *status quo* of policing in ways that may, in fact, be beneficial. If these events accelerate the movement toward community policing, rightly understood, then there is a benefit for the public and for the enterprise of policing. If these occurrences, and the reaction they generate in the public's mind, produce meaningful change in policing, then they will have contributed to real advancement. Of course, proactive police leaders are well aware of the threat posed by an erosion of public confidence in the public police. These leaders are part of the vanguard in Canadian policing moving along the continuum of change toward community policing, and have been highlighted throughout this text, and their organizations featured in Chapter Eleven.

THE RCMP

On a national scale, the Royal Canadian Mounted Police (RCMP) has a monumental task in remaking itself to reflect both its traditional roots as well as its adherence to the principles of community policing. With divisions in every province, serving as the provincial police force in every province (except Ontario and Quebec), and under contract with many jurisdictions as the provider of municipal police services, the RCMP has a scope of operation that poses an enormous challenge in terms of change management. However, this organization has made an effort to reconfigure its mission, vision, core values, and commitment to its communities in ways that are indicative of a high-order adherence to quality policing. In fact, during a five-week period in 1996, 4300 RCMP employees across Canada addressed issues revolving around how they wanted the organization to be run. This is reflected in the RCMP "H" Division (1998) document on shared leadership:

Mission—
The R.C.M.P. is Canada's national police service. Proud of our traditions and confident in meeting future challenges, we commit to preserve the peace, uphold the law and provide quality service in partnership with our communities.

Vision—
We will:

- Be a progressive, proactive and innovative organization;
- Provide the highest quality service through dynamic leadership, education and technology in partnership with the diverse communities we serve;
- Be accountable and efficient through shared decision-making;
- Ensure a healthy work environment that encourages team building, open communication and mutual respect;
- Promote safe communities;
- Demonstrate leadership in the pursuit of excellence.

Core Values of the RCMP—
Recognizing the dedication of all employees, we will create and maintain an environment of individual safety, well-being and development. We are guided by:

- Integrity
- Honesty
- Professionalism
- Compassion
- Respect
- Accountability

Commitment to our Communities—
The employees of the RCMP are committed to our communities through:

- Unbiased and respectful treatment of all people
- Accountability
- Mutual problem solving
- Cultural sensitivity
- Enhancement of public safety
- Partnerships and consultation
- Open and honest communication
- Effective and efficient use of resources
- Quality and timely services

Indeed, the "H" Division (1998) command staff realized that their continued existence could be threatened if meaningful changes were not undertaken:

> It was clear that we had to change our ways if we wanted to stay in business as Nova Scotia's police force of choice. The times were changing and we had to change with them. The only way to do it effectively was to set concrete goals, empower our employees and continuously monitor our progress. "H" Division embraced a proactive approach to the philosophy of Shared Leadership, resulting in a recognized leadership role within the Force. (p. 2)

RESTORATIVE JUSTICE

The reactive mode of policing can no longer subsist without a balancing effort in the realm of proactive policing that leads to the transactive approaches outlined in Chapter One of this text. Equally, the adversarial model of justice cannot subsist without the balancing, and more benevolent, effects of a restorative approach (Zehr, 1990). Again, the Nova Scotia Department of Justice (1998) summarizes some of the key distinctions between these two modes:

Adversarial System—

- Crime defined as a violation of rules, and a harm to the state
- Victim is inhibited from speaking about his/her real losses and needs
- Offender, victim and community remain passive and have little responsibility for a resolution
- Community role is limited
- Restitution is rare
- Controlled and operated by the state
- Offender is blamed, stigmatized and punished
- Repentance and forgiveness are rarely considered
- Assumes win-loss outcome

Restorative Justice—

- Crime is seen as a harm done to victims and community
- Victim is central to the process of defining the harm and how it might be repaired
- Offender, victim and community are active and participate in the resolution resulting from the restorative forum
- Community is actively involved in holding offenders accountable, supporting victims, and ensuring opportunities for offenders to make amends
- Restitution is normal
- Overseen by the state, but usually driven by communities
- The long-term protection of the public mandates a focus on the methods of problem solving that include the reintegration of the offender into the community, and the preservation of his/her dignity
- Repentance and forgiveness are encouraged
- Makes possible win-win outcome

PREVENTING CRIME—PROMOTING COMMUNITIES

The focus on crime prevention, as noted in Chapter Two, has been in existence for a considerable period of time and continues to offer opportunities for promoting safer communities. Crime prevention continues to be consistent with the overall goals of community policing, and operates in ways that allow the public to participate in the co-production of order. In reviewing the development of crime prevention, the 1967 U.S. President's Commission on Law Enforcement and the Administration of Justice continues to serve as a watershed for valuable ideas that have helped to shape the future of policing. Moore (1997) has re-examined the Commission's report, entitled *The challenge of crime in a free society*, and draws attention to some important elements of this study:

> It is significant, I think, that the first chapter of the Crime Commission's report was about prevention. But what I think is quite interesting is that we have a very broad juxtaposition of different systems of prevention. According to the prevention theory associated with criminal justice operations, we can prevent crime (if not the first, at least many subsequent crimes) through deterrence, incapacitation and rehabilitation. We might even prevent an offender's first offense if general deterrence works.
>
> At the other extreme is a social justice theory of prevention. It holds that if we extended great opportunities to both children and adults to live profitably and well and if we lived up to the promise of equality of opportunity, fewer people would wish to become offenders, would choose to become offenders, or would feel motivated to become offenders. So we have criminal justice prevention on one hand (reactive, with extensive use of state authority) and the social justice theory of prevention on the other (anticipatory, hopeful, accepting of our broader responsibilities to society at large, holding open the possibility of realizing a society we'd all like to be part of). These are the two ideas of prevention presented in the Commission report. (p. 28)

Both kinds of prevention, the criminal justice theory and the social justice theory, have a legitimate role to play in the transformation to community policing principles and practices in Canadian policing. The concept of doing whatever is necessary in working toward stronger, safer, and more secure communities is at work in this thinking. The magnitude of this undertaking is such that police organizations cannot hope to be the only, nor necessarily the key, players in realizing this goal. Therefore, ways must be found to ensure that communities of interest can function effectively in making the goal of safer communities possible. The City of Vancouver, British Columbia has been engaged in some interesting efforts that move in this direction.

NEIGHBOURHOOD INTEGRATED SERVICE TEAMS AND THE MAYOR'S URBAN SAFETY COMMISSION (VANCOUVER, B.C.)

In 1996, the City of Vancouver introduced a program of Neighbourhood Integrated Service Teams (NIST) that were made up of city staff as well as school and health officials. The teams members would collaborate across departmental and agency boundaries in order to help communities solve local problems that required an integrated approach. Initially, fifteen teams were established to serve the entire city. City staff team members were drawn from the following departments: police, fire & rescue services, engineering, social planning, permits & licences, library, parks, and planning. The teams would function together in order to deal with issues such as garbage, noise complaints, illegal activities, traffic issues, and problem premises. By operating as a team, these matters that frequently bogged down in bureaucratic delay could be handled decisively and directly because the NIST members could bring their departmental expertise to bear on the problem immediately and were empowered to collaborate with other team members. The principles that were formulated to guide the work of the NIST are included in Figure 12–1 below, and are certainly pertinent to any engagement of the community.

FIGURE 12–1

INTEGRATED SERVICE DELIVERY PRINCIPLES

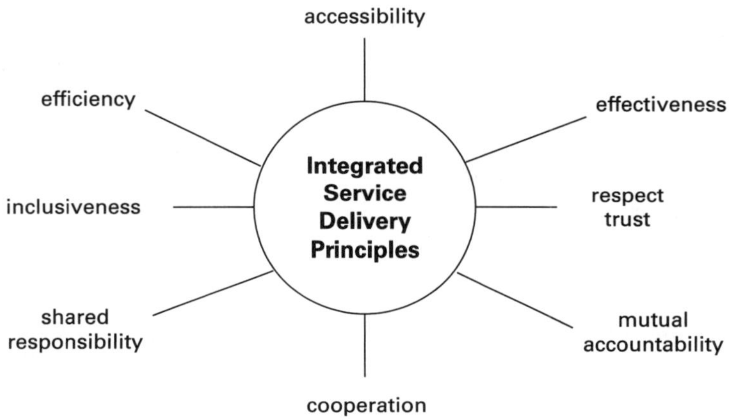

Source: City of Vancouver (1995). *Integrated service teams: orientation session, October 1995*. Used with permission.

Another initiative that reveals the commitment of the City of Vancouver to meaningful activity consistent with crime prevention and community policing, is the creation of the Mayor's Urban Safety Commission. In 1996, Mayor Owen introduced a proposal to create an Urban Safety Commission with the following mandate:

- To identify key issues affecting the safety and quality of life of Vancouver residents and visitors;
- To identify limitations in present laws and processes that restrict the city's ability to address these issues;
- To make legislators and government agencies aware of these issues, and the impact of legislation and government operations on our ability to manage our community;
- To propose and lobby for changes in legislation and government programs.

The Commission is intended to consult widely with experts, including police officers, lawyers, business people, and sociologists in order to accomplish their mandate. Figure 12–2, below, shows the extent to which the Urban Safety Commission has been encouraged to go in creating significant linkages within the wider community.

FIGURE 12–2

CITY OF VANCOUVER

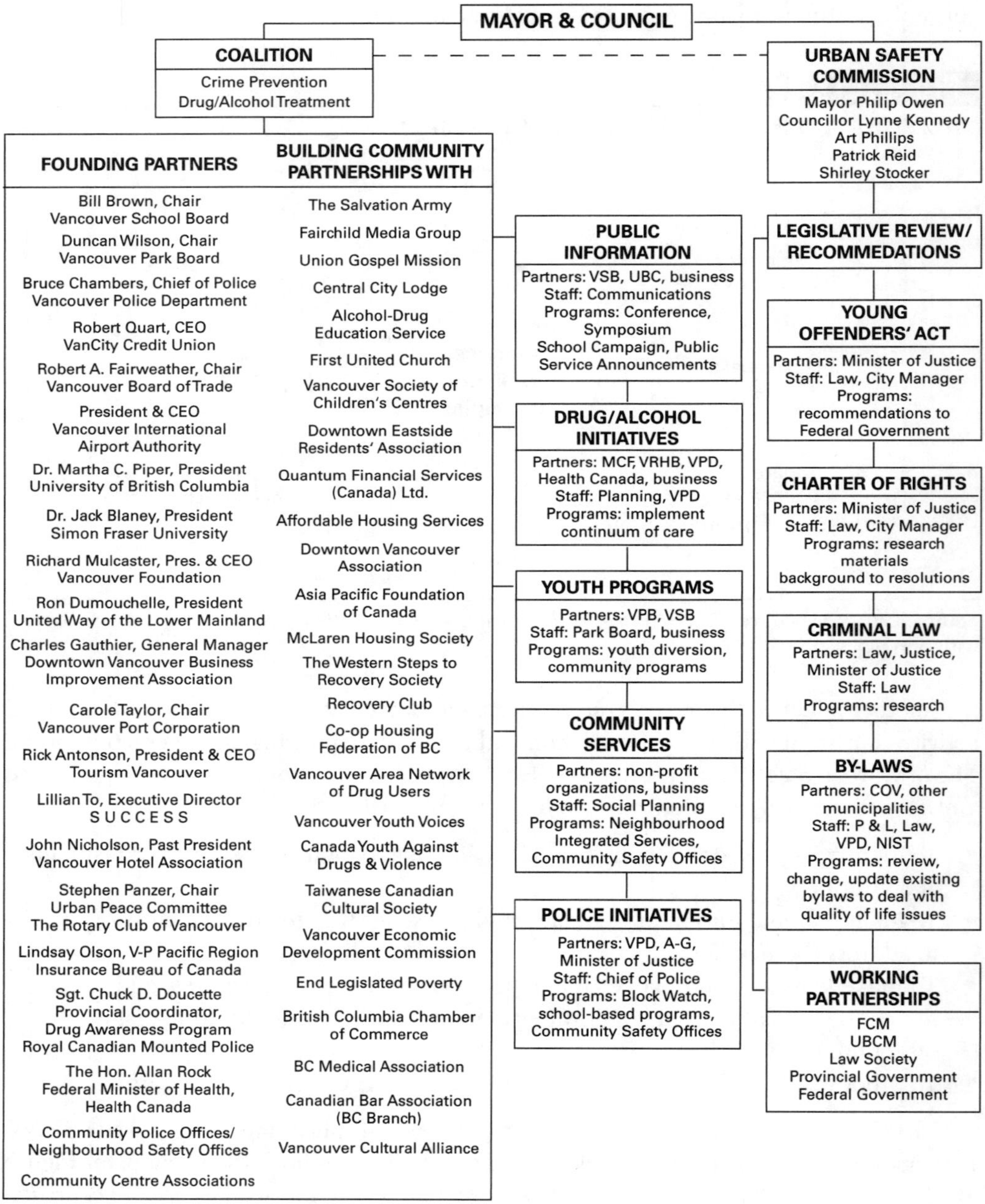

Source: City of Vancouver (1995). Used with permission.

In acknowledging the work already accomplished in Vancouver, Philips (1996) lists current city council, police department, and Safer Cities initiatives:

- Implementing a community-based policing strategy;
- Increasing the annual Vancouver Police budget 33 per cent since 1990—from $75 million to over $100 million per year;
- Intensifying enforcement against problem premises;
- Passing new pawn shop legislation to limit the number of pawn shops and the fencing of stolen goods;
- Facilitating the safe ride program to streamline a support system for addicted people that eliminates duplication of services by police, fire, and ambulance;
- Introducing guidelines for crime prevention through environmental design that are applicable to new commercial and multi-family developments;
- Establishing neighbourhood integrated service teams to provide a broad range of city services to citizens in their communities and neighbourhoods; and
- Establishing a program to reduce graffiti.

In October 1997, the Urban Safety Commission launched Vancouver's Coalition for Crime Prevention and Drug Treatment. The Mayor, the Chief Constable, and the Urban Safety Commission all pledged to work together in fighting drug abuse and property crime in the city. Announcements included:

- Investment by the city to assemble land for lease to an agency for drug treatment programs, detox, and drug and alcohol programs;
- A commitment by the Chief Constable, through restructuring and reallocation, to put 75–100 more police officers on the street;
- A joint youth education program aimed at drug awareness, supported by the Vancouver School Board, Vancouver Police Department, and Royal Canadian Mounted Police; and
- A continuing investment by the Vancouver Board of Parks & Recreation for youth programs.

MOVING TO BLEND PUBLIC AND PRIVATE POLICING

To fully appreciate the direction of community policing in Canada, it is helpful to look elsewhere for examples where it has been applied. By looking at a broader context, it is possible to see that certain assumptions are in place when we assign responsibility for the design and delivery of community policing to our public police agencies. It has been suggested earlier that the public police are not as adept at meshing with their communities as private police organizations (Stenning, 1989). It has been further suggested that the assumption that the state should have a monopoly on public safety and security is no longer compelling (Shearing, 1995). Indeed, Shearing (1995), as a result of his work in (post-Apartheid) South Africa, formulated an aim that may be contemplated in light of community policing in Canada:

> Our aim [in South Africa] has been to explore the possibility of creating a form of "community policing" that shifts both the rowing and much of the steering of policing to communities. To do this we are developing a conception of community policing for poor people that takes its lead from private policing, not state policing. In doing so, we are directly challenging the forms of community policing that seek to institutionalize "rule at a distance" strategies that have been the hallmarks of most community policing initiatives in established democracies that have had poor people as their focus. (p. 34)

If police organizations in Canada wish to function in the best interests of their communities, the thrust of Shearing's commentary should be taken seriously. Certainly, an effective system of community policing that is designed for the benefit of "poor people" would be suitable for all people, if it is theoretically sound. Much has been written regarding the continuing sense that the police are still in the "driver's seat" when community policing is being done, and the questions discussed in Chapter Nine frequently focus on this reality. Shearing's South African experience is enlightening because it has taken place in a context in which there was an opportunity to build a new foundation for policing. The new regime permitted new modes and orders to be introduced that would permit *transformative* change to take place. As Shearing (1995) indicates:

> In challenging the established vision of community policing that police in established democracies are promoting and that they are exporting to emerging democracies, we are seeking to provide an alternative to the neoconservative agenda within policing and to use this as the basis for building a conceptual and normative platform for promoting forms of neoliberal governance. (p. 34)

While it is far from certain that Shearing's efforts at "reinventing" policing will succeed in giving Black South Africans greater security, they will offer an instructive counterpoint to the approach to the community policing now being pursued in North America, under the guidance of a democratic governance that *does* believe in state control.

When police undertake community policing initiatives, they routinely omit any substantial linkages with the private sector. While this may be based on longstanding preconceptions regarding private policing, it is also certainly based, in part, on a desire for organizational self-preservation. According to Brodeur (1998):

> The neglect of private policing agencies in discussions of community and problem-oriented policing is both striking and relatively surprising. Although the development of partnerships between the police and other agencies is strongly advocated in such discussions, community and problem-oriented policing is considered almost entirely in the context of public policing. "The community" is very loosely conceived as the general public, in which only nonprivate groups such as ethnic minorities, elderly citizens, and sections of the underclass are considered as distinct. In the same way "interagency cooperation" usually means coordinating the actions of different public departments and rarely addresses the question of integrating parts of the public and the private sectors. (p. 222)

The challenge offered by many academic observers of community policing is framed around the need for the public police to clearly demonstrate that they are capable of working effectively at promoting local problem-solving, crime prevention, and other similar initiatives in ways that give communities a meaningful role to play in the process. It is no longer feasible for the public police to provide all of the myriad services expected of the community. It is certainly clear that the public police have skills, knowledge, and abilities that are essential for the advancement of safety and security. However, they are not the only possessors of these attributes. As argued by Eggers and O'Leary (1995):

> Until neighborhoods are safe again, they will not thrive economically or socially. Waiting for the government to make it all better is a losing strategy. People have to become more involved in ensuring their own security.
>
> There is a great deal that government can and should do to improve public safety, but first it must recognize that it needs help. Restoring public safety demands a renewed partnership between the police and the community. Police must reacquaint themselves with the people in the communities they serve, and communities must recognize that the brunt of the task of policing a free society does not lie with the police, but with citizens themselves.

The ideological considerations that have been raised by those who critique the liberal democratic basis of public policing in Canada are complex and well beyond the scope of this publication. However, the most radical theorists, when seeking to utterly transform the existing justice system, feel compelled to acknowledge the value of highly-skilled police officers in a future system. For example, Bianchi (1994) is quite clear when he informs his reader that:

> We will most certainly still need police, but their tasks may be far more satisfying. Under the new assensus system of control, the police not only will conduct investigations but also will assume the important task of facilitating dispute settlement. They will need to refer both plaintiff and defendant to dispute settlement localities such as neighborhood justice centers, arbitration committees and the like.
>
> We should not forget that in the not-so-distant past, constables set up that sort of dispute settlement in neighborhoods. Its was a time-honored police task before the rise of the modern over-technological society. Even today, in the Netherlands, young police recruits are told during training that dispute settlement in neighborhoods is one of their most important tasks. (p. 106)

With the world of policing being impacted by so many outside forces, it is little wonder that there has been much thought given to the need for a model that would allow the individuals who comprise police departments to operate more effectively. The frustrating lack of success of the reactive, crime control model of policing was sufficient to drive police leaders to seek out approaches that would allow their organizations to share the burden of responsibility with the communities that hold the police accountable for their actions and their policies. Community policing is the umbrella term for the approaches that have resulted from that search. Combining problem-solving, partnerships, organizational redesign, and crime prevention has taken many police departments to a point where they are beginning to see some positive impacts on crime and disorder. Governments at all levels are lending their support to the preventative mindset, and this has resulted in some significant advances in dealing with the root causes of crime and disorder, and is seen to be effective in early evaluations. Much of the effort on the prevention front is aimed at children and youth, for several reasons. This is where the preventative effort may pay the largest dividends by diverting individual youngsters from any involvement with the justice system. This is where the restorative justice approach may take a benevolent hand in guiding young adults toward a healthier life. The attention of police organizations has been drawn to children and youth in many ways through community policing and crime prevention efforts. This is particularly true at the present moment in Canadian policing history.

CHIEFS OF POLICE AND THE FUTURE

As the world of policing in Canada changes, it has been essential to find appropriate individuals to lead police organizations through frequently turbulent times. As the complexity of policing has evolved to include significant matters of law, social order, politics, jurisdiction, technology, human relations, finance, and accountability, the requirements for police executives have been transformed accordingly. As Figure 12-3 and 12-4 reveal, the qualities and qualifications of those who will lead Canadian police services into the next millennium often include specific references to community policing, and explicitly acknowledge the importance of meeting community needs.

FIGURE 12-3

Commissioner

Police Services Leadership

The Ontario Provincial Police (OPP) seeks a future-oriented leader of proven integrity who is able to lead a large policing organization in times of significant change and can plan for and accomplish long range goals of a major policing organization, while working with other organizations to identify and respond to public safety issues of provincial interest.

The Mission of the OPP is to provide a community-oriented service that promotes an environment in Ontario in which all people will be secure in their pursuit and enjoyment of all lawful activities.

The Commissioner maintains effective relationships within the OPP, with other police and public safety organizations, the community and the media, and requires knowledge of key policing issues.

The successful applicant will foster a community service orientation which entails developing strategic relationships and partnerships with the community, based on in-depth knowledge and understanding of the community's policing needs.

The Ontario Provincial Police provides generalist policing services to over 2.3 million people (3.6 million in summer) throughout the Province of Ontario. The OPP area of responsibility encompasses approximately 993,000 square kilometres of land and 174,000 square kilometres of Ontario waterways.

Qualifications:

- Demonstrated experience leading a large organization in times of change and communicating a compelling vision;
- A proven track record of meeting diverse community needs and establishing a position of trust;
- Highly strategic orientation to future directions of policing and impact on/opportunities for OPP;
- Exceptional judgment, decision making, communication skills and ability to maintain stamina and professionalism in demanding situations;
- Successful experience in persuading and influencing others to support the organization's goals in a complex environment;
- Proven track record of comparing performance against standards and holding people accountable for results;
- Knowledge of the Ministry and Government;
- Excellent knowledge of policing issues in Ontario;
- Experience in the justice system, such as police management, public safety or related areas.

Applicants must be prepared to be the subject of extensive investigations and assessment.

This is a three (3) year Order in Council appointment, with the possibility of extension.

The Commissioner is a Senior Manager Group 3.

Ontario

The Province of Ontario is an equal opportunity employer.

Please forward your résumé in strict confidence by February 6, 1998 to Doug Colling at:

KPMG Executive Search, P.O. Box 31, Stn. Commerce Court, Toronto, Ontario M5L 1B2, FAX (416) 777-8070, quoting File #284-98.

Source: *The Globe and Mail*, Friday, January 23, 1998. Used with permission.

FIGURE 12-4

Chief of Police

Realize the Vision through Leadership

The Regional Municipality of Durham, which includes the Towns of Ajax, Pickering, Whitby, the Townships of Scugog, Uxbridge and Brock, the Municipality of Clarington and the City of Oshawa, with a combined population of over 474,000, is seeking a seasoned professional, forthright and principaled leader as the new Chief of Police.

Due to a pending retirement, a new Chief is required to provide overall leadership and management to the Durham Region Police Services which has an authorized strength of 759.

As the new Chief, you will report to the members of the Police Services Board. With detailed knowledge of all aspects of municipal policing, outstanding administrative and technical skills combined with a vision for the future, you will bring an understanding of the pending changes that will impact policing throughout Ontario.

You are comfortable in dealing with elected officials, uniformed and civilian members of the police service at all levels and responding to the needs of the residents and businesses within the Region.

You understand the implications of fiscal constraint, have provided innovative alternatives to meet taxpayer demands, and recognize the need to establish new partnerships for the alternative delivery of police services.

You've developed a broad base of knowledge in the delivery of modern and effective municipal policing; have a track record of successfully delivering high levels of customer service, developing public policy and implementing Community Based Policing Programs.

You are a creative individual. You will enhance the vision and strategic direction through the Board and with your senior officers, build a winning team and create a challenging environment for all members of the police service.

KPMG
Executive Search

Interested candidates are asked to apply, in confidence to:

Doug Colling, KPMG Executive Search, P.O. Box 31, Stn. Commerce Court, Toronto, Ontario M5L 1B2, FAX (416) 777-8070 or e-mail executivesearch@kpmg.ca, quoting File #09135.

Source: *The Globe and Mail*, Monday, June 23, 1997. Reproduced with the permission of the Durham Regional Police Services Board.

In August 1999, the annual conference of the Canadian Association of Chiefs of Police (CACP) focused its attention on the theme of "Youth Crime—A Canadian Strategy." In addressing the growing concerns over youth crime and disorder, the CACP identified several topics that have an impact on Canadian communities, schools, and urban areas. The CACP recognized that youth crime constitutes a complex challenge for police leaders, and places an onus on them to find appropriate approaches to prevent crime and to work toward safer communities. The planned introduction by the federal Justice Minister of a new youth criminal justice statute, to replace the existing *Young Offenders Act*, made this a timely focus for the CACP. Among the topics discussed at this important session were:

- Current trends in youth crime and police charging practices;
- Demographics and its impact on future crime trends of youth;
- New youth gangs and effective prevention strategies;
- Overview of the proposed youth justice statute, with an emphasis on prevention and alternatives to formal court process, and changes to the transfer process to adult court;
- Examples of effective responses to youth crime such as: restorative justice, including family-group conferencing; programming for children under 12 years of age who commit offences; Aboriginal justice; keeping schools safe; street level prevention initiatives; and

- Implications of the proposed legislative changes for police executives and their communities. Developing funding and training needs and sources.

SUMMARY

Looking to the future, it is unlikely that community policing, even in its promising, problem-oriented form, will dissolve crime in Canada. It would be simplistic to presume that any style of policing would be sufficient in addressing problems that have their roots in areas far beyond the immediate reach of police organizations. However, it is becoming increasingly apparent that police leaders, at all levels, are turning their serious attention to different ways of delivering their services. They are looking to policy-makers, scholars, communities, and to each other for ways that may contribute to stronger neighbourhoods, more secure jurisdictions, and an overall healthier quality of life.

Under the banner of community policing, a great deal has been accomplished in reshaping the vision of policing in Canada. The work that remains is not only complex and controversial, it is critical to the stability of our society.

QUESTIONS FOR CONSIDERATION

1. In what ways could high enforcement tactics undermine a community policing philosophy?
2. How is it possible to strike a balance between reactive and proactive policing initiatives?
3. How could police leaders create a sense of urgency around the need to move toward a community policing model?
4. What are some areas that should be addressed **before** a police department undertakes major organizational change?
5. What are the differences between the social justice theory and the criminal justice theory of prevention identified in the 1967 U.S. President's Commission on Crime?
6. How can police organizations in Canada help individual employees overcome their fear of change?
7. How does the integrated service delivery model introduced in Vancouver, British Columbia contribute to community policing?
8. What are the advantages of the restorative justice approach? What are its disadvantages?
9. How can problem-solving, community policing, or crime prevention be applied in the detective function of a police department?
10. Are private police organizations more successful at linking with their communities? If yes, why are they are more successful?
11. Why have police organizations and crime prevention agencies focused so much attention on youth crime and disorder?

REFERENCES

Alpert, G., D. Kenney, and T. Oettmeier (1998). Facilitating organizational change: shaping philosophies through individual and organizational evaluations. Washington, D.C.: National Institute of Justice.

Anderson, T. et al. (1999). Every officer is a leader: leadership transformation in police and criminal justice. Boca Raton, Fl.: CRC Press.

Bianchi, Herman (1994). Justice as sanctuary: toward a new system of crime control. Bloomington, Ind.: Indiana University Press.

Brodeur, Jean-Paul (1998). "The assessment of police performance: conclusions." In Brodeur, Jean-Paul (ed.). How to recognize good policing: problems and issues. Thousand Oaks, Calif.: Sage Publications and Police Executive Research Forum.

Conly, Catherine H. (1993). Street gangs: current knowledge and strategies. Washington, D.C.: Department of Justice, National Institute of Justice.

Diamond, D. (1993). Community policing and regional training for small and rural jurisdictions. Washington, D.C.: Police Executive Research Forum.

Drucker, Peter F. (1988). "The coming of the new organization." Harvard Business Review, Vol. 66, no.1 (January-February), pp. 45–53.

Eck, John E. and Nancy G. La Vigne (1993). A police guide to supplying citizens and their environment. Washington, D.C.: U.S. Department of Justice, Office of Justice Programs, Bureau of Justice Assistance.

Eggers, William D. and John O'Leary (1995). "The beat generation: community policing at its best." Policy Review, Vol. 4, no. 74 (Fall).

Fraser, Craig B. et al. (1998). Challenge to change: the 21st century policing project. Washington, D.C.: Police Executive Research Forum.

Kelling, George L. and Catherine M. Coles (1996). Fixing broken windows: restoring order and reducing crime in our communities. New York: The Free Press.

Kenney, Dennis J. (1999). "Policing as a scientific method: increasing the focus on community problem solving." Subject to Debate, Vol. 13, no. 3 (March), pp. 1, 12–15.

Kenney, Dennis J. and T. Steuart Watson (1998). Crime in the schools: using student problem solving to reduce fear and disorder. Washington, D.C.: Police Executive Research Forum.

Kidder, Tracy (1999). "Small-town cop." The Atlantic Monthly, Vol. 283, no. 4 (April), pp. 47–64.

Lunney, Robert F. (1999). "The awful truth about POP." Subject to Debate, Vol. 13, no. 3 (March), pp. 6–7.

Macpherson, Sir William (1999). The Stephen Lawrence inquiry: report of an inquiry by Sir William Macpherson of Cluny advised by Tom Cook, the Right Reverend Dr. John Sentamu, Dr. Richard Stone, presented to Parliament by the Secretary of State for the Home Department by Command of Her Majesty. London: The Stationery Office.

McPherson, Nancy (1999). "The status of problem-oriented policing today and its challenges." Subject to Debate, Vol. 13, no. 3 (March), p. 8.

Moore, Mark (1997). "Looking backward to look forward: the 1967 Crime Commission report in retrospect." NIJ Journal, Issue No. 234 (December), pp. 24–30.

Nova Scotia. Department of Justice (1998). Restorative justice: a program for Nova Scotia. Halifax: Department of Justice.

Owen, Philip (1996). Memorandum from Mayor Philip Owen to members of Council, Mayor's Urban Safety Commission, 29 July 1996. [Vancouver, B.C.: City of Vancouver].

Royal Canadian Mounted Police. "H" Division (1998). Shared leadership in the RCMP "H" division. [Halifax, N.S.: RCMP "H" Division].

San Diego Police Department (1999). Community responsibilities for making neighborhood policing work. San Diego: San Diego Police Department.

Shearing, Clifford (1995). "Toward democratic policing: rethinking strategies of transformation." In U.S. Department of Justice. National Institute of Justice (1997). Policing in emerging democracies: workshop papers and highlights, Washington, D.C., December 14-15, 1995. Washington, D.C.: National Institute of Justice.

Shelley, Tara O'Connor and Anne C. Grant (eds.) (1998). Problem-oriented policing: crime specific problems, critical issues and making POP work. Washington, D.C.: Police Executive Research Forum.

Sherman, L. (1989). "Repeat calls for service: policing the hot spots." In Kenney, D.J. (ed.) Police and policing: contemporary issues. New York: Praeger.

Spergel, Irving A. et al. (1994). Gang suppression and intervention: community models. Washington, D.C.: U.S. Department of Justice, Office of Juvenile Justice and Delinquency Prevention.

Stenning, Philip C. (1989). "Private police and public police: toward a redefinition of the police role." In Loree, Donald J. (ed.) Future issues in policing: symposium proceedings. Ottawa: Canadian Police College.

Stuart, B. (1997). Building community justice partnerships: community peacemaking circles. Ottawa: Department of Justice Canada.

The Circle, Inc. (1991). The future by design: a community framework for preventing alcohol and other drug problems through a systems approach. Rockville, Maryland : U.S. Department of Health and Human Services, Office for Substance Abuse Prevention, Alcohol, Drug Abuse, and Mental Health Administration.

Trojanowicz, Robert and David Carter (1988). The philosophy and role of community policing. East Lansing, Mich.: National Neighborhood Foot Patrol Center, School of Criminal Justice, Michigan State University.

U.S. Department of Justice. Bureau of Justice Assistance (1997). Addressing community gang problems: a model for problem solving. Washington, D.C.: Bureau of Justice Assistance.

U.S. Department of Justice. Bureau of Justice Assistance (1998). Addressing community gang problems: a practical guide. Washington, D.C. : Bureau of Justice Assistance.

U.S. Department of Justice. National Institute of Justice (1997). *Policing in emerging democracies: workshop papers and highlights, Washington, D.C., December 14–15, 1995*. Washington, D.C.: National Institute of Justice.

Weick, K.E. (1984). "Small wins: redefining the scale of social problems." *American Psychologist*, Vol. 39, no. 1, pp. 40–49.

Welter, John (1999). "The transition from traditional policing to problem-oriented policing: not as easy as it looks." *Subject to Debate*, Vol. 13, no. 3 (March), pp. 1, 10–11.

Wycoff, Mary Ann (1999). "Investigations in a community policing context." *Subject to Debate*, Vol. 13, no. 3 (March), p. 9.

Zehr, H. (1990). *Changing lenses: a new focus for crime and justice*. Waterloo: Herald Press.

APPENDIX I

SAMPLE PUBLIC OPINION/ SATISFACTION SURVEYS

Gloucester Police Force
Sûreté de Gloucester

LESTER N. THOMPSON *Chief of Police / Chef de Police*

P.O. BOX (C.P.) 8333
GLOUCESTER, ONTARIO
K1G 3V5
(613) 822-2916

Dear Gloucester Resident:

As Chief of Police, I am constantly aware of the need to provide progressive and effective police services in our community.

With the support of the Board of Commissioners of Police for the City of Gloucester, we have embarked on a Five Year Strategic Plan.

One part of this process is to receive public input. With the assistance of Professor D. Forcese and his staff at Carleton University, a Public Needs Survey has been developed. By random sampling, your household has been selected to receive this survey.

As a service to your community, we would appreciate your completing the questionnaire and returning it without delay in the addressed, pre-paid envelope.

The results of this study will be prepared at Carleton University for the Gloucester Police Force and will be available in the Fall, 1989. We expect, at that time, to share the results with you.

Should you have any questions, please do not hesitate to contact Cst. A. Mackie or Sgt. G. Savage at the Gloucester Police Force (telephone no. 745-7000).

Thank you for your assistance in this matter.

Yours truly,

Lester N. Thompson

Lester N. Thompson
Chief of Police

GLOUCESTER POLICE FORCE PUBLIC NEEDS SURVEY

QUESTIONNAIRE

Your answers are ***confidential and anonymous***, and will be presented only in statistical form, under the direction of researchers at Carleton University. Research results will be made public for your information but no individual can be identified.

INSTRUCTIONS:

Please indicate the most appropriate answer. If questions do not apply, please proceed. Additional written comments are very welcome.

1. How long have you lived in Gloucester? ____________________

2. In which neighborhood or community of Gloucester do you live?

1. Orleans___	2. Blackburn Hamlet___	3. Cyrville___
4. Beacon Hill___	5. Carson Grove___	6. Blossom Park___
7. Windsor Park___	8. Gloucester South___	9. Leitrim___

10. Carlsbad Springs/Ramsayville___
11. Other (identify)____________________________________

3. How serious would you say that the following problems are in the City of Gloucester? (circle the appropriate number)

	don't know	not serious	somewhat serious	very serious
a. stealing from homes	0	1	2	3
b. stealing from stores	0	1	2	3
c. personal assaults	0	1	2	3
d. damage to property	0	1	2	3
e. family disputes	0	1	2	3
f. traffic control	0	1	2	3
g. violent crimes	0	1	2	3
h. drug abuse	0	1	2	3
i. other (please identify) ____________	0	1	2	3

4. What kind of job would you say that the Gloucester Police Force does in your community responding to the following? (circle the appropriate number)

	no opinion	very poor	adequate	very good
a. stealing from homes	0	1	2	3

b. stealing from stores	0	1	2	3
c. personal assaults	0	1	2	3
d. damage to property	0	1	2	3
e. family disputes	0	1	2	3
f. traffic control	0	1	2	3
g. crime prevention	0	1	2	3
h. violent crimes	0	1	2	3
i. drug abuse	0	1	2	3
j. noise and disturbances	0	1	2	3
k. calls for information	0	1	2	3
l. pet control	0	1	2	3
m. community relations	0	1	2	3
n. visible minority relations	0	1	2	3

5. While you may not have had direct experience of the following Gloucester Police Force programs, please indicate how important each is to you as a resident of Gloucester. (circle the appropriate number)

	not important	somewhat important	very important
a. responses to citizen requests for information	1	2	3
b. Neighborhood Watch	1	2	3
c. Block Parents	1	2	3
d. Impaired driving (RIDE)	1	2	3
e. School liaison/education	1	2	3
f. drug enforcement	1	2	3
g. traffic enforcement	1	2	3
h. drug education	1	2	3
i. foot patrol	1	2	3
j. crisis intervention (social work program)	1	2	3
k. bilingual services	1	2	3
l. seniors' programs	1	2	3
m. visible minorities' programs	1	2	3
n. response to domestic violence	1	2	3
o. motor patrol	1	2	3
p. other (please identify) ________________	1	2	3

6. Have you had face-to-face contact (non-social) with the Gloucester Police?

a. NO, never___

b. In the last 12 months Yes___ No___

c. Not in 12 months, but in last 5 years Yes___ No___
d. Not in last 5 years, but previously Yes___ No___

7. If "NO, never," please proceed to question 9. If YES, please indicate the contact(s) and your satisfaction with the contact(s). (circle the appropriate numbers)

	No Contact	Not Satisfied	Somewhat Satisfied	Very Satisfied
a. Traffic enforcement	0	1	2	3
b. Traffic accident	0	1	2	3
c. Request for information	0	1	2	3
d. Casual conversation with on-duty officer	0	1	2	3
e. Reported a crime	0	1	2	3
f. Reported a problem in the neighborhood	0	1	2	3
g. Other (please identify) ________________	0	1	2	3

8. How well did the Gloucester Police person with whom you dealt exhibit the following qualities? (circle the appropriate number)

	not at all	somewhat	very much
a. fairness	1	2	3
b. courtesy	1	2	3
c. honesty	1	2	3
d. good appearance (uniforms, grooming)	1	2	3
e. knowledge of their job	1	2	3
f. concern for the public	1	2	3
g. effectiveness	1	2	3

9. Have you telephoned a request to the Gloucester Police Force?

a. NO, never
b. In the last 12 months Yes___ No___
c. Not in 12 months, but in last 5 years Yes___ No___
d. Not in last 5 years, but previously Yes___ No___

10. If "NO, never," please proceed to question 14. If YES, did the person responding for the Gloucester Police Force identify her/himself?

Yes___ No___

11. If your call was to request police assistance, please indicate the nature of your request.

a. request for information or advice
b. report a crime committed against yourself or a member of your household
c. report a crime committed against some other person
d. report a neighborhood problem (egs., noise, suspicious persons)
e. Other (please specify)

12. If you reported a crime, please identify the type of crime, the time between the crime and your report, and the time taken for a police response.

Crime ______________________________
Time from crime occurrence to your report ______________________
Time from your report to police response ______________________

13. Please indicate your level of satisfaction with the response. (circle the appropriate number) I was satisfied with

	not at all	somewhat	very much
a. ability of the person(s) answering the call to answer in French or English	1	2	3
b. the time taken to answer the call	1	2	3
c. the time taken to respond to my request/question	1	2	3
d. the politeness of the response	1	2	3
e. person's understanding of my question/request	1	2	3
f. the police action resulting	1	2	3

14. Have you ever been the victim of a crime? Yes___ No___

15. Have you ever been the victim of a crime in the city of Gloucester?

Yes___ No___

If NO, please proceed to question 16. If YES, please indicate the nature of the crime(s). (check appropriate spaces)

a. Theft of personal belongings from Home___ Car___ Garage___
Elsewhere (describe location) ______________________
Were the police called? Yes___ No___
If "no," why? (eg., reported to another agency instead) ______________
____

b. Theft from business___
Were the police called? Yes___ No___
If "no," why? (eg., reported to another agency instead) ___________
____

c. Damage or vandalism to property at Home___ Car___ Garage___
Elsewhere (describe location) ___________
Were the police called? Yes___ No___
If "no," why? (eg., reported to another agency instead) ___________
____

d. Spousal assault or threat of physical harm at home ___
Elsewhere (describe location) ___________
Were the police called? Yes___ No___
If "no," why? (eg., reported to another agency instead) ___________
____

e. Non-spousal assault or threat of physical harm at home___
Elsewhere (describe location) ___________
Were the police called? Yes___ No___
If "no," why? (eg., reported to another agency instead) ___________
____

f. Theft or illegal use of your car from home ___
Elsewhere (describe location) ___________
Were the police called? Yes___ No___
If "no," why? (eg., reported to another agency instead) ___________
____

16. Are there areas of Gloucester in which you feel unsafe? Yes___ No___

17. How safe do you feel during daylight hours in Gloucester? (circle the appropriate number if applicable)

	very unsafe	somewhat unsafe	somewhat safe	very safe
a. in shopping centres	1	2	3	4
b. in your residence	1	2	3	4
c. in your place of work	1	2	3	4
d. in your neighborhood	1	2	3	4
e. in parks	1	2	3	4
f. while driving	1	2	3	4

18. How safe do you feel at night in Gloucester? (circle the appropriate number if applicable)

	very unsafe	somewhat unsafe	somewhat safe	very safe
a. in shopping centres	1	2	3	4
b. in your residence	1	2	3	4
c. in your place of work	1	2	3	4
d. in your neighborhood	1	2	3	4
e. in parks	1	2	3	4
f. while driving	1	2	3	4

19. If you have felt unsafe, describe the circumstances and location.

____

____

____

20. Overall, how well does the Gloucester Police Force exhibit the following qualities? (circle the appropriate number)

	don't know	not at all	somewhat	very much
a. fairness	0	1	2	3
b. courtesy	0	1	2	3
c. honesty	0	1	2	3
d. good appearance (uniforms, grooming)	0	1	2	3
e. knowledge of their job	0	1	2	3
f. concern for the public	0	1	2	3
g. effectiveness	0	1	2	3

21. Are there changes that you would like to see in the services of the Gloucester Police? No___ Yes___ Please describe ________________________

____

____

____

____

____

22. To assist in estimating different community needs, please indicate the following BACKGROUND INFORMATION:

a. In what year were you born?

b. Sex: Female___ Male___

c. Marital status: Single___ Married___ Live-in partner___ Widowed___ Divorced or separated___

d. How many people, including yourself, live in your household?___

e. What is your highest completed education?
Some high school___ High school diploma___
Some community college___ Some university___
Completed community college___ Completed university degree___

f. Please indicate your employment status:
Full-time student___
Full-time homemaker___
Full-time income-earning employee___
Part-time income-earning employee___
Self-employed income-earner___
Retired___

g. What is the TOTAL gross family income for 1988?
Under $20,000___ 20,001-40,000___
40,001- 60,000___ 60,001-80,000___
80,001-100,000___ Above 100,000___

h. Most people can identify an ethnic background, such as English, French, Italian, East-Indian, etc. What do you consider your ethnic background to be?___________

i. What is the language usually spoken in your home?___________

j. What type of housing to you live in? Rent___ Own___
Single (detached) dwelling___
Double (semi-detached) dwelling___
Small apartment building (fewer than 3 levels)___
Large apartment building (more than 3 levels)___
Row-house (town-house)___
Trailer___

23. We welcome any comments that you might have about the Gloucester Police Force. <u>THANK YOU</u> for your help. ______________________________

Peter J. Campbell
Chief of Police

Address All Correspondence
to the Chief of Police
Please Reply Attention of

HALTON REGIONAL POLICE SERVICE

Box 2700, 1151 Bronte Road, OAKVILLE, Ontario L6J 5C7
(905) 825-4777 / 878-5511
Fax: (905) 825-9416

Se necessita da ajuda de um interprete para completar este questinario contacte Staff Sergeant Dan Okuloski (825-4713).

Se richiedi i servizi di un perfetto e capace interprete per favore mettersi in contatto Staff Sergeant Dan Okuloski (825-4713).

Jeśli potrzebujesz pomocy tłumacza w wypełnieniu powyzszej adkiety skontaktuj się z Staff Sergeant Dan Okuloski (825-4713)

ਜੇਕਰ ਤੁਹਾਨੂੰ ਇਸ ਸਰਵੇਖਣ ਨੂੰ ਪੂਰਾ ਕਰਨ ਲਈ ਅਨੁਵਾਦਕ ਦੀ ਮਦਦ ਚਾਹੀਦੀ ਹੈ ਤਾਂ ਤੁਸੀਂ ਇਸ ਨੰਬਰ ਤੇ ਤਾਲ ਮੇਲ ਕਰੋ S/Sgt. Dan Okuloski (825-4713)

如果你需要在这个服务方面的解释请与我们联系
Staff Sergeant Dan Okuloski (825-4713).

Dear Resident:

On behalf of the Halton Regional Police Service, I invite you to participate in a community survey by taking a few minutes to complete and return the enclosed questionnaire.

The survey is designed to identify the perceptions and expectations of Halton citizens in relation to the role of the police. It will also enable us to understand your experiences and perceptions of crime. Your responses and opinions will allow the Halton Regional Police Service to gain a better understanding of our community needs.

Two thousand residents have been invited to participate. I assure you that your individual responses will remain confidential.

If you have any questions regarding the community survey, please feel free to contact the Halton Regional Police Service at 825-4777 / 878-5511 ext. 4713.

We would appreciate your assistance in forwarding your response in the postage paid envelope by FEBRUARY 15, 1995.

Your cooperation and participation in our community policing survey is much valued and appreciated.

Yours truly,

Peter J. Campbell
CHIEF OF POLICE

HALTON REGIONAL POLICE SERVICE

This survey is to be completed by any adult in the household 18 years of age or over. Please take a moment to complete it by circling the appropriate response:

	yes	no
1) Have you, in the last five years had occasion to speak with or contact a member of the Halton Regional Police Service?	1	2

If NO skip to question 2.
circle as many choices as you wish

a) Was the contact with:

a uniform police officer	1
a plainclothes detective	2
at a public presentation or meeting	3
records bureau staff/firearms clerk	4
police dispatcher/station clerk	5
victim services representative	6

b) Was the contact:

on the telephone only	1
at the police station in person	2
at home/office	3
in person on the street (eg. traffic stop)	4
in person at a meeting or as a committee member	5
other (example; written correspondence)	6

c) With regards to your contact were you:

a victim of crime	1
a witness to an incident	2
stopped for a traffic violation	3
charged with a criminal offence	4
at a public presentation	5
at a R.I.D.E. or seat belt program	6
at a program event or committee meeting	7
other ______________________	8

d) Were you satisfied with the response of the Halton Regional Police Service in the following areas:

	Yes	No
Telephone response	1	2
Time it took to respond	1	2
Concern of member(s) involved	1	2
Investigation conducted	1	2

2) Based on your own experience, or what you have read or heard, please tell us whether you agree or disagree with the following:

The Halton Regional Police:	Agree	Disagree	Don't Know
b) are honest and fair	1	2	3
c) are courteous and polite	1	2	3
d) have a caring attitude	1	2	3
e) respect human rights	1	2	3

Please answer the following questions as they relate specifically to **YOUR** city/town/rural area, not society in general:

3) How do you perceive the following problems in your City/Town/Rural area?

	major problem	minor problem	not a problem	don't know
a) vandalism	1	2	3	4
b) theft of cars	1	2	3	4
c) school violence	1	2	3	4
d) armed hold-ups	1	2	3	4
e) youth loitering	1	2	3	4
f) noisy parties	1	2	3	4
g) assaults	1	2	3	4
h) thefts from cars	1	2	3	4
i) family violence	1	2	3	4
j) drugs	1	2	3	4
k) fraud	1	2	3	4
l) weapons	1	2	3	4
m) littering	1	2	3	4
n) kidnapping/abduction	1	2	3	4
o) traffic offences	1	2	3	4
p) sexual assaults	1	2	3	4
q) liquor offences	1	2	3	4
r) home break in	1	2	3	4
s) impaired driving	1	2	3	4
t) business break in	1	2	3	4
u) other __________	1	2	3	4

4) How safe do you feel during the daylight hours in your City/Town/Rural Area?

	very safe	somewhat safe	somewhat unsafe	very unsafe
a) at shopping centres (local mall, plaza, or store)	1	2	3	4
b) in your residence	1	2	3	4

c) in your place of work	1	2	3	4
d) in public parks	1	2	3	4
e) while driving	1	2	3	4
f) walking in your neighbourhood	1	2	3	4
g) using public transit	1	2	3	4
h) other __________	1	2	3	4

5) How safe do you feel during the night time in your City/Town/Rural Area?

	very safe	somewhat safe	somewhat unsafe	very unsafe
a) at shopping centres (local mall, plaza, or store)	1	2	3	4
b) in your residence	1	2	3	4
c) in your place of work	1	2	3	4
d) in public parks	1	2	3	4
e) while driving	1	2	3	4
f) walking in your neighbourhood	1	2	3	4
g) using public transit	1	2	3	4
h) other __________	1	2	3	4

6) Compared to areas outside of HALTON (Burlington, Oakville, Milton, Halton Hills), would you say that your City/Town/Rural Area has:

more crime	less crime	about the same
1	2	3

7) During the past five (5) years do you think that crime in your City/Town/Rural Area has:

increased	decreased	about the same
1	2	3

8) If the police have information of a general nature, crime prevention programs, or public education material, what is the best method to relay this information to you? (you may circle more than one response)

1 television
2 radio
3 newspapers
4 flyers/newsletters
5 citizen advisory committees
6 politicians
7 word of mouth
8 neighbourhood/rural watch system

9) What are the best ways for you to communicate information of a general nature, or your concerns, interests and priorities to the police? This is not for information of a serious nature where immediate assistance is required. (you may circle more than one response)

1	citizen advisory committees
2	public surveys
3	politicians
4	letters to the police service
5	calling the police by telephone
6	in person (meeting with police after calling)
7	Crime Stoppers program

10) Do you think the police should spend more time, the same amount of time, or less time dealing with the following matters in your City/Town/Rural Area?

	more	same amount	less
a) enforcing traffic laws	1	2	3
b) organizing crime prevention activities	1	2	3
c) trying to catch criminals	1	2	3
d) controlling illegal drug use	1	2	3
e) responding in person to calls of a minor nature	1	2	3
f) working with the public to solve problems	1	2	3
g) foot/bicycle patrol when practical	1	2	3
h) patrolling your community in marked police cars	1	2	3
i) patrolling your community in unmarked police cars	1	2	3
j) providing educational programs at schools	1	2	3

11) Do you believe the public and the police have a shared responsibility in providing a safe community?

yes	no
1	2

12) Would you be willing to report the following minor offenses to the police in a manner other than sending a patrol car? (i.e. over the phone, attending the station, making an appointment, via the mail). Your report will still be dealt with.

	yes	no
Minor Motor Vehicle Collision (no injury & damage under $ 4,000.00)	1	2
Fraud/theft	1	2
Municipal Bylaws (noise complaints, animal complaints)	1	2
Vandalism/Mischief to property	1	2

13) How many officers do you believe are available at any given time during the **DAY**, to respond to calls for service in your City/Town/Rural Area?

1-5   6-10   11-19   20-29   30+   don't know

14) How many officers do you believe are available at any given time during the **NIGHT**, to respond to calls for service in your City/Town/Rural Area?

1-5 6-10 11-19 20-29 30+ don't know

		Yes	No
15) a)	Are you aware that some Police Stations in Halton have limited public access hours?	1	2

b) How concerned are you that public access to Halton Regional Police Service offices is not available 24 hours a day?

very concerned	somewhat concerned	don't have an opinion	somewhat unconcerned	not concerned
1	2	3	4	5

Please explain why?

____

____

____

16) If you call the Halton Regional Police Service, and there is no immediate danger to anyone, how long should it take for officers to respond to in the following situations:

	less than 30 minutes	30-90 minutes	more than 90 minutes	don't have an opinion
residential burglary	1	2	3	4
motor vehicle accidents (no injuries)	1	2	3	4
thefts	1	2	3	4
parking complaints (not fire route)	1	2	3	4
alarms	1	2	3	4
vandalism	1	2	3	4
municipal by-laws	1	2	3	4

17) Do you believe that residents or businesses should be charged a fee for certain police services.(eg., a fee for police responding to a false alarm)

Yes 1 No 2

18) I reside in:

Urban Burlington	1	Rural Burlington	2
Urban Oakville	3	Rural Oakville	4
Urban Milton	5	Rural Milton	6
Halton Hills-Acton	7	Halton Hills-Georgetown	8
Halton Hills Rural	9		

19) How long have you lived in this area?

less than 3 years	3-10 years	more than 10 years
1	2	3

20) Your gender male 1 female 2

21) Your age

18-30	31-44	45-64	65+
1	2	3	4

22) Which describes your current work status.

working full time	1	working part time	2	not working	3
student	4	home maker	5	retired	6

23) Which best describes where you live.

single family dwelling	townhouse or row house	high rise building
1	2	3
low rise building less than 5 floors	duplex, triplex, fourplex	farm
4	5	6

24) From an overall perspective, how satisfied are you with the work of the Halton Regional Police Service in your City/Town/Rural Area?

Very Satisfied	Reasonably Satisfied	Don't Have an Opinion	Reasonably Dissatisfied	Very Dissatisfied
1	2	3	4	5

Why do you feel this way? Please explain in as much detail as possible. **Please also use this space to make any additional comments you wish.**

____

Thank You for Taking the Time to Complete the Survey, We Value Your Input.

APPENDIX II

NEIGHBOURHOOD CRIME PREVENTION SAFETY AUDIT GUIDE*

"BUILDING PARTNERSHIPS FOR SAFER NEIGHBOURHOODS"

SAFETY AUDIT WORKING GROUP
SAFER CITIES INITIATIVES OFFICE

Updated: July, 1996

"It is efficient to confine the initial survey to bare essentials and gather special data later in the design process as new questions arise."
Kevin Lynch

Neighbourhood Crime Prevention Safety Audits are a practical way for people to create a ***Safer City***. It is hoped that citizens, volunteer groups, and businesses will use this tool to reduce opportunities for crime by identifying and addressing specific safety concerns in their area. By working together, people in communities and the City of Edmonton can help to prevent crime.

Use the Neighbourhood Crime Prevention Safety Audit Guide:

- TAKE IT TO YOUR FRIENDS
- TAKE IT TO YOUR COMMUNITY GROUP
- TAKE IT TO YOUR LOCAL MERCHANT
- TAKE IT TO YOUR NEIGHBOURHOOD SCHOOL
- TAKE IT . . .

and help build a Safer Edmonton for yourself, your neighbours, and for us all.

We wish to acknowledge and thank METRAC [Metro Action Committee on Public Violence Against Women and Children], a Toronto-based group, for their pioneering work on Safety Audits. We used their resource materials extensively. This guide has been developed from their Women's Safety Audit Guide, adapting it to our situation in Edmonton.

Crime Prevention Safety Audits focus on working together as a neighbourhood to increase people's safety in public and semi-public places like:

---

* Source: City of Edmonton. Safer Cities Initiatives Advisory Committee (1996). Reproduced with permission.

- parks
- bus stops
- streets
- the workplace
- colleges and universities
- underground parking garages
- school yards
- washrooms in shopping malls
- the transit system
- pedways
- laundry rooms
- parking lots
- recreation centres
- and anywhere you feel unsafe

TABLE OF CONTENTS

THE NEIGHBOURHOOD CRIME PREVENTION SAFETY AUDIT CONCEPT

A Neighbourhood Crime Prevention Safety Audit is a partnership among residents, community groups, local businesses, neighbourhood institutions and government. It helps your neighbourhood feel like a safer place by enabling you to assess how safe or unsafe you and your neighbours currently feel in your surroundings. It facilitates communication with your neighbours, local merchants, neighbourhood schools, Community and Family Services Centre, Police Neighbourhood Foot Patrol, and other stakeholders to address safety problems and bring about solutions that will make your neighbourhood a safer place for you and your family.

Government alone cannot make us feel safer nor can they create safer neighbourhoods. Our feelings of safety are our own and each of us must work for the solutions that satisfy our own feelings.

INTRODUCTION

"What can we do to make our communities safer? This question is asked more and more by Canadians across the country. For many, the obvious answer is more cops, courts and corrections. But it is becoming increasingly clear that this response cannot work alone. If we are to succeed in improving the quality of life in our communities, we will need to do much more to prevent crime, and to deal with the underlying social problems that cause it." *Toward A Safer Edmonton For All, Final Report, Mayor's Task Force on Safer Cities, May, 1992.*

It is our belief that "the community is the focal point of crime prevention." Government and citizens must work together to foster community-based efforts to increase neighbourhood safety. What follows is *one* tool that can be used to assist in this process. This tool was developed with two important principles in mind:

1. The reduction of fear and vulnerability leads to enhanced safety and quality of life for us all.
2. Crime Prevention Through Environmental Design [CPTED] offers perspectives that help in the identification of problems and leads to solutions for a safer neighbourhood.

We trust that this guide will aid you in addressing questions you might have, inspire you to get involved, and challenge you to work for change in parts of your neighbourhood where you or your neighbours feel unsafe or uncomfortable. By working together, we can make Edmonton a ***Safer City*** for us all.

MAYOR'S TASK FORCE ON SAFER CITIES

The starting point for Neighbourhood Crime Prevention Safety Audits in Edmonton was the Mayor's Task Force on Safer Cities. The Task Force first met in October 1990 and built upon the work being done across the country connecting crime prevention with social development and environmental design.

Several recommendations put forward by the Mayor's Task Force encouraged the development and use of Neighbourhood Crime Prevention Safety Audits. This Neighbourhood Crime Prevention Safety Audit Guide was developed in response to these recommendations and it is our hope that it will be one step forward in creating a ***Safer City***.

WHAT IS A CRIME PREVENTION SAFETY AUDIT?

A Crime Prevention Safety Audit is a tool that can be used to note, for future corrective action, what evokes uncomfortable or unsafe feelings in your surroundings. It is an inventory of the features in an area, (or building, path, street) which you feel affect your safety and allows you to take action to correct them. Whether there is sufficient lighting, whether you would be heard if you called for help, whether there are people who can help, or what improvements you'd like to see to enhance safety are questions whose answers help determine the appropriate action to take.

The main goal of a Crime Prevention Safety Audit is to work together to create safe places. The result will be reduced opportunities for violence and crime in the area in which you live.

Crime Prevention Safety Audits are simple but powerful tools. The strength of the Audit lies in people's experience. We each have a wealth of knowledge from living in or frequenting an area. We are the experts on our own neighbourhoods. By sharing what we know and what we feel, and by working together, we can make change happen.

HOW TO BEGIN

Now that you have obtained a copy of this Neighbourhood Crime Prevention Safety Audit Guide you are interested in how to go about conducting an Audit. If you are a member of a neighbourhood group (e.g., community league, church, neighbourhood association) you may want to work with and involve them and perhaps request that they sponsor the Neighbourhood Crime Prevention Safety Audit. While this is not essential to a successful Audit, such collaboration can bring the resources of the group to the Audit and help build upon its strengths.

To begin your Crime Prevention Safety Audit, contact the Citizen Action Centre (telephone 496-8200), the Police Service, Community and Family Services, or Parks and Recreation office in your area to obtain more information and to establish a liaison to aid you in conducting the Audit. These Departments will be able to answer many of the questions you have regarding Safety Audits. If you or your group require assistance in organizing or orienting volunteers for the Crime Prevention Safety Audit or other area of the Audit, Community Services Department, Parks and Recreation, Police Service or Fire Department staff will provide you with the necessary aid.

The steps you need to follow to do your Safety Audit are outlined in the next five sections. These sections show you:

1. How to invite others to join you.
2. How to prepare for an Audit
3. How to do an Audit.
4. What you need to do after you've finished the Audit.
5. Some ideas for special types of Audits.

A. TALKING TO OTHERS

Neighbourhood Crime Prevention Safety Audits do not have to involve a lot of people, but often the very places you want to audit are those where you feel unsafe. Since the best time to audit is at night, it is recommended that you do the Audit with other people.

You can start by talking to people or groups you know. For example, you could contact:

- friends and neighbours
- your community league
- a tenants' or residents' association
- your church group or other organization

If you are doing the Audit at work, invite co-workers and your union or staff representative. If you are auditing your university or college campus, invite student union members and staff.

INVITING OTHERS TO COME ON THE CRIME PREVENTION SAFETY AUDIT:

You may want to invite people to join you on the Audit who can help you get improvements made, such as your city councillor or school trustee, reporters from a community newspaper, or the person in charge of a place that concerns you.

B. DEFINING THE AUDIT

HOW BIG AN AREA DO YOU WANT TO COVER?

Sometimes it is not clear what size of area you want to audit — a building, a street, an area surrounding an LRT station, a neighbourhood, or an entire city. You may want to start small. For example, if your concern is the whole neighbourhood, you could do things in steps or stages:

- do a full Audit of a typical street;
- audit the whole neighbourhood from the point of view of one or two factors like lighting and signs or the social connections of residents; or
- audit your route to and from the Transit zone, the store, the community league hall or the school.

At work or school, you may want to concentrate on the places that most concern you. These may be washrooms, parking lots, stairways and tunnels, or any isolated areas.

A survey of local residents and others who frequent the area of concern could be taken around the neighbourhood or in a specific building. This is one way to find out where and in what situations people in the area feel least safe or least comfortable.

Survey questions could include:

1. How safe do you feel in your building / on the street / waiting at the bus stop?
   ❒ very safe ❒ safe ❒ unsafe ❒ very unsafe
2. Have you limited what you do because you don't feel safe? ❒ yes ❒ no
3. Have you ever felt at risk of sexual assault in this area? ❒ yes ❒ no
4. Please list five specific places where you feel the least safe.

   ____

   ____
5. Do you have friends and/or neighbours you can count on if in trouble?
   ❒ yes ❒ no
6. Please comment on what would help you to feel more safe.

   ____

   ____

Audits have been done in different ways. For example:

- Two adjacent cities invited women to participate; in one night they did an Audit of the areas in the city that were of most concern to them.

- Over a 5-month period, every building and outdoor area was audited on a major University campus.
- A neighbourhood group encouraged residents to do Audits on their own street and to hand in the results to the residents' association.
- Employees in an office building audited their workplace, the parking garage, and bus stops in both the summer and winter.

HOW MANY PEOPLE ON AN AUDIT TEAM?

The best size for an Audit team is three to seven people. This size allows for different points of view, but is still small enough to enable much discussion. If you want to cover a large space, like a neighbourhood, you may need more than one team.

THE AUDIT GROUP:

The Audit group should reflect the needs and opinions of people in the whole neighbourhood. It should, if possible, include isolated seniors, those with disabilities, and other vulnerable people. If this is not possible, try to be aware of the unique points of view of other people who might be:

Remember, when scheduling time for an Audit, consider potential child care needs. This consideration will make it possible for more people to participate.

- using a wheelchair
- hard of hearing
- blind
- mentally challenged
- very young
- elderly
- poor
- shift workers
- travelling with young children
- carrying parcels
- members of a minority culture or group
- unable to read
- not familiar with English

TIME FRAME:

The ideal time to allow for a "first Audit" is approximately two to three hours. You will need about 1/2–1 hour to talk about the Audit and decide on the location and size, 1–1 1/2 hour to complete the Audit, and 1/2–1 hour to discuss the findings and plan to write the recommendations. If the same group continues to work together, subsequent Audits may not require as much time.

C. DOING THE AUDIT

WHAT YOU NEED FOR THE AUDIT:

- Take a flashlight along.
- Use a red or black pen or marker instead of a blue one. Your notes will photocopy better.

- Take a camera or video camera if possible. If you're going out at night, use a high speed film—not less than 400 ASA (high speed films can be used to take pictures indoors). It is also a good idea to write down the number of each photo and note the location from which it was taken.
- Take notes or use your camera to document ***positive*** features as well as problem areas. It can be very powerful to contrast both good and bad examples of the same factor—for example, a well-lit street and a poorly-lit street.
- It is important to talk to people you meet during the Audit. Introduce yourself. Tell them that you are looking at safety in the area and would like to know what they think. You might ask how often and why they are there, whether they have had any bad experiences, and what changes they'd like to see.
- If it's difficult for you to take notes, use a tape-recorder.
- Make arrangements for a place to meet before and after the Audit.
- Ensure each participant has safe transportation home from the Audit.

WHEN TO AUDIT:

- *Time of Day—*

It is recommended to do outside Audits after dark. It's the only way to know if there is a problem with lighting—one of the most important safety features. Night time is also the time of day when people are most isolated and feel least safe.

Sometimes a place is more deserted and threatening early in the morning or in the middle of the afternoon when everyone else is at work. Only you will know which part of each 24 hours is of most concern TO YOU.

- *Time of Year—*

Seasons also affect how safe a place feels. For example, safety concerns are different when tree branches and bushes are bare in winter. Trees and plants in full bloom in summer potentially hide an attacker or blocking out light. Dry parking lots are different when they are full of puddles, covered in ice, or surrounded by snow banks which block the view.

You may want to go back to the same place more than once to note differences at alternative times of day, week, or year.

USING A CHECKLIST:

You've decided to do an Audit because you feel unsafe in a part of the city and you want to do something about it. The idea, then, is to gather the information that will help you press for changes.

The questions you are trying to answer are:

- "Why don't I like this place?"
- "When and why do I feel uncomfortable here?"
- "What changes would make me feel safer?"

Experiences from other cities' Audits indicate that taking the time to think about the questions on a checklist enables the auditor to get a clearer picture of the area.

Stopping to take a closer look at different factors also allows people to share feelings about a place.

Sometimes one person will remember something, and that story will trigger another memory in someone else. Each of these stories helps us determine why a place does or doesn't feel safe.

Finally, if you are working with other people who do not understand the problems as well as you, the checklist can help you work together. For example, your building superintendent may rush past an area, but the checklist will remind you to take the time to look at the lighting there before you move on.

- What if you were walking along here late at night?
- What if you had to wait for someone to pick you up?
- Is this doorway a possible entrapment site?
- Does it feel safe in winter? In the rain?
- Are there fewer people around at different times of the day, week, or year?

Going through a checklist gives you an outline of the kinds of questions to ask. It takes the whole picture and breaks it down into parts. Remember, this checklist only contains suggested questions. There may be additional ones you want to answer that are not included.

You can photocopy and enlarge the checklist on-the following pages to make it easier to write on

CRIME PREVENTION SAFETY AUDIT CHECKLIST

❐ Outdoors ❐ Indoors

General Area:____

Specific Location:______________________________________

Date:____

Day:____

Time:____

Audited By:____

Reason for Auditing the Area:_____________________________

TIPS ON TAKING NOTES:

- Regardless of how sharp your memory is, you will not remember everything. Therefore it is important to take good notes. Our experience shows that using a checklist and writing notes on it will make it easier to organize your ideas and suggestions later on.

- Write down any questions that you have (even if you don't have the time to find the answers).
- Sometimes a place is so poorly designed that there aren't any real solutions beyond a temporary band-aid. It is still important to note the problem. Identifying and naming a problem is the beginning of changing your surroundings and the way new buildings and spaces are designed.
- Take notes on everything, including your comments on the process of the Audit itself.
- Look over your notes a day or two later to see if they still make sense. Would someone who wasn't on the Audit understand what you mean? If not, try to make your notes clearer.

1. General Impressions

a] Describe your first initial reactions to the site:

____

____

____

b] What 5 words best describe the place?

________________________ ________________________

________________________ ________________________

____

2. Lighting [good lighting allows you to see the area and what's happening in it]

a] What is your impression of the lighting?

❐ very poor ❐ poor ❐ too dark
❐ satisfactory ❐ good ❐ too bright
❐ very good

b] Is the lighting consistent throughout the space?

❐ yes ❐ no

c] Are any lights out?____________________

d] If so, how many?____________________

e] What proportion of lights are out?____________

[e.g. Maybe only two bulbs in your hallway are burned out, but if there are only three bulbs to start with, then a more powerful way to say this is that two-thirds of the lights are out.]

f] Are you able to see and identify a face 25 metres (75 feet) away?

____

g] Do you know where/whom to call if lights are out, broken, not yet turned on, etc.?
❐ yes ❐ no

h] Outdoors: Is the lighting obscured by trees or bushes?
❐ yes ❐ no

i] How well does the lighting illuminate pedestrian walkways and sidewalks?
❐ very poor ❐ poor ❐ satisfactory
❐ good ❐ very good

3. SIGNAGE [GOOD SIGNAGE LETS YOU KNOW WHERE YOU ARE, WHAT RESOURCES ARE AVAILABLE, AND HELPS YOU DEVELOP SOME FAMILIARITY WITH THE LOCATION]

a] Is there a sign (i.e. room no., building name) identifying where you are?
❐ yes ❐ no

b] If no, are there directional signs or maps nearby which can help you identify where you are?
❐ yes ❐ no

c] Are there signs which show you where to get emergency assistance if needed?
❐ yes ❐ no

d] Are there signs which direct you to wheelchair access?
❐ yes ❐ no

e] Do exit doors identify where they exit to?
❐ yes ❐ no

f] Is there information posted describing the hours the building is legitimately open?
❐ yes ❐ no

g] What is your impression of the overall signage?
❐ very poor ❐ poor ❐ satisfactory
❐ good ❐ very good

4. SIGHTLINES [CLEAR SIGHTLINES ARE IMPORTANT AS THEY LET YOU SEE, WITHOUT INTERFERENCE, WHAT LIES AHEAD]

a] Can you clearly see what is up ahead?
❐ yes ❐ no

b] If no, the reasons may be:

Indoors: ❐ sharp corners ❐ walls
❐ pillars
❐ others____________________

Outdoors: ❐ bushes ❐ fences
❐ hill
❐ others____________________

c] Are there places someone could be hiding?
❐ yes ❐ no

d] If yes, where?

e] What would make it easier to see? e.g.:
❐ transparent materials like glass ❐ vehicles moved ❐ angled corners
❐ security mirrors ❐ trimmed bushes ❐ snow cleared

Other comments?

___

___

5. Isolation—Eye Distance [this lets you assess how far away things are from the location and if someone would see you if you were in trouble]

a] At the time of your Audit, did the area feel isolated?
❐ yes ❐ no

b] How many people are likely to be around?

- In the early morning:
❐ none ❐ a few ❐ several ❐ many
- During the day:
❐ none ❐ a few ❐ several ❐ many
- In the evening:
❐ none ❐ a few ❐ several ❐ many
- Late at night (after 10 pm):
❐ none ❐ a few ❐ several ❐ many

c] Is it easy to predict when people will be around?
❐ yes ❐ no

d] Is there a monitor or surveillance system?
❐ yes ❐ no ❐ don't know

e] Other comments?

___

___

6. Isolation — Ear Distance [lets you assess if you could be heard in an emergency]

a] How far away is the nearest person to hear a call for help?

___

b] How far away is the nearest emergency service such as an alarm, security personnel, crisis telephone? ___ ❐ don't know

c] Can you see a telephone, or a sign directing you to emergency assistance?
❒ yes ❒ no

d] Is the area patrolled?
❒ yes ❒ no ❒ don't know

e] If yes, how frequently?
❒ every hour ❒ once per afternoon/evening ❒ don't know

Other comments?

____

____

7. Movement Predictors [a predictable or unchangeable route or path; this allows you to assess whether or not you can determine the way or direction people will move]

a] How easy is it to predict people's movements? (e.g. their routes)
❒ very easy ❒ somewhat obvious ❒ no way of knowing

b] Is there an alternative well-lit route or path available?
❒ yes ❒ no ❒ don't know

c] Is there an alternative frequently travelled route or path available?
❒ yes ❒ no ❒ don't know

d] Can you tell what is at the other end of the path, tunnel, or walk?
❒ yes ❒ no ❒ don't know

e] Are there corners, alcoves, or bushes where someone could hide and wait for you?
❒ yes ❒ no

f] Other comments?

____

____

8. Possible Entrapment Sites [lets you assess whether or not there are locations which are of special concern]

Indoors:

a] Are there empty rooms that should be locked?
❒ yes ❒ no

b] Are there small, well-defined areas? e.g.:
❒ stairwells
❒ recessed doorways or lockers
❒ unlocked closets
❒ elevators
❒ others:____________________

Outdoors:

c] Are there small, confined areas where you would be hidden from view? e.g.:
- ❒ unlocked equipment or utility shed
- ❒ alley or lane
- ❒ recessed doorway
- ❒ construction site
- ❒ others:____________________

9. Escape Routes [lets you assess whether or not there are ways to escape should there be an incident]

a] How easy would it be for an offender to disappear?

❒ very easy ❒ quite easy ❒ don't know

b] How difficult would it be for you to escape to safety if you had to?

❒ very difficult ❒ quite difficult ❒ don't know

10. Nearby Land Uses [lets you assess the impact of how the land is used as it relates to your comfort and safety]

a] What is the surrounding or nearby land used for?

❒ stores ❒ offices ❒ restaurants
❒ factories ❒ heavily treed/wooded area ❒ busy traffic
❒ parking lots ❒ residential houses and streets ❒ river bank
❒ don't know ❒ other:____________________

b] Can you identify who owns or maintains nearby land?

❒ yes ❒ no

c] What are your impressions of nearby land use?

❒ very poor ❒ poor ❒ satisfactory
❒ good ❒ very good

d] Is the land use in the area changing?

____

____

e] Does its new use make you feel more or less comfortable than its old use?

____

____

____

f] What about the land use change makes you feel more or less comfortable?

____

____

____

11. Factors That Make the Place More Human [these questions let you assess whether or not the location is used or abused by people]

a] Does the place feel cared for?
❐ yes ❐ no

b] Does the place feel abandoned?
❐ yes ❐ no

c] What gives you that feeling?

____

____

____

d] Is there graffiti on the walls?
❐ yes ❐ no

e] In your opinion, are there racist or sexist slogans/signs/images on the walls?
❐ yes ❐ no

f] Are there signs of vandalism?
❐ yes ❐ no

g] Would other materials, tones, textures or colours improve your sense of safety?
❐ yes ❐ no

h] Other comments?

____

____

____

12. Maintenance [these questions help you tell if the area is well looked after and well used by people]

a] What are your impressions of maintenance?
❐ very poor ❐ poor ❐ satisfactory
❐ good ❐ very good

b] Is there litter lying around?
❐ yes ❐ no

c] Is there need for major repair?
❐ yes ❐ no

d] Do you know to whom maintenance concerns should be reported?
❐ yes ❐ no

e] From your experience, how long do repairs generally take?
❐ yes ❐ no

13. Overall Design [lets you express your overall feeling after you have looked at the site in detail]

a] Describe your impressions of the overall design:
❒ very poor ❒ poor ❒ satisfactory
❒ good ❒ very good

b] If you weren't familiar with the place, would it be easy to find your way around?
❒ yes ❒ no

c] Is the entry visible and well defined?
❒ yes ❒ no

d] Are public areas visually protected?
❒ yes ❒ no

e] Does the place make sense?
❒ yes ❒ no

f] Is the place too spread out?
❒ yes ❒ no

g] Are there a confusing number of levels?
❒ yes ❒ no

h] Other comments?

____

____

14. Social Concerns [this lets you assess whether or not there are groups and organizations in place which add to your feelings of comfort and safety]

a] Are there cultural and social activities occurring in the neighbourhood?
❒ yes ❒ no

b] Describe how this makes you feel more or less comfortable:

____

____

____

c] Are there organizations and groups in the neighbourhood which are concerned about the neighbourhood and its people?
❒ yes ❒ no

d] Describe how this makes you feel more or less comfortable:

____

____

____

e] Do you have friends or neighbours in the area you could count on in an emergency?
❒ yes ❒ no

f] Is the population of the area changing?
❒ yes ❒ no

g] Describe how more or less safe this makes you feel:

h] Are there people with special needs in the area whose needs are not being met?
❒ yes ❒ no

i] Describe who they are:

j] Are there institutions in the area which make you feel more or less comfortable?
❒ yes ❒ no

k] Describe how they make you feel more or less comfortable:

From the Crime Prevention Safety Audit questions that you just completed, identify what can be done to make you feel safer about this location

What improvements would you like to see?

Do you have any specific recommendations?

What skills or resources could you contribute to making these improvements?

D. AFTER THE AUDIT

ORGANIZING YOUR FINDINGS:

After you have completed the Audit, you will have a lot of information about problem areas and many ideas about changes you'd like to see.

One method of organizing the information is grouping all the points on one factor, such as "lighting" together. As well, comments about how far a person has to go to get help, whether there are enough telephones nearby, and how likely it is that other people might see an assault, can be brought together under "isolation" factors.

Another way of organizing the information from the checklist(s) is by *type* of space. For example, safety factors common to all parking lots could be grouped together.

When you have finished organizing the findings, ensure that no part of the area has been overlooked. If it has, consider collecting more information using a mini-Audit or by surveying people in the area.

SHARING THE RESULTS:

Whether they were part of the Audit or not, you may want to get support, information, ideas, and feedback from people living or working in the area. Consider holding a small meeting where those who did not participate can talk about their concerns and help with the recommendations.

This will give you more information concerning the problems, encourage ideas for improvement, and create support for the changes you want to see.

MAKING RECOMMENDATIONS:

The first step is to look at the checklist(s) and determine what are the most important concerns. List these concerns in order of importance, most urgent first.

The second step is to make recommendations aimed at alleviating the problems. For example, if the Audit shows that buildings are hard to identify and find, the recommendation might be to put up signs. Signs need to be readable at night and by those who may be blind (for example, lit signs with symbols instead of words or braille signs). Staff of the Police Service, Community and Family Services, Parks and Recreation and Fire Departments can aid you in developing recommendations and in helping ensure they are specific. Audit recommendations should be prioritized to aid those who need to act in understanding what is most important to you.

WORKING FOR CHANGE:

Once your Neighbourhood Crime Prevention Safety Audit is complete and you have made the necessary recommendations, contact the City staff person assigned to your Audit or the Citizen Action Centre (telephone 496-8200).

City staff will refer the recommendations to the appropriate City Department(s) for action. A meeting(s) will be organized between you and the appropriate City Department(s) to discuss your recommendations.

Some recommendations can be implemented easily and relatively quickly, while others may take time to effect. Some negotiations will be necessary as there may be many requests for changes. Respective Departments will advise your organization of the action it is taking or is making on your behalf.

Some recommendations will require action be taken by other stakeholders for example, local businesses, schools or the Provincial government. The City cannot ensure that these other stakeholders will comply with the recommendations. The City staff assigned to your Audit can advise your group as to strategies that have been effective in getting attention and action from them. One strategy that may be useful in grabbing the attention of property owners is sending a follow-up letter. This letter will alert the stakeholder to areas of concern. An example of one such letter can be found in the appendix.

Implementation of recommendations of Neighbourhood Crime Prevention Safety Audits will take time. The strength and support of your group will make the work seem easier and the benefits of having a ***Safer City*** more worthwhile.

REVIEW AND ONGOING ASSESSMENT:

Your group will need to review progress in the implementation of recommendations periodically. Has the situation changed? Have changes made as a result of recommendations alleviated or reduced the need for others? Are other changes more urgent? After a year or two, you and your group may want to evaluate the effectiveness of the Audit. Is there a more positive feeling about safety in the neighbourhood now, as opposed to prior to the Audit? In what areas do people feel safer? If you conducted a survey, this could be repeated and the results compared. You could conduct a second Neighbourhood Crime Prevention Safety Audit and compare the situation now with that during the first Audit. An evaluation may support resident improved feelings of safety. An evaluation could point out additional areas needing change.

CONNECTING WITH YOUR COMMUNITY:

Recommendations and actions to address recommendations which stem from the Safety Audit may affect the entire neighbourhood. Your group should develop ways to inform the community of the Audit, the Audit Recommendations and of the actions that are taking place. As well, the community should also be advised of the reasons some recommendations will not be implemented. Your local community league newsletter or school newsletter may be useful vehicles to make local residents aware of the audit results.

E. SPECIAL AUDITS

LARGE AUDITS:

If you want to do an Audit of a large area of a city, you will need to do a lot of extra planning. We are unable to predict everything you will need to do, but here are some things to think about:

- How many teams do you need?

- Do you have maps and any other important information?
- Can you arrange for safe transportation to and from the Audit areas?
- What about child care?
- How will you organize the volunteers, train the team leaders, collect the checklists, write a summary report and put forth recommendations?
- How will you present the findings?

If the place being audited is large or complex, the Audit should focus on one part at a time. For example, when a subway system was audited, three to fifteen copies of the checklist were filled in for each subway station, depending on its size and complexity. One checklist might concern the station platform, another the tunnel, one for the passenger pick-up area, or any area where lighting, availability of help, and other factors were considered.

JOINT AUDITS:

If there are many safety problems in your area (or you want to do a large area) and if you have a lot of energy, you could pursue others for a joint Audit team before you go out on the Audit.

Here is a list of some of the people you may wish to contact to be partners in a joint Audit:

- Neighbourhood businesses (shops, restaurants, gas stations, 24 hour stores)
- Business Associations
- Planning and Development Department (City of Edmonton)
- Residents Associations
- Parks and Recreation Department (City of Edmonton)
- Community Leagues
- Local churches
- Community & Family Services Department (City of Edmonton)
- Edmonton Transit System
- Neighbourhood Watch
- Edmonton Police Service
- Block Parent
- Emergency Response Department (City of Edmonton)
- Neighbourhood Centres
- Elected Representatives
- Sexual Assault Centre
- School Representatives

Who you want to involve is up to you.

AUDITING THE TRANSIT SYSTEM:

Whenever you are doing an Audit of a street area that includes a bus stop, or an LRT station, be sure to look carefully at where people wait for a bus, and how they get to and from the stops.

Here are some extra questions for auditing the transit system:

- How good is the lighting inside the shelter or LRT station?
- How far away is the nearest public phone? The nearest private phone (commercial or residential)?
- Is there visible information for passengers about what to do in an emergency?

SAFETY IN RELATIONSHIPS

While you have been reading the Neighbourhood Crime Prevention Safety Audit Guide or conducting the Audit of your neighbourhood, concerns for your personal safety or issues about your relationships may have come to mind. Should any of these concerns trouble you, there are agencies which you can turn to for help. Some of the major service agencies you can contact are:

- Sexual Assault Centre of Edmonton, 423-4102, [24 hour crisis line 423-4121]
- The Family Centre, 423-2831 or 448-1622
- Community Services, 496-5801

APPENDIX A

SAMPLE FOLLOW-UP LETTER:

After an Audit has been completed, a follow-up letter may be sent to anyone your group wants to inform of the results. The letter will encourage implementation of recommendations by alerting the person in charge of the area to the existence of safety concerns. The letter will also serve as a record of action taken.

SAMPLE LETTER

Date:

From:

To:

On _____ (date of Audit), we conducted a Neighbourhood Crime Prevention Safety Audit of _____ (specify location). The results of the Audit indicate the following safety concerns:

-
- (List concerns)
-

We have brought our safety concerns to your attention in an effort to work together to increase the safety in this area.

Thank you for your time. We look forward to working together towards a safer community.

Sincerely,

(Group Name)

INDEX